D1042353

AD HOC NETWORKING TOWARDS SEAMLESS COMMUNICATIONS

Ad Hoc Networking Towards Seamless Communications

LILJANA GAVRILOVSKA
University "Ss Cyril and Methodius" — Skopje, Macedonia

and

RAMJEE PRASAD
Aalborg University, Denmark

A C.I.P. Catalogue record for this book is available from the Library of Congress.

ISBN-10 1-4020-5065-8 (HB)
ISBN-13 978-1-4020-5065-7 (HB)
ISBN-10 1-4020-5066-6 (ebook)
ISBN-13 978-1-4020-5066-4 (ebook)

Published by Springer,
P.O. Box 17, 3300 AA Dordrecht, The Netherlands.

www.springer.com

Printed on acid-free paper

To my husband, Miroslav to our daughter, Ada, and to our son, Alek
— Liljana Gavrilovska

To my wife, Jyoti, to our daughter, Neeli, to our sons, Anand and Rajeev, to our granddaughters, Sneha and Ruchika, and to our grandson, Akash
— Ramjee Prasad

There is nothing that old that you can not say something new about it
— Dostojevski

TABLE OF CONTENTS

ACKNOWLEDGEMENTS

This book was a result of several years of active involvement in research and EU projects covering the area of ad hoc networking. The authors would like to express their deepest and sincere appreciation to the many people with whom they have shared their careers through these years.

Special thanks goes to Vladimir Atanasovski (Faculty of Electrical Engineering, Skopje, Republic of Macedonia) for his dedicated support on many aspects of this book.

The authors would like to thanks Junko Prasad for her patience and cooperation in freeing us from the enormous editorial burden.

PREFACE

मुक्तसंगोऽनहंवादी धृत्युत्साहसमन्वितः ।
सिद्ध्यसिद्ध्योर्निर्विकारः कर्ता सात्त्विक उच्यते ॥२६॥

mukta-sango 'naham-vadi
dhrty-utsaha-samanvitah
siddhy-asiddhyor nirvikarah
karta sattvika ucyate

The worker who is free from all material attachments and false ego, who is enthusiastic and resolute and who is indifferent to success or failure, is a worker in the mode of goodness.

– The Bhagvad Gita (18.26)

Future wireless systems (3G, B3G and 4G) envision higher data rates, mobility support and seamless communications, provided through a common platform which unifies evolving access technologies, seamless inter-working and adaptive multimode terminals. Ad hoc wireless networks, as an access networking paradigm, is considered to be an inevitable part of the future wireless systems enabling broadband access.

Ad hoc wireless networks are foreseen to experience the widespread use within the following years. The salient features of these networks are that they can freely and dynamically self-organize into arbitrary and temporary network topologies, and that they can operate without any infrastructure and under different channel and network operating conditions, providing seamless communications. Their quick deployment and easy establishment of communications, as well as huge number of new wearable wireless terminals, stipulate numerous applications varying from military applications on the battlefield to a pervasive communications in sensor networks. The expected degree of penetration of these networks will depend on the successful resolution of the key features.

The evolution of the future ad hoc networks will contribute to realization of all IP networking concept enabling connectivity of the smart appliances to the Internet. Last mile solutions, either ad hoc or mesh-based, will provide the broadband access capabilities and adapt to future multimedia applications.

Ad hoc wireless networking paradigm confronts numerous problems in front of the network designers. Enlarged opportunities impose additional requirements and research problems concerning power efficiency, device discovery and topology control, routing, addressing, security, etc. Many solutions have been offered within the academic, industrial and standardization institutions worldwide.

This book aims to reveal the state-of-the-art in ad hoc wireless networking and to address the features that previously have not been extensively addressed. It presents in a comprehensive and systematic manner areas such as basic concepts (topology, capacity and self-organization), service discovery, mobility, and cross-layer optimizations in ad hoc wireless networks. It also discusses and updates the key research topics that are expected to promote and accelerate the commercial application of these networks (e.g. PHY/MAC, routing, QoS, security).

The book is designed to be used as a reference work and as a textbook. As a reference book it is primarily intended to ad hoc network technologists and engineers, and scientists in academia, industry and research labs. As a textbook, it is intended to students completing graduate and research oriented courses in ad hoc and wireless networking. The book can serve as a source to comprehensive reference material on wireless ad hoc networks. The material in this book is also intended to those who want to learn more about this field.

The reader is assumed to have some knowledge of communications networks principles and concepts, and wireless communications.

ABOUT THE AUTHORS

Liljana Gavrilovska received her B.Sc., M.Sc. and Ph.D. from University "Ss. Cyril and Methodius", Skopje (76), University of Belgrade (85) and University "Ss. Cyril and Methodius", Skopje (95) respectively.

She has worked with the R&D department at the Macedonian Telecom Company in the field of telephone network planning. She joined the Faculty of Electrical Engineering, University "Ss. Cyril and Methodius", Skopje, Republic of Macedonia, in 1977. Dr. Gavrilovska currently holds a position of Professor at the Institute of Telecommunications and chief of Telecommunications Laboratory and CWMC (Center for Wireless and Mobile Communications), working in the area of telecommunication networks and wireless and mobile communications.

In 1982 receiving Carl-Duisburg grant, she was involved in a research project in the area of digital data transmission at University of Erlangen, Germany. During 1992/1993, as a recipient of bilateral Yugoslavian-Canadian government grant, she was involved with University of Toronto, Canada, in research in ATM switches. Dr. Gavrilovska joined the Centre for PersonKommunikation, Aalborg University, Denmark, in January 2000, where she was holding first a position of invited professor and later as Associate Research Professor (2001–2002). From 2003–2005 she was involved with CTiF (Center for TeleInfrastructure) with the same position on a partly basis and participated in several EU and national/international projects.

Dr. Gavrilovska is author/co-author of more than 100 research journal and conference publications, and technical papers, and has held several tutorials in the field of wireless ad hoc networks and future wireless systems. She has participated in ACTS project ASAP working on the books Towards a Global 3G System: Advanced Mobile Communications in Europe, Vol. I and II (Artech House, 2001, Boston).

Her major research was concentrated on ad hoc networking, wireless and personal area networks, cross-layer optimization, future mobile systems, traffic analysis and admission techniques.

She is a senior member of IEEE and serves as a Chair of Macedonian Communication Chapter.

Ramjee Prasad was born in Babhnaur (Gaya), India, on July 1, 1946. He is now a Dutch citizen. He received his B.Sc. (eng.) from the Bihar Institute of Technology, Sindri, India, and his M. Sc. (eng.) and Ph. D. from Birla Institute of Technology (BIT), Ranchi, India, in 1968, 1970, and 1979, respectively.

He joined BIT as a senior research fellow in 1970 and became an associate professor in 1980. While he was with BIT, he supervised a number of research projects in the area of microwave and plasma engineering. From 1983 to 1988, he was with the University of Dar es Salaam (UDSM), Tanzania, where he became a professor of telecommunications in the Department of Electrical Engineering in 1986. At UDSM, he was responsible for the collaborative project Satellite Communications for Rural Zones with Eindhoven University of Technology, The Netherlands. From February 1988 through May 1999, he was with the Telecommunications and Traffic Control Systems Group at Delft University of Technology (DUT), where he was actively involved in the area of wireless personal and multimedia communications (WPMC). He was the founding head and program director of the Center for Wireless and Personal Communications (CWPC) of International Research Center for Telecommunications – Transmission and Radar (IRCTR). Since June 1999, Dr. Prasad has been with Aalborg University, where currently he is Founding Director of Center for Teleinfrastruktur (CTIF), and holds the chair of wireless information and multimedia communication. He was involved in the European ACTS project FRAMES (Future Radio Wideband Multiple Access Systems) as a DUT project leader. He is a project leader of several international, industrially funded projects. He is the project coordinator of the European sixth framework integrated project "My Personal Adaptive Global NET (MAGNET)". He has published over 500 technical papers, contributed to several books, and has authored, coauthored, and edited nineteen books: *CDMA for Wireless Personal Communications, Universal Wireless Personal Communications, Wideband CDMA for Third Generation Mobile Communications, OFDM for Wireless Multimedia Communications, Third Generation Mobile Communication Systems, WCDMA: Towards IP Mobility and Mobile Internet, Towards a Global 3G System: Advanced Mobile Communications in Europe, Volumes 1 & 2, IP/ATM Mobile Satellite Networks, Simulation and Software Radio for Mobile Communications, Wireless IP and Building the Mobile Internet, WLANs and WPANs towards 4G Wireless, Technology Trends in Wireless Communications, Multicarrier Techniques for 4G Mobile Communications, OFDM for Wireless Communication Systems, Applied Satellite Navigation Using GPS, GALILEO, and Augmentation Systems, From WPANs to Personal Networks: Technologies and Applications, Towards the Wireless Information Society: Heterogeneous Networks, Towards the Wireless Information Society: Systems,*

Services, and Applications, and *4G Roadmap and Emerging Communication Technologies,* all published by Artech House. His current research interests lie in Wireless networks, packet communications, multiple-access protocols, advanced radio techniques, and multimedia communications.

Dr. Prasad has served as a member of the advisory and program committees of several IEEE international conferences. He has also presented keynote speeches, and delivered papers and tutorials on WPMC at various universities, technical institutions, and IEEE conferences. He was also a member of the European cooperation in the scientific and technical research (COST-231) project dealing with the evolution of land mobile radio (including personal) communications as an expert for The Netherlands, and he was a member of the COST-259 project. He was the founder and chairman of the IEEE Vehicular Technology/Communications Society Joint Chapter, Benelux Section, and is now the honorary chairman. In addition, Dr. Prasad is the founder of the IEEE Symposium on Communications and Vehicular Technology (SCVT) in the Benelux, and he was the symposium chairman of SCVT'93.

In addition, Dr. Prasad is the coordinating editor and editor-in-chief of the *Kluwer International Journal on Wireless Personal Communications* and a member of the editorial board of other international journals, including the *IEEE Communications Magazine* and *IEE Electronics Communication Engineering Journal.* He was the technical program chairman of the PIMRC'94 International Symposium held in The Hague, The Netherlands, from September 19–23, 1994 and also of the Third Communication Theory Mini-Conference in Conjunction with GLOBECOM'94, held in San Francisco, California, from November 27–30, 1994. He was the conference chairman of the fiftieth IEEE Vehicular Technology Conference and the steering committee chairman of the second International Symposium WPMC, both held in Amsterdam, The Netherlands, from September 19–23, 1999. He is the general chairman of WPMC'01 which was held in Aalborg, Denmark, from September 9–12, 2001.

Dr. Prasad was also the founding chairman of the European Center of Excellence in Telecommunications, known as HERMES. He is now HERMES honorable chair. He is a fellow of IEE, a fellow of IETE, a senior member of IEEE, a member of The Netherlands Electronics and Radio Society (NERG), and a member of IDA (Engineering Society in Denmark). Dr. Prasad is adviser to several multinational company. He has received several international award; the latest being the Telenor Nordic 2005 Research Prise.

LIST OF ABBREVIATIONS

2G	2nd Generation
3G	3rd Generation
4G	4th Generation
7DS	Seven Degrees of Separation
A/SHF-CDMA	Adaptive/Slow Frequency Hopping CDMA
AAA	Authentication, Authorization and Accounting
ABC	Always Best Connected
ABR	Associativity Based Routing
AC	Access Category
ACK	ACKnowledgement
ADMR	Adaptive Demand-Driven Multicast Routing
ADV	Adaptive Distance Vector
AEDCF	Adaptive Enhanced Distributed Coordination Function
AF	Assured Forwarding
AFEDCF	Adaptive Fair Enhanced Distributed Coordination Function
AIFS	Arbitration InterFrame Space
AIFSN	Arbitration InterFrame Space Number
ALM	Application Layer Metrics
AODV	Ad-hoc On demand Distance Vector
AP	Access Point
AQR	Active QoS Routing
ARAN	Authenticated Routing for Ad hoc Networks
ARQ	Automatic Repeat reQuest
ASAP	Adaptive Reservation and Pre-allocation Protocol
ATM	Asynchronous Transfer Mode
AUSNet	Autonomous Undersea System Network
B3G	Beyond 3rd Generation
BA	Broker Agent

BAN	Body Area Network
BE	Best Effort
BER	Bit Error Rate
BI	Busy Indication
BIP	Broadcasting Incremental Power
BNDMA	Blind Network Diversity Multiple Access
BSAR	Bootstrapping Security Associations for Routing
BSS	Basic Service Set
BT	Backoff Timer
BTMA	Busy Tone Multiple Access
BWA	Broadband Wireless Access
CA	Certificate Authority
CAC	Connection Admission Control
CAMA	Cellular-Assisted Mobile wireless Ad hoc network
CAN	Content-Addressable Network
CATA	Collision-Avoidance Time Allocation
CBR	Constant Bit Rate
CBSD	Clustering Based Service Discovery
CBTC	Cone Based Topology Control
CCK	Complementary Code Keying
CDMA	Code Division Multiple Access
CDS	Connected dominating set
CEDAR	Core Extraction Distributed Ad hoc Routing
CEPT	Conference of European Postal and Telecommunications
CGA	Cryptographically Generated Address
CGSR	Clustered-Gateway Switch Routing
COMPASS	COMPASS selected routing
CP	Content Provider
CP	Control Point
CRL	Certificate Revocation List
CSI	Channel State Information
CSMA/CA	Carrier Sense Multiple Access/Collision Avoidance
CSMA/CD	Carrier Sense Multiple Access/Collision Detection
CSP	Class-based Service Propagation
CTS/BI	CTS/Bussy Indication
CW	Contention Window
DA	Directory Agent
DAD	Detection of Duplicated Addresses
DARPA	Defence Advanced Research Project Agency
DCF	Distributed Coordination Function
DDR	Dynamic Distributed Routing
DFCP	DiffServ Code Point
DHCP	Dynamic Host Configuration Protocol
DHT	Distributed Hash Table
DiffServ	Differentiated Services

DIFS	Distributed InterFrame Space
DLC	Data Link Control
DLPS	Distributed Laxity-based Priority Scheduling
DM	Direct mode
DoS	Denial-of-Service
D-PRMA	Distributed Packet Reservation Multiple Access
DPS	Distributed Priority Scheduling
DREAM	Distance routing effect algorithm for mobility
DS	Direct Sequence
DS	Dominating Set
DSCR	Dual Stage Contention Resolution
DS-DEDCA	Dual Stage Dynamic EDCA
DSDV	Destination Sequenced Distance Vector
DSDV-SQ	DSDV SeQuence number
DSR	Domain Space Resolver
DSR	Dynamic Source Routing
DSSS	Direct Sequence Spread Spectrum
DT	Distance Table
DV	Distance Vector
DWOP	Distributed Wireless Ordering Protocol
EDCA	Enhanced Distributed Channel Access
EF	Expedited Forwarding
ETSI	European Telecommunications Standards Institute
FA	Foreign Agent
FAMA	Floor Acquisition Multiple Access
FC	Factor of Correction
FCC	Federal Communications Commission
FDMA	Frequency Division Multiple Access
FEC	Forward Error Correction
FH	frequency hopping
FHSS	Frequency Hopping Spread Spectrum
FIFA	Foundation for Intelligent Physical Agents
FIFO	First In First Out
FORP	Flow Oriented Routing Protocol
FPQ	Fixed Priority Queueing
FPRP	Five-Phase Reservation Protocol
FQMM	Flexible Quality of Service Model
FSR	Fisheye State Routing
FU	Functional Unit
GDH	Generalized Diffie-Hellman
GEDIR	GEographic DIstance Routing
GG	Gabriel graph
GPRS	General Packet Radio System
GPS	Global Positioning System
GRS	Greedy Routing Scheme

GSD	Group Service Discovery
HA	Home Agent
HC	Header Comrepssion
HIPERLAN	HIgh-PErformance Radio Local Area Network
HMAC	Hash Message Authentication Code
HP	Heterogeneity level
HRMA	Hop-Reservation Multiple Access
HSR	Hierarchical State Routing
HTTP	Hyper Text Transfer Protocol
IARA	Interference Aware Routign Adaptation
IARP	IntrAzone Routing Protocol
IBSS	Independent Basic Service Set
iCAR	integrated Cellular Ad hoc Relay
ICI	Inter-Carrier Interference
ICMP	Internet Control Message Protocol
ICPK	Implicitly Certified Public Key
ICSMA	Interleaved Carrier Sense Multiple Access
IDS	Intrusion Detection System
IEEE	Institute of Electrical and Electronic Engineers
IERP	IntErzone Routing Protocol
IETF	Internet Engineering Task Force
INORA	INSIGNIA+TORA
INS	Intentional Naming System
INSIGNIA	INband SIGNallIng for mobile Ad hoc networking
INTMOD	INTernational mobility MODel
IntServ	Integrated Services
IP	Internet Protocol
IR	InfraRed
IrDA	Infrared Data Association
ISI	Inter-Symbol Interference
ISM	Industrial, Scientific and Medical
ISO	International Standards Organization
ITU	International Telecommunication Union
JVM	Java Virtual Machine
KDC	Key Distributed Center
KDS	Kinetic Data Structures
KEK	Key Encrypting Key
KR	Knowledge Range
LAN	Local Area Network
LAR	Location-Aided Routing
LCA	Linked Cluster Algorithm
LCT	Link Cost Table
LET	Link Expiration Time
LIDS	Local Intrusion Detection System
LLA	Link Layer Adaptation

LMST	Local Minimal Spanning Tree
LORA	Least Overhead Routing Algorithm
LOS	Line Of Sight
LPR	Low cost Packet Radio
LS	Link State
MAC	Medium Access Control
MAC	Message Authentication Code
MACA	Multiple Access Collision Avoidance
MACA/PR	Multiple Access Collision Avoidance with Piggyback Reservation
MACA-BI	MACA By Invitation
MACAW	MACA for Wireless LANs
MAGNET	My personal Adaptive Global NET
MANET	Mobile Ad hoc NETwork
MBN	Mobile Backbone Node
MC-CDMA	Multi Carrier CDMA
MC-CSMA	Multi-Channel CSMA
MCDS	Minimum Connected Dominating Set
MCN	Multi-hop Cellular Network
MDS	Minimal Dominating Set
MEA	Mesh network Enabled Architecture
METMOD	METropolitan mobility MODel
MFR	Most Forward within Radius
MGCP	Mulitmedia Gateway Control Protocol
MIMO	Multiple Input Multiple Output
MLM	MAC Layer Metrics
MMAC	Multimedia Mobile Access Communication Systems
MN	Mobile Node
MP3	MPeg layer 3
MPR	MultiPle Relay
MRL	Message Retransmission List
MSDP	MAGNET Service Discovery Protocol
MSDU	MAC Service Data Unit
MST	Minimum Spanning Tree
MuPAC	Multi-Power Architecture for Cellular networks
NATMOD	NATional mobility MODel
NBS	Number of Busy Slots
ND	Node Degree
NDM	Non-Disclosure Method
NDMA	Network Diversity Multiple Access
NDP	Network Discovery Protocol
NEWCOM	Network of Excellence on Wireless COMmunications
NFP	Nearest Forward Progress
NLM	Network Layer Metrics
NLOS	Non-Line-of-Sight

NoSs	Number-of-Sending stations
NSF	National Science Foundation
ODMRP	On Demand Multicast Routing Protocol
OFDM	Orthogonal Frequency Division Multiplexing
OFT	One-way Function Tree
OLMQR	On demand, Link-state, Multi-path QoS Routing protocol
OLSR	Optimized Link State Routing
OM	Obstacle Mobility model
ORA	Optimum Routing Algorithm
OSI	Open System for Interconnection
OSPF	Open Shortest Path First
OTH	Over-The-Horizon
P2P	Peer-to-peer
PAN	Personal Area Network
PAP	Peak-to-Average-Power
PCF	Point Coordination Function
PCM	Power Controlled Multiple access
PDA	Personal Digital Assistant
PER	Packet Error Rate
PF	Persistence Factor
PGP	Pretty Good Privacy
PHY	PHYsical
PIN	Personal Identification Number
PKI	Public Key Infrastructure
PLBQR	Predictive Location-Based QoS Routing protocol
PLBR	Preferred Link-Based Routing
PN	Personal Network
PPM	Pulse Position Modulation
PRNET	Packet Radio NETwork
PRTMAC	Proactive Real-Time MAC
PSTN	Public Switched Telephone Network
PTKF	Partial Topology Knowledge Forwarding
QAM	Quadrature Amplitude Modulation
QoS	Quality of Service
QPART	QoS Protocol for Ad hoc Realtime Traffic
QPSK	Quadrature Phase Shift Keying
RA	Rate Adaptation
RAN	Radio Access Network
RAND	RANDom number
RASD	Routing Assisted Service Discovery
RBAR	Receiver Based AutoRate
RD	Relative Direction
RDMAR	Relative Distance Micro-discovery Ad hoc Routing
RET	Route Expire Time

RFC	Request For Comments
RFID	Radio Frequency IDentification
RIP	Routing Information Protocol
RLM	Random LandMarking
RLS	Reactive Location Service
RNG	Relative Neighborhood Graph
RNS-OFDM	Residue Number System–OFDM
RP	Randezvous Point
RPC	Remote Procedure Call
RPGM	Reference Point Group Mobility model
RREP	Route REPly
RREQ	Route REQuest
RRM	Radio Resource Management
RS	Reed Solomon
RSVP	ReSource reserVation Protocol
RT	Real Time
RT	Routing Table
RTMAC	Real-Time MAC
RTP	Real Time Protocol
RTS/BI	RTS/Busy Indication
RTS/CTS	Request To Send/Clear To Send
RWMM	Random Walk Mobility Model
RWP	Random WayPoint mobility model
SA	Security Association
SA	Service Agent
S-AODV	Security-aware AODV
SAP	Service Access Point
SAR	Security Aware ad hoc Routing protocol
SARA	Scalable Adaptable Reservation Architecture
SBRP	Secure Bootstrapping and Routing protocol
SCTP	Session Control Transport Protocol
SDMA	Spatial Division Multiple Access
SDP	Service Discovery Protocol
SDR	Software Defined Radio
SEAD	Secure Efficient Ad hoc Distance vector
SIFS	Short InterFrame Space
SIG	Special Interest Group
SINR	Signal-to-Interference plus Noise Ratio
SIP	Session Intitiation Protocol
SKDS	Soft Kinetic Data Structures
SLA	Service Level Agreement
SLM	Salutation Location Manager
SLP	Service Location Protocol
SMN	Service Management Node
SMS	Short Message Service

SMT	Secure Message Transmission
SNIR	Signal-to-Noise-Interference Ratio
SNMP	Simple Network Management Protocol
SNR	Signal-to-Noise Ratio
SoC	System on a Chip
SOPRANO	Self-Organizing Packet Radio Ad hoc Network with Overlay
SPOT	Secure PosiTioning
SPT	Shortest Path Tree
SQUAWB	Secure QoS-enhanced UltrA WideBand
SR	Speed Ratio
SRMA/PA	Soft Reservation Multiple Access with Priority Assignment
SRP	Secure Routing Protocol
SSDS	Secure Service Discovery Service
SSH	Secure SHell program
SSL	Secure Socket Layer
SSR	Signal Stability Routing
STAR	Source Tree Adaptive Routing
STDMA	Spatial reuse Time Division Multiple Access
STRAW	STreet RAndom Waypoint mobility model
SU	Slot Utilization
SUCV	Statistically Unique and Cryptographically Verifiable identifier
SURAN	SURvivable Adaptive radio Network
SWAN	Service differentiation in stateless Wireless Ad hoc Networks
SWG	Small World Graph
TBP	Ticket Based Probing
TBR	Trajectory Based Routing
TBRPF	Topology dissemination Based on Reverse-Path-Forwarding
TC	Topology Control
TCP	Transport Control Protocol
TCP/IP	Transport Control Protocol/Internet Protocol
TCR	Topology Change Rate
TD-CDMA	Time Division CDMA
TDD	Time Division Duplex
TDMA	Time Division Multiple Access
TDR	Time Dependent Routing
TGDH	Tree-based Group Diffie-Hellman scheme
TI	Tactical Internet
TIARA	Techniques for Intrusion-resistant Ad hoc Routing Algorithms

TIGER	Topologically Integrated Geographic Encoding and Referencing
TMPO	Topology Management by Priority Ordering
TORA	Temporally-Ordered Routing Algorithm
TTP	Trusted Third Party
UA	User Agent
UAV	Unmanned Aerial Vehicle
UDDI	Universal Description, Discovery and Integration
UDG	Unit Disk Graph
UDP	User Datagram Protocol
UDP/IP	User Datagram Protocol/Internet Protocol
UMP	User Mobility Profile
UMTS	Universal Mobile Telecommunications System
U-NII	Unlicenced-National Information Infrastructure
UPnP	Universal Plug and Play
URI	Unified Resource Identity
URI/URL	Uniform Resource Identifier/Uniform Resource Locator
URL	Unified Resource Locator
UWB	Ultra Wide Band
VANET	Vehicular Ad hoc NETwork
VID	Virtual Identifier
VoIP	Voice over Internet Protocol
VTT	Virtual Antenna Arrays
WAN	Wide Area Network
WBAN	Wireless Body Area Network
WB-PAN	Wireless Broadband Personal Area Network
WCA	Weighted Clustering Algorithm
WEP	Wired Equivalent Privacy
WFPQ	Weighted Fair Priority Queueing
Wi-Fi	Wireless Fidelity
WING	Wireless Internet Gateways
WLAN	Wireless Local Area Network
WMAN	Wireless Metropolitan Area Network
WPAN	Wireless Personal Area Network
WRP	Wireless Routing Protocol
WSN	Wireless Sensor Network
WWP	Weighted WayPoint mobility model
WWRF	Wireless World Research Forum
WWW	World Wide Web
XML	eXtended Markup Language
ZHLS	Zone-based Hierarchical Link State routing
ZRP	Zone Routing Protocol

1

Introduction

1.1 What is Ad Hoc Networking?

Ad hoc networks are a key factor in the evolution of wireless communications envisioned as cornerstones of future generation wireless networking technologies (B3G and 4G) [1–4]. They are infrastructureless networks formed on-the-fly (anytime, anywhere, for virtually any application) with limited life of existence. Ad hoc networks can be established as a stand alone group of mobile terminals, which communicate autonomously in a self-organized manner or are connected to a pre-existing infrastructure and use it to communicate with outside networks [5]. Ad hoc terminals (i.e., nodes) communicate wirelessly and share the same media (radio, infrared, etc.) and are free to move while communicating with other nodes.

Ad hoc networks inherit the traditional problems of wireless and mobile communications due to the specifics of the wireless transmission media, such as bandwidth optimization, power control and QoS issues. In addition, the possible lack of a fixed infrastructure introduces new problems regarding network configuration, topology maintenance, discovery procedures (device/neighbor/link discovery), ad hoc addressing and self-routing. Ad hoc networks are self-configuring networks with nodes capable of overtaking the routing functionalities. This allows multi-hop connectivity, extending the ad hoc networks domain over the connectivity range.

Motivations behind focusing interest in ad hoc wireless networks lie in the expectation that in near future there will be high proliferation of wireless devices. They fit into these developments by self-configuration and independence of existing infrastructure [5]. Ad hoc networks are realized in a variety of networks ranging from body area networks (BAN) [6] to vehicular ad hoc networks (VANET) [7] and autonomous undersea system networks (AUSNet) [8]. The multi-hop communication capabilities enlarge the capacity limitations due to the limited range of the wireless devices. Independent wireless nodes can be connected to a fixed-backbone via dedicated gateway devices enabling the IP networking services in areas that the Internet otherwise does not reach. Cooperativeness, self-organization and capability of neighborhood discovery enhance the interoperability among different wireless technologies [9]. Embedded options to provide location information are a strong contribution to the overall network management. All these make ad hoc networks fit in the future vision of the Wireless World of "continued connectivity".

1.2 Services and Applications

Ad hoc wireless networks recently have attracted considerable attentions due to the new developments in wireless technology and standards. There are numerous possibilities for applications of the ad hoc concept in the networking world. Ad hoc networks hide great commercial potential and advantage of flexible set up anywhere, at any time, without infrastructure and centralized administration [10].

Variety of services and applications were developed ranging from tactical military networks, through different commercial and educational applications, sensor networks, to location-aware services [11]. Fast deployment and easy establishment of functionalities, autonomous or relayed communications, cooperativeness and emerging areas of nomadic and ubiquitous computing, as well as improved IP-based networking in dynamic autonomous wireless environments promote ad hoc networks as a desirable wireless access technology. Ad hoc networking also gets momentum in emergency communication in catastrophic disaster areas and during terrorist attacks [12, 13]. They participate in collaborative and distributed computing and mesh (infrastructure relayed) and hybrid (integrated cellular and ad hoc) wireless networks. Table 1.1 gives an overview of existing applications and examples of services they provide.

Devices enabling wireless ad hoc networking are becoming smaller, cheaper, wearable, powerful and with lots of embedded capabilities delivering services seamlessly to end users. They comprise laptops, PDAs, camcorders, mobile phones, MP3 players, game stations, medical sensors, etc. Their number grows exponentially supporting the ad hoc networking expansion.

1.3 A Glance of History

The idea of ad hoc networking is a result of convergence of different concepts behind the need to establish survivable, infrastructureless communications between entities which were originally out of reachable range from each other. The concept of ad hoc communication based on *multi-hop wireless relaying of messages* has started in the ancient world. Darius I, the King of Persia, 500 BC, has introduced the principle of multi-hop relaying using line of shooting man and providing a much faster method of spreading information than

Table 1.1. Applications of wireless ad hoc networks

Applications	Descriptions/Services
Tactical networks	Military communication, operations Automated Battlefields
Emergency services	Search-and-rescue operations Disaster recovery, e.g., early retrieval and transmission of patient data (record, status, diagnosis) from/to the hospital Replacement of a fixed infrastructure in case of earthquakes, hurricanes, fire, etc.
Commercial environments	E-Commerce, e.g., electronic payments from anywhere (i.e., in a taxi). Business: – dynamic access to customer files stored in a central location on the fly provide consistent databases for all agents – mobile office Vehicular Services: – transmission of news, road conditions, weather, music – local ad hoc network with nearby vehicles for road/accident guidance – Unmanned Aerial Vehicles (UAV) — remotely piloted or self-piloted aircrafts that can carry cameras, sensors, communications equipment or other payloads
Home and enterprise networking	Home/office wireless networking (WLAN), e.g., shared whiteboard application, use PDA to print anywhere, trade shows Personal Area Network (PAN)
Educational applications	Set up virtual classrooms or conference rooms Set up ad hoc communication during conferences, meetings or lectures
Entertainment	Multiuser games Robotic pets Outdoor Internet access
Sensor networks	Home security and tracing Indoor/outdoor environmental monitoring Disaster prevention Health and wellness monitoring Power monitoring Location awareness Factory and process automation Military applications
Mesh networks	Residential zones (broadband Internet) Highway communication facilities for moving vehicles Business zones, important civilian regions, university campuses
Hybrid networks	Enhancements of cell coverage and connectivity of holes

any available method at that time. In 1970, Norman Abramson proposed ALOHAnet aiming to connect distant parts of the Hawaii Island. Aloha was based on single hop wireless packet switching with multiple accesses and was extended to multiple relaying in PRNET (packet radio network) project. PRNET was supported by DARPA (Defence Advanced Research Project Agency) and demonstrated the feasibility and efficiency of infrastructureless networks for military and civilian applications. After PRNET, Survivable Radio Networks (SURAN) was developed in 1983 as its extension. It provided more efficiency in terms of network scalability, security, processing capabilities and energy management [14]. As a next step, the low cost packet radio (LPR) showed up in 1987 [15]. LRP was based on DS spread spectrum technology combined with microprocessor-based packet switching. Further developments included advanced network management protocols, hierarchical network topology and dynamic clustering and security [16]. Wireless Internet Gateways (WINGs) deployed a flat peer-to-peer network architecture [17]. At the beginning of the 1980s, the military ad hoc applications were developed around the globe. These developments continued in 1997 with the largest implementation of mobile wireless multi-hop packet radio that implemented Internet connectivity (TI — Tactical Internet) [14].

The 1990s saw an explosive growth of wearable devices, notebook computers, open-source software and radio equipment placing the idea of infrastructureless gathering of mobile hosts in the wireless arena. As a result, the IEEE 802.11 committee [18], for the first time, accepted the term "*ad hoc network*" [19]. After that, a constant interest in the field of commercial, non-military, use of ad hoc networks emerged and resulted in approval of the first WLAN standard in 1997. IEEE developed a collision avoidance based medium access scheme that enables efficient ad hoc network formation.

In the meantime, the Internet Engineering Task Force (IETF) has created a working group (WG) called *mobile ad hoc networks (MANET)*, primarily aiming to solve and standardize the routing functionality in mobile ad hoc environment [20]. In 1999 another MANET deployment demonstrated over-the-horizon (OTH) communications from object at the sea to objects on land via an aerial relay. Primarily developed for military purposes, the MANET applications today are mainly focused on commercial and public sector, attracting attentions of international standardization bodies. European Telecommunications Standards Institute (ETSI) delivered another solution for ad hoc networking called HIPERLAN [21]. The Bluetooth Special Interest Group (SIG) introduced the Bluetooth technology [22].

The effort to enable efficient ad hoc networking continues. The following subsections describe the concept of ad hoc networks and ad hoc network types, as well as the major representatives of the existing technologies that support ad hoc connectivity.

1.4 Definition and Enabling Technologies

Ad hoc networks are organized spontaneously by nodes wishing to communicate without infrastructure support. Ad hoc communication can take place in different scenarios and is independent of any specific device, wireless transmission technology, network or protocol. Nodes can freely enter or leave the network at any time. Ad hoc networks can significantly vary in size from small size PANs comprising few devices, to few hundreds of sensors in wireless sensor networks. However, network functionality and protocols are distributed among all network participants.

There are several types of ad hoc networks, such as Wireless Body Area Networks (WBANs), Wireless Personal Area Networks (WPANs), Wireless Local Area Networks (WLANs), sensor networks, etc., that differ in their implementation domains and other features. However, most of them share the common feature of limited energy resources and capability to communicate using one or more wireless technologies. The Bluetooth technology, WLAN 802.11 and Ultra Wide Band (UWB) are the most frequently considered technologies in various ad hoc scenarios.

1.4.1 Characteristics of Ad Hoc Networks

Design, performances and successful deployment of wireless ad hoc systems disclose major issues and challenges in front of scientific and industrial communities [23]. Ad hoc specific characteristics include: wireless shared media, ad hoc based temporary and dynamic topology, autonomous operation in a distributed peer-to-peer mode without predefined infrastructure, multi-hop routing, mobility and limited energy resources. Designers of wireless ad hoc networks need to solve the distributed arbitration for the shared channel through appropriate MAC schemes providing solutions to different issues (synchronization, hidden/exposed terminal, throughput, access delay, real time support, etc.). The responsibilities of the routing protocols include aspects of mobility, location dependencies, loop-free communication, minimum control overhead, scalability, QoS, security, etc. QoS provisioning and pricing schemes are one of the key issues responsible for successful deployment. Self-organization aspects are recently gaining attention. Security is also very important and it includes understanding of threats and attacks and offering appropriate solution. Other important areas are addressing and service discovery, energy management and scalability. All these issues are differently addressed in various wireless networks appropriate for ad hoc design.

1.4.2 Network Types and Corresponding Enabling Technologies

Ad hoc networks are defined as a category of wireless networks that utilize multi-hop radio relaying and are capable of operating without support of any infrastructure. Different networks may exhibit ad hoc capabilities even though not all of them are ad hoc networks by their nature (e.g., WLAN). Each of them is addressing a specific range of commercial applications.

Wireless LANs (WLANs) were originally introduced as an alternative to wired LANs, within a building or campuses, mostly for user and data communications. They offer wireless connectivity and allow mobility of users. WLANs operate in the unlicensed ISM (Industrial, Scientific and Medical) frequency bands (902 to 928 MHz, 2.4 to 2.4853 GHz and 5.725 to 5.85 GHz). Typical communication range is 100 to 500 m and data rates vary from 1 to 54 Mbps, depending on the underlying technology. Wireless LANs can use different physical media for transmission (infrared-IR LAN, microwave and radio frequencies) [24]. The most popular enabling technologies are the IEEE's 802.11 [18] family of protocols and ETSI's HiperLAN [20]. The former uses spread spectrum technology, which can be either direct sequence spread spectrum (DSSS) or frequency-hopping spread spectrum (FHSS), and OFDM. The latter operates in the U-NII (Unlicenced-National Information Infrastructure) 5 GHz and 17 GHz frequency bands, with different physical layer [25]. Both networks can enable ad hoc configurations: IEEE 802.11 being in infrastructureless mode (see Fig. 1.1) and HiperLAN using the direct mode (DM) of operation [24].

The most prominent *de facto* standard for WLAN is *IEEE 802.11* [18], which was the first wireless standard aiming to provide high data rates in small-scale environments while providing support for mobility of nodes. When acting in an *ad hoc* mode, the mobile nodes can communicate directly to each other without using access points (APs). The absence of a central entity (AP) is overcome with *self-organization* and *multi-hop* relaying properties. When two or more stations communicate together in an ad hoc mode, they form an *independent basic service set (IBSS)*, otherwise they form a *basic service set (BSS)*. IEEE 802.11 MAC layer provides a basic access mechanism for peer-to-peer ad hoc networking in contention-based (DCF — distributed coordination function) and contention-free (PCF — point coordination function) medium

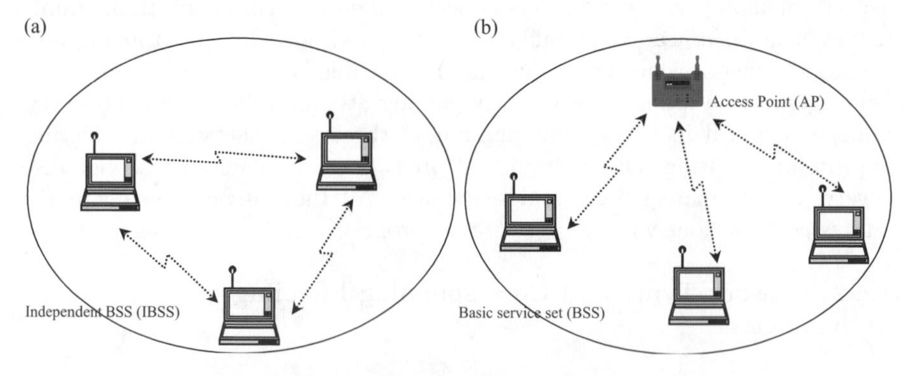

Fig. 1.1. A typical deployment scenario of IEEE 802.11: (a) ad hoc mode; (b) infrastructure mode

access. DCF incorporates a CSMA/CA (carrier sensing multiple access with collision avoidance) protocol as a basic medium access mechanism. RTS/CTS handshaking mechanism is introduced in order to solve the problem of *hidden/exposed* terminals (see Chapter 3).

After the first version of IEEE 802.11 (1997) and 802.11b, two other versions with different solutions for physical layer showed up: IEEE 802.11a and IEEE 802.11g. Other enhancements include: IEEE 802.11e (QoS), IEEE 802.11i (security), IEEE 802.11n (high throughput), IEEE 802.11p (vehicular environment), IEEE 802.11u (interworking with External Networks), IEEE 802.11v (wireless networks management), IEEE 802.11w (contention based protocol). The enhancement process is still active.

Recently, the concept of *hot spots (Wi-Fi — wireless fidelity)* has become frequently addressed, referring to an area covered by one or more WLAN access points providing Internet connectivity. The penetration of WLAN technology is growing exponentially threatening to overtake the cellular users.

Wireless personal area networks (WPANs) are short-range person-centered networking paradigm [1–3, 26]. They consist of networked devices which belong to a personal space surrounding a person [27]. Typically, a wireless personal area network uses some technology that permits communication within about 10 m. WPANs do not necessarily require an infrastructure. They form single-hop networks in which two or more devices are connected in a point-to-multipoint "star" manner. Proposed operating frequencies are around 2.4 GHz in digital modes. The objective is to facilitate seamless operation among home or business devices and systems.

Every device in a WPAN will be able to plug in to any other device in the same WPAN, provided they are within each other's physical range. Enabling technologies are Bluetooth, IrDA, UWB, HomeRF, etc. Bluetooth technology provides significantly lower data rates than wireless LANs. However, inherent support for ad hoc connectivity, low power consumption and wide availability make the Bluetooth technology a main candidate for PAN communication technology provider today. It was used as a basis for a new standard, i.e., IEEE 802.15.1.

Bluetooth technology is an industry wireless specification standard for use in various devices for short-range communications. As a radio-based wireless PAN technology, it allows devices which are in proximity to share information. Bluetooth nodes can operate in a contention-free token-based multi-access mode implementing a polling scheme. Nodes are using frequency hopping scheme with 79 frequencies. Bluetooth networks are classified into two different network topologies: *piconet* and *scatternet* (as depicted in Fig. 1.2). Nodes in piconets are organized in a *master–slave* manner. Each master device may cooperate with mostly seven other active nodes (slaves). A slave can communicate with the different piconets it belongs only in a time-multiplexing mode. Independent piconets that have common coverage may form a scatternet architecture. Scatternet exists when a node is active in more than one piconet at the same time.

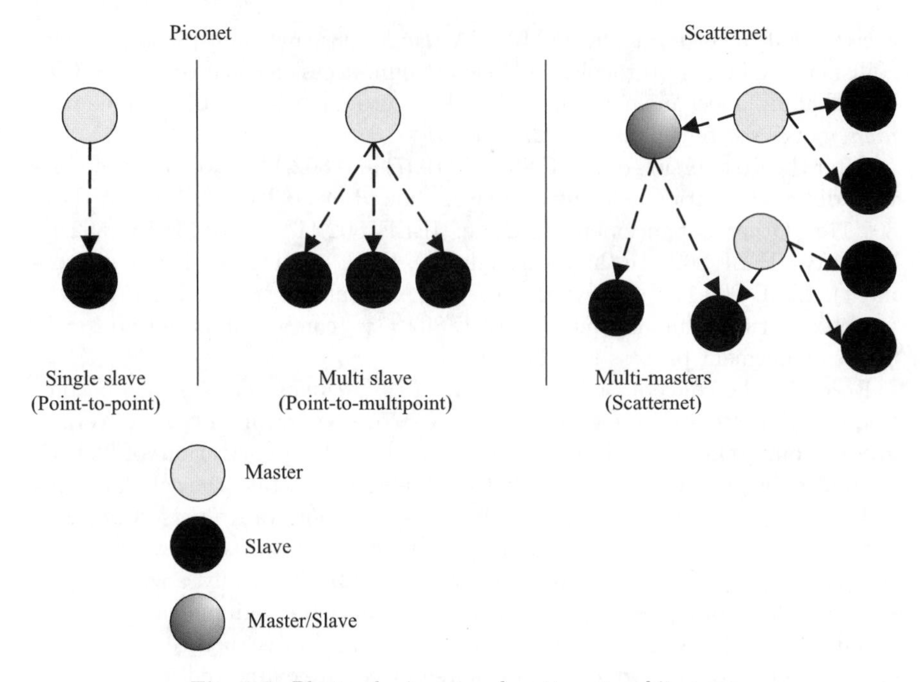

Fig. 1.2. Bluetooth piconet and scatternet architectures

Bluetooth enables mobile nodes (computers, mobile phones and portable handhelds) to communicate without cables. Some of the benefits are: enhances users experience, connecting devices without the need for cables, reduced power consumption. Bluetooth is becoming more affordable for everyone and is increasingly integrated in laptops, mobile phones, handhelds and many other devices.

Standardization efforts started in 1999 with the approval of the IEEE 802.15 committee. This committee comprises four working groups, each dedicated to particular areas (e.g., 802.15.3 — high data rates). Table 1.2 gives an overview of the major characteristics of different technologies and standards relevant to WLAN/WBAN design.

Wireless body area networks (WBANs) are another networks foreseen for short-range communications of small user devices [27]. WBANs provide a basic infrastructure element for control and monitoring of medical sensors and actors (e.g., implants). WBANs are also used in context-aware applications (e.g., automatic exchange of business cards with handshaking).

Wireless sensor networks (WSNs) are special category of ad hoc networks which provide wireless communication infrastructure among sensors deployed in a particular region and for a particular purpose. Sensors are tiny devices which can sense certain physical parameters (e.g., humidity, temperature, radiation, border intrusion), process data and communicate over the network to the monitoring station in a periodic or a sporadic fashion (see Fig. 1.3).

Table 1.2. Examples of ad hoc network types and corresponding enabling technologies and standards

Standard	Enabling technology	Network type	Frequency range	Physical layer	Max. throughput	MAC
802.11	802.11	WLAN*	2.4 GHz	Infrared, DSSS, FHSS	2 Mbps	CSMA/CA, polling
802.11a	802.11	WLAN*	5 GHz	OFDM	54 Mbps	CSMA/CA, polling
802.11b	802.11	WLAN* (Wi-Fi)	2.4 GHz	DSSS	11 Mbps	CSMA/CA, polling
802.11g	802.11	WLAN*	2.4 GHz	OFDM	54 Mbps	CSMA/CA, polling
HiperLAN/2	HiperLAN	WLAN	5 GHz	OFDM	54 Mbps	TDMA/TDD
802.15.1	e.g., Bluetooth	WPAN, WBAN, WSN	2.4 GHz	FHSS	1 Mbps	Polling
802.15.3	802.15.3	WPAN	2.4 GHz	QPSK, QAM	55 Mbps	TDMA
802.15.3a	802.15.3a	WPAN	3.1–10.6 GHz	UWB	110 Mbps	TDMA with QoS guarantees
802.15.4	e.g., ZigBee	WPAN, WBAN, WSN	2.4 GHz, 915 and 868 MHz	DSSS	250 Kbps	CSMA/CA
HomeRF	Radio	WPAN	2.4 GHz	FH combined with FSK	10 Mbps	Combination of CSMA/CA and TDMA
IrDA	IrDA	WPAN	875 ± 30 nm	Infrared combined with pulse modulation	4 Mbps	–
802.16/ 802.16e	802.16	WMAN (Wi-Max)	10–66 GHz, 2–11 GHz	OFDM	70 Mbps	TDMA, FDMA
HiperMAN	HiperMAN	WMAN	2–11 GHz	OFDM	70 Mbps	TDMA, FDMA

Note:
*The 802.11 technology lately becomes a candidate for technology enabler in WPANs.

They have certain specifics: mobility of nodes is not a mandatory requirement, usually they consist of a large number of nodes, density of nodes varies depending on the application, power constraints are more rigorous than in other ad hoc networks, etc. WSNs operate with low data rates (tens of kbps). They are a giant leap towards "proactive computing": a paradigm where computers anticipate human needs and, if necessary, act on its behalf. Examples of enabling technologies implemented in sensor networks are ZigBee [28], radio-frequency identification (RFID) and UWB.

Even though ad hoc networks are expected to be able to work without any infrastructure, recently there are interesting solutions revealed that enable

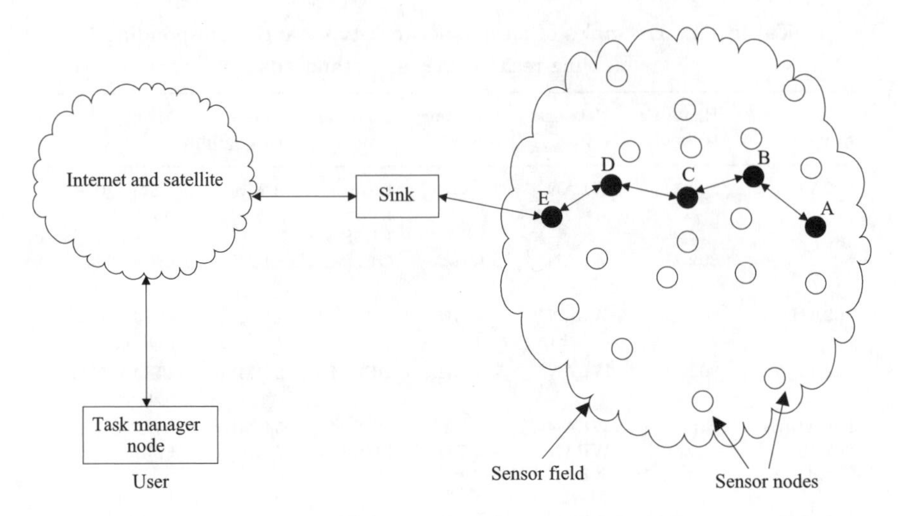

Fig. 1.3. Illustration of a sensor network

the mobile ad hoc nodes to function in the presence of an infrastructure. Examples of these solutions comprise WMANs, wireless mesh and hybrid networks.

Wireless metropolitan area networks (WMANs). The IEEE in December 2005 announced the approval of IEEE 802.16e, the mobile WirelessMAN standard that will facilitate the global development of mobile *broadband wireless access* (BWA) systems [29]. The 802.16 WirelessMAN standard, which addresses WMANs for broadband wireless access, previously supported only fixed (stationary) terminals. The amended standard specifies a system for combined fixed and mobile BWA supporting subscriber stations moving at vehicular speeds in licensed bands under 6 GHz. The 802.16e standards development project begun in late 2002, whereas the final draft was completed in October 2005. ETSI's counterpart to 802.16 WirelessMAN is HiperMAN [30]. *Wireless mesh networks* provide an alternate communication infrastructure for mobile or fixed nodes [31]. The infrastructure is formed by small radio relaying devices, which are fixed on marked places (e.g., roof of a house in residential areas, lamp posts by the highway), as depicted in Fig. 1.4. Mesh networks offer much cheaper coverage than cellular networks without one point of failure. They operate in the licence free ISM bands around 2.4 GHz and 5 GHz with data rates of 2 Mbps to 60 Mbps. Enabling technologies for wireless mesh networks are IEEE 802.11 and Wi-MAX [32] (IEEE 802.16).

The *hybrid architectures* [23] combine benefits of cellular and ad hoc approaches incorporating properties of multi-hop relaying with the support of existing infrastructure (see Fig. 1.5). Such hybrid networks increase the capacity, flexibility and reliability of routing (no one point of failure), and provide coverage and connectivity in holes (areas without cellular coverage). Several

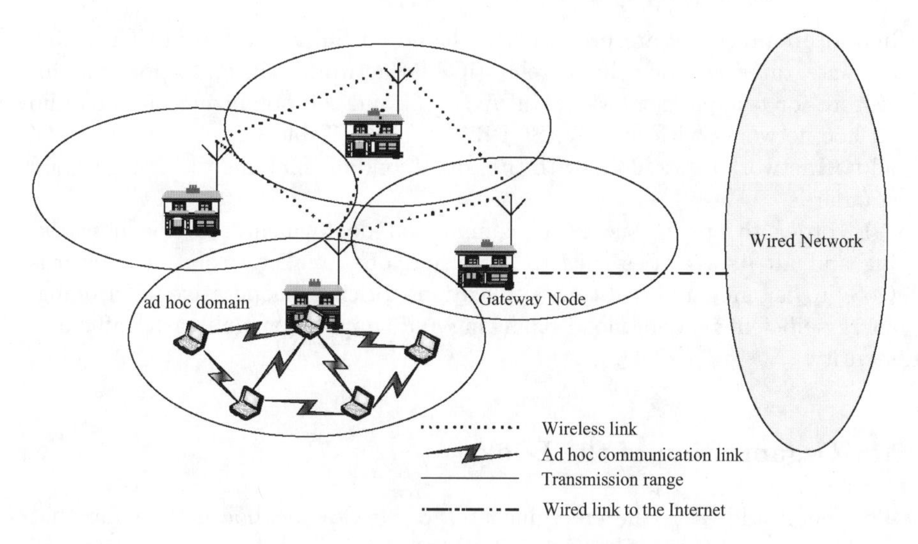

Fig. 1.4. Illustration of a mesh architecture

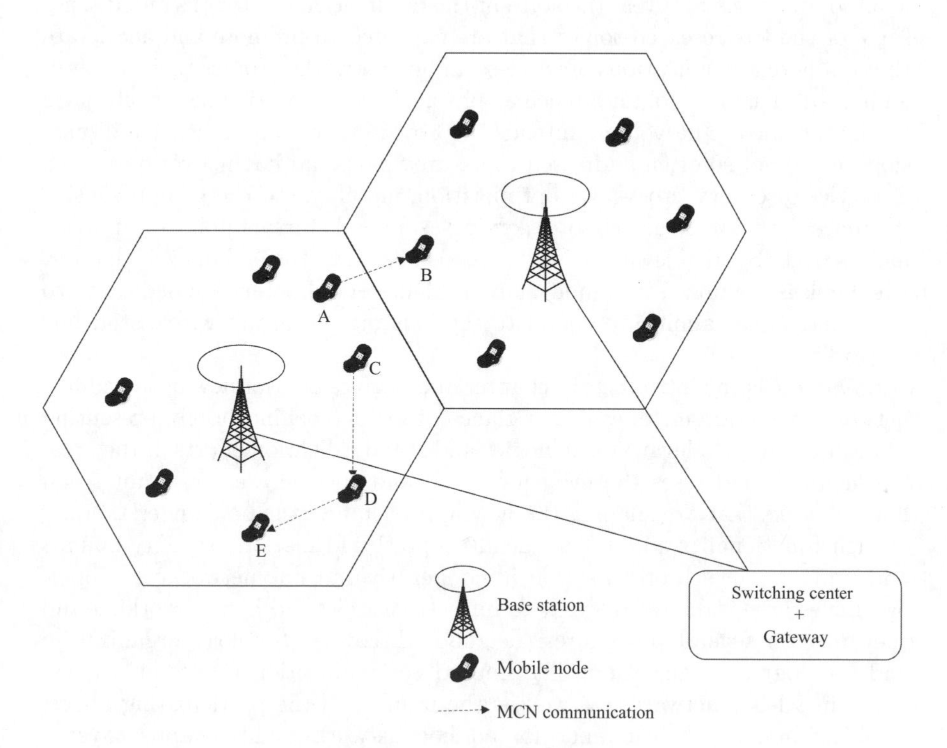

Fig. 1.5. Illustration of a hybrid architecture

hybrid architectures are proposed: multi-hop cellular networks (MCNs) [33, 34], integrated ad hoc cellular relay (iCAR) network [35], multi-power architecture for cellular networks (MuPAC) [23] and self-organizing packet radio ad hoc networks with overlay (SOPRANO) [36]. Table 1.2 gives examples of ad hoc network types along with the corresponding enabling technologies and standards.

In order to ensure successful commercial deployment, ad hoc networking and all its varieties, has to solve and offer realistic solutions towards QoS provisioning and real-time applications, pricing, cooperative functioning, energy-efficient relaying, load balancing and support for multicast traffic and security.

1.5 Organization of the Book

This book addresses the emerging area of wireless ad hoc networking that lately attracts growing interest in both academia and industry. Wireless ad hoc networks pose many complex and open problems to researchers. It is almost impossible to give a deep insight into all relevant topics in one volume. This volume aims to reveal the state-of-the-art in ad hoc networks and discuss some of the key research topics that are expected to promote and accelerate the commercial applications of wireless ad hoc networks. Topics that are often addressed in similar volumes (such as physical, MAC, routing issues, etc.) are briefly outlined. The volume intends to provide more comprehensive discussion on the not so often addressed topics in wireless ad hoc networking (such as service discovery, cross-layer optimisation, mobility, etc.). A comprehensive reference list follows each chapter serving as a good starting point for individual research. Some relevant projects are also listed at the end of each chapter. The book is organized into nine chapters. Different chapters are dedicated to particular issues aiming to complete the existing picture of wireless ad hoc networking.

Chapter 1 is the introductory chapter and it gives an overview of the ad hoc networking paradigm, as well as a glance of history behind briefly presenting some of the most relevant technologies and standardization efforts in the area.

Chapter 2 addresses the most advanced and specific areas relevant to ad hoc networks that are shaping the new approach and the new understanding of their functionality and design such as topology formation, capacity bounds and problems of self-organization in cooperative ad hoc networking. Topology issues are extremely relevant to infrastructureless ad hoc networking and they rely on several procedures: neighbor discovery, topology organization and topology reorganization. The limited communication ranges of wireless nodes in ad hoc networks, as well as the number of the participating nodes, establish new capacity bounds for ad hoc networks. This chapter gives a brief overview on the ongoing research and emerging solutions necessary to understand network planning and design. It also surveys self-organization and cooperativeness of ad hoc networks which are gaining momentum in understanding of the ad hoc phenomena and providing new possibilities in areas such as routing, security, management, etc.

Chapter 3 outlines the most relevant aspects and specifics of physical and MAC layer approaches in ad hoc networks. It also gives a classification of routing protocols briefing their main characteristics. The chapter aims to give a broad understanding of this important part of ad hoc networks design through which a variety of system characteristics and performances can be created.

Chapter 4 enlightens the cross-layering paradigm. Cross-layering (CL) is proven as a successful design tool in ad hoc networking which can be beneficiary to network performances. The chapter presents the major ideas behind different cross-layering approaches and offers some layer-related CL solutions. It also takes in consideration cautionary perspectives on cross-layer system design.

Chapter 5 outlines the general concepts in achieving QoS guarantees in wireless ad hoc networks. It covers issues such as definitions of novel QoS frameworks, which unify different techniques under one umbrella building a system approach that can adopt to network dynamism and unstable resources in a satisfactory manner. Examples of MAC layer QoS solutions as enhancements, as well as link-layer based solutions for QoS support, illustrate these efforts.

Chapter 6 deals with problems and solutions related to service discovery in ad hoc wireless networks, a topic which lately attracts a lot of attention imposing new challenges in front of network designers. Evolving service discovery protocols have to manage a large scale environment and resource constraints of present and future devices. Efficient and flexible service discovery is a crucial feature of ad hoc networks towards seamless communications and their integration into future wireless systems. This chapter gives an overview and definitions of basic entities and functionalities of service discovery in ad hoc mobile networks. It presents possible architectures and explains implemented mechanisms in different service discovery protocols.

Chapter 7 covers mobility, which is another demanding and challenging issue in wireless ad hoc networking. Mobility influences network dynamism, topology and connectivity. The chapter gives a survey of the most important existing mobility models (e.g., traces or synthetic models) that can mimic the movements of real mobile nodes. It presents several relevant protocol independent and protocol dependent mobility metrics. It also explains the impact of mobility on different aspects of wireless ad hoc networking, such as capacity and cooperation, energy conservation, security and query resolution. Mobility is an important issue which is in focus of the worldwide research community paving the path towards seamless communications.

Chapter 8 gives an overview of the security aspects in wireless ad hoc networking. Security is gaining attention and has become a primary concern in attempt to provide secure communication in a hostile wireless ad hoc environment. This chapter gives an introduction to security peculiarities, challenges and needs in a wireless ad hoc domain. It presents possible threats and attacks, introduces the most important security mechanisms and explains several security architectures.

Chapter 9 is the concluding chapter aiming to highlight the role of ad hoc networks in the future wireless systems. This chapter discusses recent advances in the area of wireless networking, addresses the foreseen all IP solutions and aspects relevant to seamless communications towards 4G.

References

[1] Wu, J. and Stojmenovic, I., "Ad Hoc Networks," *IEEE Computer Magazine*, February 2004.

[2] Prasad, R. and Gavrilovska, L., "Personal Area Networks," keynote speech, *Proceedings of EUROCON*, Bratislava, Slovakia, 2001.

[3] Gavrilovska, L. and Prasad, R., "B-PAN — a New Network Paradigm," in *Proceedings of the 4th International Symposium on Wireless Personal Multimedia Communications — WPMC 2001*, Aalborg, Denmark, 9–12 September 2002.

[4] Prasad, R. and Gavrilovska, L., "Research Challenges for Wireless Personal Area Networks," keynote speech, *Proceedings of the 3rd International Conference on Information, Communications and Signal Processing (ICICS)*, Singapore, 2001.

[5] Norismaa, M., "Wireless World Research Forum: WG3 and WG4," *Research Seminar on Middleware for Mobile Computing*, Helsinki University of Technology, Spring 2002.

[6] Van Dam, K., Pitchers, S. and Barnard, M., "From PAN to BAN: Why Body Area Networks," *WWRF meeting*, Helsinki, Finland, May 2001.

[7] Mohapatra, P. and Krishnamurthy, S., "Ad Hoc Networks: Technologies and Protocols," *Springer Science and Business Media, Inc.*, Chapter 1, 2004.

[8] Benton, C., Kenney, J., Nitzel, R., Blidberg, R., Chappell, S. and Mupparapu, S., "Autonomous Undersea Systems Network (AUSNet) — Protocols to Support Ad-Hoc AUV Communications," *IEEE/OES AUV2004: A Workshop on Multiple Autonomous Underwater Vehicle Operations*, Sebasco Estates, Maine, June 2004.

[9] Srinivasan, V., Nuggehalli, P., Chiasserini, C. F. and Rao, R. R., "Cooperation in Wireless Ad Hoc Networks," *IEEE INFOCOM 2003*, San Francisco, 2003.

[10] Basagni, S., Conti, M., Giordano, S. and Stojmenovic, I., *Mobile Ad Hoc Networking*, IEEE Press, Wiley-Interscience, 2004.

[11] Gavrilovska, L. and Atanasovski, V., "Ad Hoc Networking Towards 4G: Challenges and QoS Solutions," *7th International Conference on Telecommunications in Modern Satellite, Cable and Broadcasting Services — TELSIKS 2005*, Nis, Serbia and Montenegro, September 28–30, 2005.

[12] Bodanese, E., Gavrilovska, L., Rakocevic, V. and Stewart, R., "Eliminating the Communication Black Spots in Future Disaster Recovery

Networks," *Proceedings of the 8th International Symposium on Wireless Personal Multemedia Communications — WPMC 2005*, Aalborg, Denmark, September 2005, 1930–1934.

[13] Utkovski, Z. and Gavrilovska, L., "On the Communication Issues in Emergency Networks," *7th National Conference with International Participance ETAI'05*, Ohrid, Macedonia, September 2005, T-18–T-23.

[14] Freebersyser, J. A. and Leiner, B., *A DoD Perspective on Mobile Ad Hoc Networks, Ad Hoc Networking*, Addison Wesley, 2001.

[15] Fifer, W. and Bruno, F., "The Low-Cost Packet Radio," *Proceedings of the IEEE*, 75(1), January 1987, pp. 33–42.

[16] Shacham, N. and Westcott, J., "Future Directions in Packet Radio Architectures and Protocols," *Proceedings of the IEEE*, 75(1), January 1987, pp. 83–98.

[17] Garcia-Luna-Aceves, J. J., Fullmer, C. L., Madruga, E., Beyer, D. and Frivold, T., "Wireless Internet Gateways (WINGS)," *IEEE MILCOM'97*, Monterey, California, November 1997.

[18] IEEE 802.11 Working Group for Wireless LANs, http://grouper.ieee.org/groups/802/11

[19] Toh, C. -K., *Ad Hoc Mobile Wireless Networks: Protocols and Systems*, Prentice Hall PTR, 2002.

[20] IETF Mobile Ad-Hoc NETworks (MANETs) Charter, http://www.ietf.org/html.charters/manet-charter.html

[21] ETSI BRAN HIPERLAN1/2 Specifications, http://portal.etsi.org/bran

[22] Bluetooth Core Specification Version 1.2, November 2003, http://www.bluetooth.org

[23] Siva Ram Murthy, C. and Manoj, B. S., *Ad Hoc Wireless Networks: Architectures and Protocols*, Prentice Hall Communications Engineering and Emerging Technologies Series, 2004.

[24] Aggelou, G., *Mobile Ad Hoc Networks: From Wireless LANs to 4G Networks*, McGraw Hill, 2005.

[25] Korhonen, J., "HIPERLAN/2", 1.11.1999, http://www.tml.tkk.fi/Studies/Tik-110.300/1999/Essays/hiperlan2.html

[26] My personal Adaptibe Global NET, MAGNET (IST-507102), http://www.ist-magnet.org

[27] Zimmerman, T. G., "Personal Area Networks (PAN): Near-Field Intra-Body Communication," M.S. thesis, MIT Media Lab., Cambridge, MA, 1995.

[28] ZigBeeTM Alliance, http://www.zigbee.com

[29] The broadband wireless access home page, http://www.broadband-wireless.org/home.htm

[30] ETSI HiperMAN Specifications, http://portal.etsi.org/radio/hiperman/hiperman.asp

[31] Akyildiz, I. F., Wang, X. and Wang, W., "Wireless Mesh Networks: A Survey," *Computer Networks Journal (Elsevier)*, March 2005, pp. 445–487.

[32] IEEE 802.16 Working Group on Broadband Wireless Access Standards, http://grouper.ieee.org/groups/802/16

[33] Lin, Y. D. and Hsu, Y. C., "Multi-Hop Cellular: A New Architecture for Wireless Communications," *IEEE INFOCOM 2000*, March 2000.

[34] Ananthapadmanabha, R., Manoj, B. S. and Siva Ram Murthy, C., "Multi-Hop Cellular Networks: The Architecture and Routing Protocol," *IEEE PIMRC 2001*, October 2001.

[35] Wu, H., Qiao, C., De, S. and Tonguz, O., "Integrated Cellular and Ad Hoc Relaying Systems: iCAR," *IEEE Journal on Selected Areas in Communications*, 19(10), October 2001.

[36] Zadeh, A. N., Jabbari, B., Pickholtz, R. and Vojcic, B., "Self-Organizing Packet Radio Ad Hoc Networks with Overlay," *IEEE Communications Magazine*, 40(6), June 2002.

2

Basic Concepts

2.1 Introduction

Although not a new topic, ad hoc wireless networks continue to provide new research horizons. According to their basic characteristics, infrastructureless and dynamic topology, sensitive wireless links and energy aware communication, they initiated many research areas that cover basic concepts, which are main building blocks in their architecture. Such areas are: design of physical/MAC layer, routing, energy awareness, scheduling, middleware services, security, resource management etc. Some of them will be covered in the following chapters. However, these areas do not completely reflect the peculiarities of ad hoc networking. Infrastructureless communication of ad hoc nodes gives them freedom to equally participate in network organization and management. The concept of self-organization was developed, in order to catch up with this specific behavior.

This chapter aims to cover the most advanced and specific areas relevant to ad hoc networking that are shaping new approach and understanding of their functionality and design: capacity of ad hoc wireless networks and self-organization issues. Topology formation, as a basis for better understanding of the ad hoc structures is also briefly overviewed. There are similar topics covered in different areas, but approached from a different viewpoint and with a different level of abstraction. It all contributes to a conclusion that networking is not only a physical design, but it also incorporates philosophy of communication.

2.2 Topology Formation

Topology issues are one of the basic aspects of ad hoc networking. The absence of fixed links, wireless connectivity and dynamism of ad hoc nodes promote topology of ad hoc wireless networks into specific paradigm to which many researchers lately dedicated their attention. Topology information, as important ad hoc networks profile, is often incorporated in different procedures such as routing, security, power saving, etc. It is also extensively used in self-organizing processes in ad hoc networking.

Ad hoc networks must be able to fulfill topology control to maintain network connectivity, optimize network lifetime and throughput and make it possible to design power efficient routing. Several procedures are relevant for topology control: neighbor discovery, topology organization and topology reorganization. Following section briefly covers major concepts concerning topology formation in ad hoc networks.

2.2.1 Neighbor Discovery

During the *neighbor discovery* phase, each node investigates its neighbors and maintains the gathered information in appropriate data structures. The investigation can be performed with periodic transmission of short packets called *beacons* ("hello" packets to advertise itself and discover other nodes) or with promiscuous snooping on the channel in order to detect the neighbors' activities. The goal of neighbor discovery at each node is to determine the set of other nodes within direct *communication range* (within one hop), called also *topology knowledge range* (KR) [1], Fig. 2.1. The source S periodically sends the neighborhood discovery packet to all the packets within its chosen knowledge range (KR). The nodes within that range *(N1–N3)* receive the neighbor discovery packet and reply with the location update packets, which provide required information to continue with the topology formation (e.g., geographical position in a geographical routing scheme). The knowledge range can be also dynamically adjusted by each node as in PRADA [1], which results in fast convergence to a near-optimal solution.

The set of potential neighbors either includes all physical neighbors within the communication range or the neighbors' graphs from origin to destination can be chosen according to some neighbor selection criterion [2]. Very often, the goal is to achieve shortest *average hop distance*, which is the average number of hops from source to destination. The neighbor selection depends upon certain factors, uncontrollable factors such as mobility, noise, weather, interference, etc., or controllable ones such as transmit power and antenna direction. The choice of the next to source *one-hop neighbor* is often combined with specific forwarding (routing) protocols (see more in Chapter 3). Fig. 2.2 shows several forwarding schemes: most forward within radius (MFR) scheme [3], which forwards the packet to the maximum progress neighbors, (e.g., node M, whose *progress* is $\overline{Sm}$); nearest forward progress (NFP), which forwards to minimum progress neighbor within the topology knowledge range of source S (e.g., node N) [4]; greedy routing scheme (GRS), which is based on the geographical distance and chooses as neighbor the node closest to the destination D [5]; compass selected routing (COMPASS), which selects node

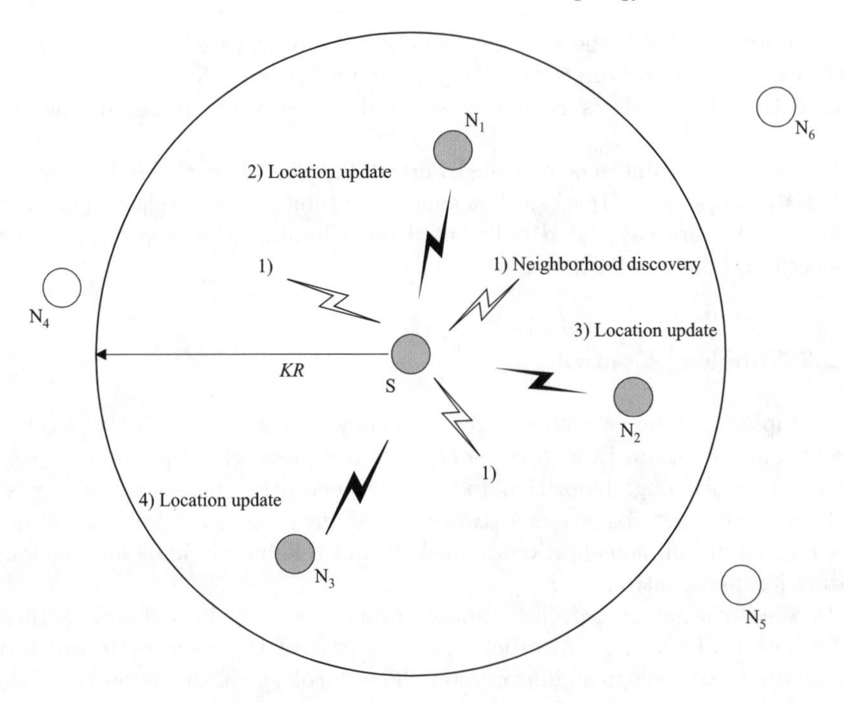

Fig. 2.1. Neighborhood discovery protocol

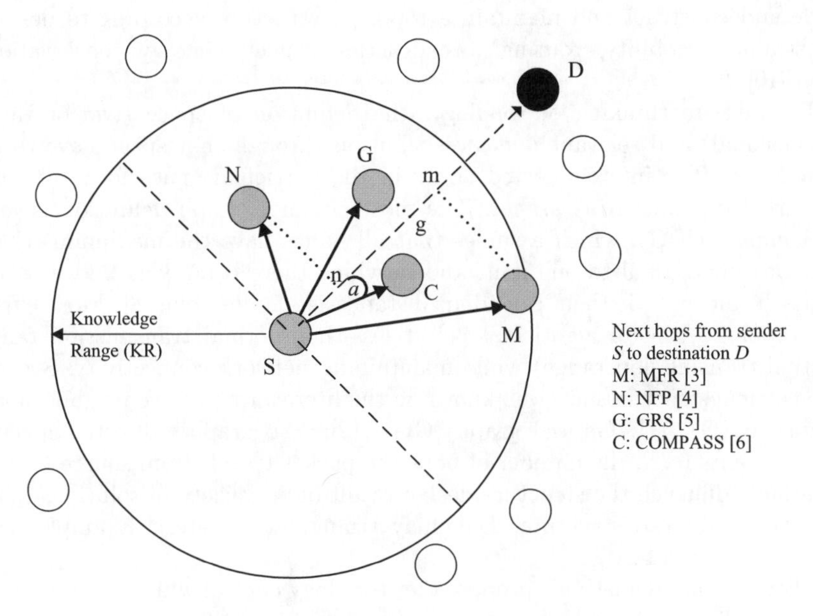

Fig. 2.2. Different forwarding schemes

C in order to achieve the smallest angle ∠CSD, providing that the direction SC is closest to the direction CD [6]; random process forwarding scheme, where the next node is randomly selected between the nodes in the KR range [7].

A novel forwarding scheme called Partial Topology Knowledge Forwarding (PTKF), proposed in [1] is based on a localized shortest path weighted routing, where routes are calculated only based on a limited local topological view concerning power link metric.

2.2.2 Topology Control

The topic of topology control or topology management lately receives significant attention. Emerging concepts of new networking paradigms (e.g., sensor networks), globalization (e.g., next generation networks, 4G, mesh networks and heterogeneous networks) and growing attention on ad hoc design, require implementation of efficient and reliable methods for topology formation and control.

In the *topology organization* phase, every node in the network gathers information about the entire network or a part of the network in order to maintain the topological information. The *topology* of the network is the union of L_i, where L_j represents the set of links discovered by a node i. The efficiency of a communication network depends significantly on its topology and can benefit of the appropriately chosen *topology control* mechanisms [8, 9]. The topology control algorithms select the transmission range of a node and construct and maintain a topology structure according to desired constraints (mobility, routing, broadcasting aspects, energy conservation, etc.) [10].

To start up thinking of topology, the definition of space (two or three dimensional) and relevant nodes distribution through that space have to be cleared up. The interconnected nodes in the particular space form geometric structures known as *graphs.* The most common graph definition is *unit disk graph* (UDG), which assumes that all nodes have the maximum transmission range equal to one unit, and in which there is an edge between two nodes if and only if their Euclidian distance is at most one [8]. Most existing topology control algorithms select less-than-normal transmission range (actual transmission range) while maintaining network connectivity. Several geometrical structures are well known in the literature: relative neighborhood graph (RNG) [11]; Gabriel graph (GG) [12]; Yao graph [13], etc. Specific requirements (e.g., the number of hops the packet travels from source to destination) influence the structure design resulting in variety of solutions (sink structure, YaoYao structure, Dalaunay triangulation, etc.). Examples can be found in [8, 14].

There are many methods proposed for topology control which can be classified according to different criteria. Considering the way how topology control is managed, the proposed algorithms can be categorized into several groups: *centralized* methods, *distributed* methods and *localized* methods.

Two most general approaches to topology control in ad hoc networks are: *hierarchical topology* (clustering) organizations [15] and *power-based control* (adjust power on a per-node base) organizations [10, 16]. Elements of previously mentioned methods are incorporated in both approaches in a *static* or a *dynamic* manner.

The following text covers the clustering and the power-based control approaches as major representatives of topology control techniques. Incorporated methods cannot be sharply distinguished in both approaches. Additional subsection is dedicated to the localized algorithms as lately emerging topology control techniques that best suite the dynamic ad hoc networking.

2.2.2.1 Clustering Approach

The clustering approach consists of *clusterheads selection* (or leader selection/reselection), definition of gateways and constitution of a connected (virtual) backbone in an optimized manner using different heuristics. The topology control mechanisms aim to weed out redundant and unnecessary topology information and to achieve better network behavior and higher stability. The clustering approach is related to methods developed in graph theory, such as the *minimum* dominating set problem and the relevant *minimum connected dominating set* (MCDS) problem [17].

The MCDS approach defines a minimum subset of nodes in the original graph such that the nodes compose a dominating set of the graph and the induced subgraph of an MCDS has the same number of connected components as the original graph. The MCDS is also a well-known NP-complete problem [18] and sub-optimum solutions must be used to approximate the optimum solution. Such sub-optimum solution, presented by a connected dominating set (CDS), is proposed in [10].

The dominating set in graph theory defines a graph where all nodes are classified into two categories: dominating set (clusterheads or dominators) and hosts (dominatees). In general, a *dominating set* (DS) of a network is a subset of nodes such that each node is either in the dominating set or is a neighbor of a node in the dominating set. In a *connected dominating set* (CDS) the dominating nodes form a connected subgraph of the network. Fig. 2.3 depicts two examples of possible dominating sets in an ad hoc network.

The dynamic topology of ad hoc networks makes computing the "minimum" dominating set impossible. This problem is known to be NP-hard, even when the complete network topology is available [10]. Alternative *minimal dominating set* (MDS) was proposed, based on various heuristics, that can guarantee a local minimum election of the dominators in a particular number of steps [10, 19].

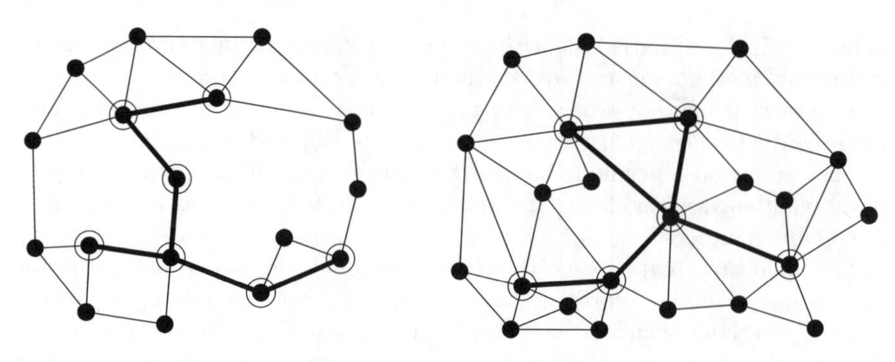

Fig. 2.3. An example of connected dominating set

Although wireless ad hoc networks have no physical backbone infrastructure, a *virtual backbone* can be formed by nodes in a connected dominating set. The MDS is usually proposed as a solution for a clusterhead selection, while the CDS method presented a union of clusterheads and gateways.

Clusterheads can be elected via non-deterministic negotiations or by applying deterministic criteria. The *non-deterministic* negotiation requires multiple increment steps in order to avoid the election jitter and attain minimum conflicts between the clusterheads in its one-hop neighborhood. Some examples of this approach are: CEDAR [20], the spanning tree algorithm [17], SPAN [21], etc. Clustering algorithms can build clusters within D-hops (D-clustering) [19] using Max-Min Leader Election and based on D-hops dominating set. They aim to achieve smaller clusterhead density and redundant backbone, good run time, better balancing among clusterheads, fairness and stability.

The *deterministic* criteria can determine the clusterheads in a single round. Different heuristics are used in *cluster election* process. Deterministic approach is implemented in the following algorithms: Lowest ID (the lowest node identifier in one-hop neighborhood including itself is used to elect MDS) [19, 22]; Max Degree (highest degree in one-hop neighborhood) [20, 23]; MOBIC (received signal strength variation from one-hop neighbors) [24], Load Balance (based on virtual identifier (VID), assigned to each node, that runs a budget reflecting the energy status) [25], etc.

Authors in [10] propose topology management by priority ordering (TMPO) algorithm that combines the MDS and CDS approaches into an algorithm that offers better load balancing capabilities and higher topology maintenance stability. Fig. 2.4 illustrates the simplified backbone topology construction using TMPO. The network graph is generated by randomly placing 100 nodes over a $1000 \times 1000\,\mathrm{m}^2$ area. In (a) all nodes are hosts, in (b) clusterheads are elected (MDS is formed, but not connected) and in (c) gateways are inserted (CDS is formed).

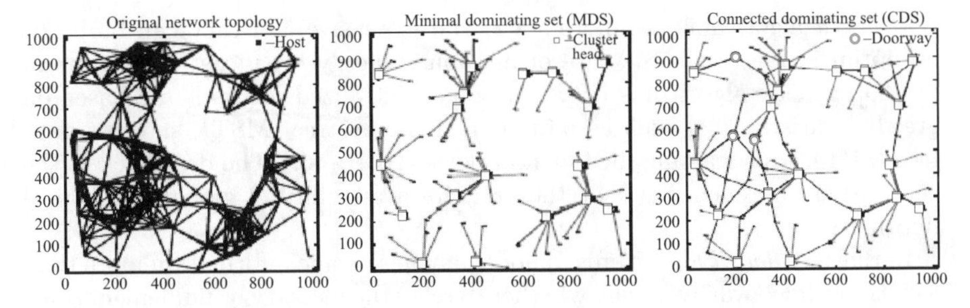

Fig. 2.4. A network topology control example [10]

The effect of clustering is obvious. The election process involves dynamic control mechanisms that incorporate the three key factors for topology management of ad hoc networks: node battery life, mobility and load balancing.

2.2.2.2 Power-Based Control

The energy conservation is extremely important in wireless ad hoc networks since each node has battery supply to power its processing capabilities [16]. Transmission from node to node requires power consumption during: source preparation to transmit, link power support during transmission (path loss) and receiver processing power to receive, store and process the signal. The *path loss* between nodes u and v is $\|uv\|^{\beta}$, where $\|uv\|$ is the Euclidian distance between u and v, and β is a real constant between 2 and 5 depending on the transmission environment. It is usually considered as a major part of power consumption and presents the power cost metric.

The power-based control usually considers some minimization objectives and connectivity constraints and can be classified in non-adjustable and adjustable methods. If the power consumption per node is *non-adjustable*, meaning that the transmission power of every node is equal and is normalized to one unit, the problem of minimizing the total power is equivalent to the MCDS problem , i.e., the minimization of the number of nodes that relay the message, since all relaying nodes form a connected dominating set (CDS) [15]. In the *adjustable* algorithms, the optimization criteria are applied to minimize the maximum of the total transmission power. Interesting dynamic power schedules are proposed in [26], where two approximation heuristics are introduced and the lifetime of a dynamically adjusted multicast connection is made several times longer than in the static assignment. A comprehensive study of centralized power aware topology control algorithms can be found in [8, 15].

In the algorithms for power based control, implementation of a centralized or a distributed approach has produced a variety of proposals. The proposed *centralized* algorithms are more static and expensive and they can be based on optimizations as in the case of minimum energy broadcast tree proposal [27–29], genetic algorithms [30], etc. Some centralized methods are based on greedy heuristics [31] such as minimum spanning tree (MST), shortest path tree (SPT), broadcasting incremental power (BIP), etc. The drawback of the centralized approaches is that they require availability of global topological information.

In the *distributed* algorithms, a node may need some information more than a constant hop away to decide whether to relay the message or implement some optimization criteria. In the localized methods, each node has to maintain the state of its local neighbors (within some constant hop). The concept of *localized* algorithms was proposed as distributed algorithms where simple local node behavior, based on local knowledge, achieved desired global objectives.

2.2.2.3 Localized Algorithms

Localized topology control does not require global topology information. Based on geometric nature of the multi hop ad hoc wireless networks it promotes the idea of localized routing, i.e., the decision to forward a packet based only on local information [32]. Elimination of the communication intensive tasks of updating the routing tables in localized routing protocols can offer mobility support. The localized energy efficient broadcast can be supported by implementation of directional antennas [33, 34]. Localized methods for topology control can also include the clustering approach and developed methods based on distributed CDS [10], localized low weight structures, area-based [35] and graph-based [36] decision, flooding, probabilistic schemes, etc. Methods for dynamic selection of dominating nodes, active scheduling and exploitation of different modes for energy consumption (active, idle, sleep periods) are also exploited in proposed algorithms [37, 38].

Existing localized topology control algorithms can be grouped as:

- *RNG-based algorithms.* The RNG (Relative Neighborhood Graph) is a geometrical graph used to remove edges (i.e., reduce the number of neighbors) while maintaining the network connectivity. In localized topology control protocols each node determines its logical neighbor set based on location information of 1-hop neighbors. Two nodes are logical neighbors if and only if edge (u, v) exists in RNG. An edge (u, v) is removed if there exists a third node w such that $\mathrm{d}(u, v) > \mathrm{d}(u, w)$ and $\mathrm{d}(u, v) > \mathrm{d}(v, w)$ where $\mathrm{d}(u, v)$ is the Euclidean distance between u and v.
- *Minimum energy algorithm.* Each node collects location information of nodes within a small *search region* to control message overhead [39, 40]. The search region is adjusted to cover the entire normal 1-hop neighborhood. Algorithms in this group reduce the number of edges using the transmission power as edge cost. They try to maintain the network connectivity preserving the minimum energy paths.

- *Cone-based algorithms.* The cone-based topology control algorithms (CBTC), [41, 42], use dynamic search regions (*cones* — k-th parts of disk centered at originating node) to reduce control overhead. Several optimizations are proposed, using different cones, reducing the number of neighbors in transmission range [34] or providing k-connectivity.
- *LMST-based algorithm.* The local minimal spanning tree (LMST) is build at each node based on 1-hop location information and selects neighbors in MST as local neighbors [43]. The maximum number of local neighbors is six.

Special attention in topology structures is dedicated to broadcasting and multicasting mechanisms. They are involved in many applications in ad hoc wireless domain. Broadcasting is often referred as one-to-all communication model. In local broadcasting, a distributed broadcast protocol is based on solely local state information. Specifics of broadcasting topology control can be found in [10, 44]. The localized energy efficient broadcast can be supported by implementation of directional antennas [33].

The dynamic nature of ad hoc wireless networks is better covered by localized algorithms. The concept of localized algorithms, which are based on local knowledge, allows scalability and achieves desired global objectives. However, not all structures can be constructed locally (e.g., MST). A comprehensive study of localized topology control algorithms can be found in [32, 34].

2.2.2.4 Probabilistic Algorithms

Probabilistic algorithms [45–47] adjust transmission range to maintain the optimal number of neighbors that balances energy consumption, contention level and connectivity. The probabilistic methods cannot provide a hard guarantee on the network connectivity [48, 49]. However, the problem of *critical power* (the universal minimum power used by all wireless nodes such that induced wireless topology is connected) and fault tolerance within a certain topology can be approached using probabilistic mechanisms.

2.2.3 Topology Reorganization

In *topology reorganization* phase, the ad hoc networks require the updating of the topology information. The topology changes in dynamic ad hoc networks can occur due to mobility of nodes, failure of nodes, complete depletion of power sources of nodes, new activated nodes, etc. The reorganization is performed in two phases: periodic or aperiodic exchange of topological information and adaptability (recovery from major topological changes in the network). The reorganization should also cover the network portioning and merging [8].

2.2.4 Summary

Topology formation is a basic topic in understanding the ad hoc network structures. Principles of topology formation, discovering and some existing topology structures are closely related to routing (see Chapter 3), power awareness (see Chapter 4) and security issues (see Chapter 8). They also play an important role in self-organization.

Specific networking technologies apply different topology formation algorithms according to appropriate networking paradigm. The Bluetooth technology concerns the piconet and scatternet topology formation as a major problem [50–54]. MANET is mainly occupied with efficient broadcasting topology structures [55]. The lately emerging concept of sensor networks also dedicates significant research attention to topology formation aspects [56].

2.3 Capacity Bounds for Ad Hoc Networks

Ad hoc networks are wireless networks without any fixed base station or wireline backbone infrastructure. The nodes use peer-to-peer packet transmission and multihop routes to communicate with each other by purely using the remaining ad hoc nodes as their relays. Since every node all over the domain shares whatever portion of the channel it is utilizing with nodes in its local neighborhood, capacity constriction occurs. The capacity constraints of wireless ad hoc networks have been a major research interest since the landmark paper of Gupta and Kumar [57].

The key reasons why the overall capacity is reduced are:

- *Interference* in the zone around the receiver node;
- "*Forwarding burden*" — as the number of nodes increases, the number of hops increases and the nodes spend a larger fraction of their capacity for relaying other nodes' traffic.

The capacity of wireless ad hoc networks also depends on *network size* (i.e., number of nodes) and *traffic patterns.* The network's per node capacity can determine the scaling to large networks. The average *distance between the source and destination* can also influence the capacity.

Gupta and Kumar have proposed a model for studying the capacity of *static* ad hoc networks, where n immobile nodes are located in a region of 1 m^2 (either on the surface S^2 of a three-dimensional sphere or in a disc in a plane) and with the use of omni-directional antennas. Each node can transmit at W bit/sec over a common wireless channel. The channel can be broken into several sub-channels of capacity W_1, W_2, $\ldots W_M$ bit/sec, where $\sum_{m=1}^{M} W_m = W$. However, this does not influence the results.

Table 2.1. Characteristics of arbitrary and random networks

Characteristics \ Network type	Arbitrary network	Random network
Node's location	Arbitrary	Random, i.e., independent and uniformly distributed
Destination for each node	Arbitrary	Random
Traffic pattern	Arbitrary	Random
Range of power level per transmission node	Arbitrary	Homogeneous (same nominal range of power)

Two capacity metrics are introduced:

- *Transport capacity*, defined as the total bit-distance product per second that can be transported by the network, and
- *Throughput capacity*, defined as the maximum common throughput that can be provided to each node with a randomly chosen destination.

The network transports one *bit-meter* when one bit has been transported for a distance of one meter towards its destination. If the area of the domain is $A\,\text{m}^2$, rather than the normalized $1\ \text{m}^2$, then all transport capacity results should be scaled by $\sqrt{A}$.

Gupta and Kumar consider two types of networks, *Random Networks* and *Arbitrary Networks*. Random Networks basically imply statistical nodes distribution (i.e., uniform and independent), while Arbitrary Networks means that the distribution of the nodes could be arranged to achieve best results in terms of capacity. The specifics of both networks are shown in Table 2.1.

2.3.1 Communication Models and Implications

Gupta and Kumar in [57] have analyzed the capacity of wireless networks in general, spreading the results towards ad hoc network design. Involving elements of information theory in their investigations they have tried to define the boundary behavior of capacity (transport and throughput capacity) of the wireless networks paving the road for many following authors' deeper research under more realistic scenarios [58, 59]. Gupta and Kumar propose two different models of communication in order to catch up the organization of the network area, the node distribution across that area, the space geometry,

Table 2.2. Upper capacity bounds under Protocol and Physical Models

Model / Network type	Protocol model	Physical model
Arbitrary network	$CW\sqrt{n}$ (bit-meters/sec)	$C'Wn^{\frac{\alpha-1}{\alpha}}$ (bit-meters/sec)
Random network	$\Theta\left(\frac{W}{\sqrt{n \log n}}\right)$ (bit/sec)	$\Theta(\frac{W}{\sqrt{n}})$ (bit/sec)

Note:
O, Θ and Ω are Knuth's operators, C and C are deterministic constants independent of n and W [57]

the communication behavior, the power range and certain limitations. These models are:

- The *Protocol Model*, which is a more restrictive model and it requires that each receiver node lies outside the interference region of every other transmitter node. As a non-interference model, it requires each transmitter transmitting to a receiver at distance r forming an interference region consisting of a disk of radius $(1 + \Delta)r$ centered around itself. The region around a node with radius $\Delta\, r/2$ is called *exclusion region.*
- The *Physical Model*, requires that the signal to interference plus noise ratio (SINR) is below predefined threshold β. The signal power is assumed to decay as $r^{-\alpha}$ with distance r, for $\alpha > 2$ [57].

The major achieved results for upper bounds of the capacity capabilities for the previously mentioned networks and models are presented in Table 2.2.

Concerning Arbitrary Networks, the listed results for a Protocol Model are achieved under optimal conditions (nodes optimally placed, traffic pattern and range of each transmission optimally chosen). In addition, if the transport capacity is equally divided between all n nodes, then each node will obtain $\Theta(W/\sqrt{n})$ bit-meters/sec. If the source-destination pair distance is about 1 m away, then each node will obtain a throughput capacity of $\Theta(W/\sqrt{n})$ bit/sec. The increase of number of nodes in order to increase the overall transport capacity leads to inconvenient implication of dramatically decreasing the throughput capacity. The results also show that transport and throughput capacity improve when α is larger (for the Physical Model), i.e., when the signal power decays more rapidly with distance.

When nodes need to communicate only with nearby nodes (at destination of $O\left(\frac{1}{\sqrt{n}}\right)$), it is obvious that all nodes can transmit data to nearby neighbors at a bit rate that does not decrease with n. Under physical model, the lower bound on capacity is the same as for the protocol model. More details about the particular capacity bounds and model definitions are available in [57]. The results in this paper are obtained under some restrictions. The nodes are not mobile and a perfect scheduling algorithm is considered. The mobility of the nodes and the unknown traffic demands can additionally influence the capacity.

There are several feasible scenarios suggesting improvements in capacity based on different features such as communication only with nearby nodes, limitations in signal power, clustering, involvement of directional antennas, construction of hybrid nodes, etc. Major ideas behind these scenarios are elaborated in the following text.

2.3.2 Capacity Improvement

Several implications follow from the capacity analysis in [57] pointing out the directions of ad hoc networks design development. Many authors propose and analyze the influence of different solutions in order to improve the capacity. This modern approach leads to design of more efficient and scalable ad hoc networks. Different mechanisms concerning nodes organization and cooperation, as well as overall system, may influence the capacity bounds. In the scenario when nodes communicate only with their neighbors, analysis in [57] shows that all nodes can transmit data to nearby nodes at a bit rate that *does not decrease with n*. Such scenario can occur in collections of "smart homes", where larger number of sensors and actuators communicate by wireless means.

Gupta and Kumar pointed out that the aggregate throughput capacity becomes $\Theta\left(\sqrt{\frac{n+m}{\log(n+m)}}W\right)$ if m additional homogeneous nodes are deployed as pure relays in random positions. Therefore, the addition of kn nodes, that would serve as pure relays, provides a less than $\sqrt{k+1}$ – fold increase in throughput [57, 60]. Note that there are no wire connections between the relay nodes. More appropriate method to increase the capacity is offered through infrastructure support, i.e., deployment of hybrid networks.

The capacity analysis shows that it could be beneficial if ad hoc nodes are grouped into clusters and each cluster designates one specific node for relaying functionality and taking care of all burden of relaying and multi-hop packets. It will reduce the transmission power consumed by vast majority of other nodes.

Cooperation between nodes in a cluster can play significant role in achieving higher throughput capacity. Authors in [61] analyzed the cooperation between the nodes in transmitting and receiving clusters. The clusters contain nodes separated with small distances which allows joint encoding (in the transmitting cluster) or decoding (in the receiving cluster) of their messages. It is shown in [61] that the cooperation between the nodes in the transmitting cluster performs significantly better and increases the throughput for more than 50% approaching MIMO (Multiple Input Multiple Output) upper bound for high power gain. The cooperation at both sides is only beneficial in very high power gains.

Several features that can improve the ad hoc wireless network capacity bounds are analyzed in the following subsections.

2.3.2.1 Improving the Network Capacity Bounds by Infrastructure Support

Hybrid architectures contain number of base stations connected by a high-bandwidth wired links [60]. The effects on capacity were investigated when the number of base stations was m and the number of ad hoc nodes n and each of them was capable of transmitting at W bit/sec over the common wireless channel. The aggregate capacity depends on relations between m and n. So, under a *deterministic* routing strategy, if m grows slower than $\sqrt{n}$, the maximum aggregate throughput capacity is $\Theta\left(\sqrt{\frac{n}{\log \frac{n}{m^2}}}W\right)$, which is insignificant growth. However, if the number of base stations m (under the k-th nearest–cell routing strategies) grows faster than $\sqrt{n}$, the maximum throughput capacity scales as $\Theta(mW)$, which increases linearly with the number of base stations [58, 60, 62]. The boundary for m growth is $\sqrt{\frac{n}{\log n}}$, under *probabilistic* routing strategies, when a transmission mode is independently chosen with certain probability for each source destination pair. If m grows slowlier, then the maximum throughput capacity is $\Theta\left(\sqrt{\frac{n}{\log n}}W\right)$, which is of the same asymptotic behavior as a pure ad hoc network without any benefit. However, if m grows faster than $\sqrt{\frac{n}{\log n}}$, the maximum throughput capacity scales as $\Theta(mW)$, which means that it increases linearly with the number of base stations. The authors in [58] underscore the importance of choosing the minimum power level of communication and suggest that simply communicating with the closest node or base station could yield good capacity. They proved that in case of a random network with n wireless nodes and kn base stations ($k > 0$), allowing nodes to perform power control, it is possible to provide a throughput of $\Theta(1)$ to a fraction $f (0 < f < 1)$ of nodes. The investment costs in hybrid networks must consider the number of base stations.

2.3.2.2 Improving the Capacity Bounds by Mobility

Grossglauser and Tse [63] consider ad hoc network of mobile nodes and show that the average long term throughput per node can stay constant. In their work the node movement process is ergodic with uniform stationary distribution over the network. The basic idea is to allow a source node to distribute packets to as many different nodes as possible. These nodes then relay the packets to the final destination whenever they get close to it. Therefore, the expected path length remains constant. However, the result depends on the movement model. Furthermore, the fixed throughput guarantee is achieved only over very long time frames. The result, nevertheless, suggests a way to take advantage of node movement when sending packets from applications that can tolerate long delays.

Several researchers have studied the delay incurred using the mobility to improve the capacity. Perevalov and Blum [64] approximately expressed the capacity as a function of the maximum allowable delay in an all-mobile network. Assuming physical model, they defined a critical value for delay (d) considering that:

- for values of d below critical the capacity does not benefit appreciably from the motion,
- for moderate values of d above critical the capacity increases as $d^{2/3}$,
- the dependence of a critical delay on the number of nodes n is a very slowly increasing function ($n^{1/14}$).

The *multiuser diversity*, based on the delay tolerance in mobile wireless networks [63], can additionaly improve the capacity. It is based on a rule that a high-speed wireless base station should provide coverage only to nodes in its proximity and not to ones which are far away. It means that the entire transmit power budget, for delay sensitive users, will be dedicated to a nearby node. The multiuser diversity differs from traditional technique of *path diversity* (based on multiple paths), because each packet is sent along only *one route* to take advantage of the closeness of the relay node. In such techniques, two hop routes are sufficient to achieve the maximum throughput capacity of the network within the interference limits. The total agregate throughput per source–destination (S–D) pair is approaching $\Theta(1)$. This supports the conclusion that throughput per user increases dramatically when nodes are mobile rather than fixed.

The multi-hop relaying protocols include the *delay* in packet delivery from source to destination node. The trade-off between the delay and capacity was addressed in several papers. Grossglauser and Tse [64] define an expected packet delay in 2-hop relaying algorithm as $\Theta(nT_p(n))$, where $T_p(n)$ is the packet duration. Sharma and Mazumdar in [65] show that any protocol that allows only nearest neighbor transmission incurs expected packet delay of $\Omega\left(T_p(n)\sqrt{n}\right)$. Using algebraic manipulations they developed general expression for capacity/delay trade-off. It shows that within the network with n mobile nodes (moving according to the random way-point mobility model (RWMM)) and n S–D pairs, each generating traffic at an expected rate of $\lambda(n)$ per time slot and considering the arbitrary control protocol, an expected paket delay of $E\{D(n)\}$ can be achieved when:

$$\frac{E\{D(n)\}}{\lambda(n)} \geq \Theta(nT_p(n)) \tag{2.1}$$

For static model [66], the optimal throughput-delay trade-off is $D(n) = \Theta(nT(n))$, where $T(n)$ and $D(n)$ are throughput and delay respectively. The mentioned results inspire questions about the fundamental limits of mobile ad hoc networks. It is challenging to investigate the mobility models which can support the mentioned trade-off.

2.3.2.3 Improving the Network Capacity Bounds Via Directional Antennas

The use of directional antennas reduces the interference area caused by each node, which can potentially increase the capacity of the network. Several works define the capacity limits, varing from theoretical to technology-based [59, 62, 67]. Yi et al. in [62] show that in a random wireless network, the use of directional antennas with beamwidth α for transmitters increases the capacity by a factor of $2\pi/\alpha$ and the use of directional antennas with beamwidth β for receivers can increase the capacity for a factor of $2\pi/\beta$. If both, transmitter and receiver have directional antennas, it can improve the capacity by a factor of $4\pi^2/\alpha\beta$. Effects of different types of direcional antennas and appropriate propagation model were investigated in [67]. Peraki and Servetto [68] show that even under conditions when interference is almost removed (transmitters generate arbitrarily narrow beams) and the transmission range is set to minimal possible to maintain connectivity, the capacity can only be improved by an order of $\Theta(\log^2(n))$.

2.3.2.4 Improving the Network Capacity with the Use of UWB

The most promising approach to improving the capacity bounds in wireless ad hoc networks is to employ unlimited bandwidth resources (spectrum), such as the ultra wide band (UWB). UWB is defined as any radio technology using a spectrum that occupies a bandwidth of least 500 MHz. UWB is most appropriate for short-range communication. By Shannon's capacity theory, UWB transmitters are capable of transmitting at rates from 100 Mbps to 500 Mbps with very low power. The capacity of UWB networks was investigated in [69] where upper and lower bounds to throughput capacity per node were obtained. However, Zhang and Hou in [70] show that under the limit case when B tends to infinity (B is the bandwidth in Hz), each node can obtain a throughput of $\Theta(n^{(\alpha-1)/2})$, where n is the density of the nodes and $\alpha > 1$ is the path loss exponent. Zhang and Hou assume that each link achieves the Shannon's capacity and use the theory of percolation to prove the lower and upper bound of the network capacity. The characteristics of UWB make it well suited for wireless sensor networks and, in addition, to wireless personal networks in "smart home" environments.

2.3.3 Summary

The capacity analysis involves more complex information theory, geometry and algebra, so the details are omitted from this text and can be found in the referenced papers. The previous brief overview of the ongoing research in capacity of ad hoc wireless networks underlines the importance of this research area and implications it may have on networking conditions. There is still a lot to be done towards more realistic scenarios. The ongoing research can reshape the understanding of wireless networks planning and design. New emerging networking paradigms, such as mesh and heterogeneous networks can build future structures upon it.

2.4 Self-Organization and Cooperative Ad Hoc Networking

Tremendous growth, variety and personalization of networking paradigm inevitably lead to new approaches in solving networking dilemmas. A fresh look in design and development of networks is offered by self-organizing approach, which minimizes the human intervention and aims to offer reliable, trustworthy, scalable and robust solutions.

Traditional approaches to ad hoc networks do not exploit certain characteristics such as cooperation and relationship between nodes. The ad hoc networking can bring a paradigm shift in the way the networks are organized and operated. Unlike many traditional solutions where the network environment is relatively stable (i.e., Internet or cellular networks), ad hoc networking solutions try to cope with unstable network situation and entrusted topology (links and nodes), because of different nodes' communication ranges, low power level of individual nodes, sensitive links, limited resources, etc. The main goal of self-organization within the area of communication and computer networks is to minimize the need for configuration, develop protocols that facilitate network operation and enable new types of communication networks, such as completely decentralized ad hoc and sensor networks.

The self-organized mobile ad hoc networks convey a promise of going a step further, entrusting users with the operational activities of the networks including resource management, security etc. Ad hoc nodes can act without any infrastructure (or providers) where all the nodes must act cooperatively and exchange the necessary information in distributive manner. The *self-organization* characteristic is a fundamental and more innovative aspect of ad hoc networks.

A prototype of large-scale self-organized mobile ad hoc networks is presented in the Terminodes project, where the nodes are called *terminodes* (terminal + node) [71]. It foresees the new solutions for infrastructureless communication, security and applications and presents the shift from personal communications (dominant so far) towards the communications between objects (sensor networks).

The following sub-section presents the idea of self organization and related small world phenomenon, explains communication and coordination aspects in ad hoc networking, and briefly looks at the design and policing issues of self-organizing networks.

2.4.1 Definition of Self-Organization

The concept of self organizing communication networks emerged in the 1990s [72]. Some relevant phenomena such as small world [73, 74] were introduced much earlier in social sciences. Until the late 1990s, random graph or network theory, which offered completely regular or random approach, were the main tools to study complex networks [75]. Today two approaches are most relevant in self-organization, one offered in statistical physics and the other in engineering [76]. However, it is difficult to declare a unique definition of self-organization.

Having in mind engineering approach, the self-organization can be defined as a property of certain dynamic mechanisms (structures, patterns and decisions) appearing on the global level of a system based on interactions among low level components. Within such system, rules that specify the interactions among systems' components are executed on the basis of local information, without reference to the global pattern. For example, the ad hoc nodes in military deployment experience the same level of cooperation, they depend on unified authority and must cooperate between each other. From the other hand, the civil ad hoc network installation can experience nodes that refuse cooperation (e.g., forwarding the received packets) and try to maximize their own benefit that they get from the network. Their activities depend on their own authority and influence the overall system characteristics [8]. The users in a self-organized system have to meet some constraints in order to allow the system to behave as users expect and to be able to correct erratic behavior. This includes definition and understanding both user and technical requirements of self-organization and potentially affects all system's layers (from physical, middleware to application). The complex system of self-organized networks is based on imposed properties or rules of artifacts or technological systems, often defined by human intervention.

2.4.1.1 The Small-World Concept

An interesting model for social relationship is represented by the *small-world phenomenon* [74, 77]. Stanley Milgram introduced this phenomenon completing experimental study in social sciences [74] in the 1960s. His experiments showed that the acquaintanceship graph connecting the entire human population has a diameter of six or less (the *six degrees of separation*). The small worlds phenomenon introduced a class of graphs – *small worlds graphs* (SWG), that have similar characteristics as small graph phenomenon: very large graphs that tend to be sparse, clustered and have a small diameter (the average path length between nodes). The behavior of a small world system differs from a behavior of a physical system. In the SWG, the local actions can have global consequences.

The small-world approach investigates nodes behavior, distinguishing between the connectivity achieved through the wireless links and the communication which is based on a relationship that does not always respect the local connectivity. A mobile ad hoc network is a community of users acting very similar to human communities and can be modeled with a SWG [75]. As a result, many novel topics in ad hoc networking consider this phenomenon in the analysis (e.g., mobility, security, QoS, etc.) (See chapter 6).

2.4.1.2 Communication in a Self-Organized Network

A mobile ad hoc network shows similarities to a social model [77] of human communities of users and collaborations and communications which can happen in the real life. The *communication* task is identified with the capability of delivering some information to the destination. The nodes' *collaboration* can involve different aspects, similar to real behavior of people. The social behavior of people or other communities (e.g., bees, ants) is often characterized by self-organization, optimization of communications and adaptation to the environment. A *community* is a structure that derives from individual's interactions in a shared environment. In all communities, a certain degree of cooperation is required to carry on all communications tasks. Cooperation can be built on a strong *hierarchy*, where the role of each member of the group is strictly assigned, or on spontaneous *altruism*.

Game theory considers the cooperation aspects in ad hoc networks [78] and can model the freedom of every node to choose cooperation or isolation. Basic unit of *game theory* is the game, consisting of three basic elements: description of strategic interactions between players, set of constraints on the actions the players can take and specification of the interests of the players. Applied in ad hoc domain, it can interpret the interactions, cooperation or even competition among nodes. Nodes try to optimize their own payoff and adopt cooperative behavior to obtain better network performance. In case of misbehavior of nodes, game theory can also provide punishment strategies.

2.4.2 Design Paradigm

The phenomenon of self-organization is pervasive in many areas (e.g., nature, statistic physics, sociology) following some general principles. In order to understand them, many researchers have observed this phenomenon in dynamic complex systems and tried to understand how and why these systems evolve from the interaction of simple entities. In complex communication network domain, in order to understand the self-organization, a constructive engineering approach requires designing the rules and protocols for interactions among the nodes. Prehofer and Bettstetter in [79] have recently proposed a design process for self-organized network function and defined four design paradigms for self-organized networking:

- Design local interactions that achieve global properties,
- Exploit implicit coordination,
- Minimize the maintained state,
- Design protocols that adapt to changes.

The problem of shaping a *localized* behavior requires design of localized rules that, if applied to all entities, automatically lead to the desired *global* behavior (or at least approximate it). The notation localized means that entities have only a local view of the network, and, interact only with their neighbors. Examples for such functionality can be found in unique addresses (solution in IPv6 without centralized server), connectivity, some topology control approaches (mentioned in previous text), clustering (using Basagni's algorithm), etc. The *divide-and-conquer* concept can be implemented when the global property can not be reduced to local one. In this approach the information is collected locally, aggregated and exchanged with other nodes (e.g., in the routing information protocol, RIP). If each node performs this process, global property can still be achieved without knowing the complete topology. Generally, even when the localized algorithms provide solutions only *close to the optimum*, they are useful, if they lead to fast convergence to a stable configuration. The locality helps the network to become stable and robust towards dynamic changes and failures, but also can introduce some inconsistency because of imperfect coordination (e.g., two nodes might have same IP addresses).

The *implicit coordination* means that the coordination information is not communicating explicitly by signaling messages, but is inferred from the local environment. The implicit coordination (together with conflict detection and resolution) provides mechanisms for better control over the available resources provided. The goal is to achieve resource and time efficient means of coordination among the nodes in self-organized networks. An extreme form of explicit coordination is *zero* coordination, which avoids coordination between nodes at all. To avoid resource conflicts it can be combined with *randomization* approaches (as in Aloha) along with timers. Many self-organized systems in nature also use randomization to initialize the system and recover from errors or deadlocks. The implicit coordination together with the local rules can achieve, in self-organized manner, the networking properties, which are usually achieved through global rules in conventional communication networks (e.g., global connectivity, global addressing, supervised resource management, etc.).

The amount of *long-lived state information*, which plays important role in traditional communication networks (e.g., configuration information, security databases, addresses of dedicated network entities) has to be minimized in self-organized networks in order to achieve a higher level of self-organization. The possible approach is employment of *discovery mechanisms*, involved to discover information about certain network entity or service. The state information achieved through discovery procedure needs to be accordingly refreshed, which makes it *not long-lived*. The on-demand routing protocols for ad hoc networking, which employ a broadcast for route search and unicast for reply, only maintain short-lived state which is updated on a regular basis (for more details see Chapter 3).

Another important design aspect is *adaptability*. The dynamic ad hoc networks can experience many unpredictable changes such as resource constraints, user requirements, node mobility and failures. They must be able to adapt efficiently without any centralized entity that could notify the nodes. Each node has to continuously monitor its local environment and react in appropriate manner. Authors in [79] define *three levels of adaptation*. Appropriate protocols must be able to deal with: failures and motilities (level 1), adapt to own parameters (e.g., value of timers, cluster size) in order to optimize the system performances (level 2) and trigger to alternative solution (e.g., switch from cluster-based routing to flooding) when employed mechanisms are no longer suitable (level 3).

Implementation of complex *learning algorithms* in dynamic ad hoc networks represent more advanced level of adaptation, which can effectively change algorithms in the nodes. Parunak and Brueckner, in [80], proposed a self-organizing approach to MANET management, based on *stigmatic learning* and self-organized systems of agents.

2.4.3 Self-Policing in Ad Hoc Networking

The behavior of self-organized networks is an evolutionary field that still needs to be completely understood and discovered. At the moment it is dominated by the policy-based approaches that additionally need the inbuilt policies and learning properties in order to change and optimize concrete user/system behavior. So, the input provided by users must be adapted to the way users think, and their intuitive concepts, and understanding of how systems should work.

In wireless ad hoc networks, nodes communicate with far-off destinations using intermediate nodes as relays. Wireless nodes are energy constrained and they might not always accept the relaying responsibilities. Nodes, which agree to relay traffic, but fail to do so, are termed as *misbehaving* nodes. Two mechanisms are proposed to handle this problem: *watchdog* and *pathrather*

[81]. The former is in charge of identifying the misbehaving nodes, and, the latter is in charge of defining the best route avoiding these nodes.

Node misbehavior due to selfish or malicious reasons (intentional) or faulty nodes (unintentional) can significantly degrade the performance of mobile ad hoc networks. The misbehavior can cause serious degradations to network performances such as complete non-functionality (selfish nodes safe power and deny to relay other node's messages), decrease of throughput and packet loss, network partitioning, jeopardized security, etc.

To cope with misbehavior in self-organized networks, nodes need to be able to automatically adapt their strategy to changing levels of cooperation. Solutions are offered through:

- *Secure routing*, which provides prevention against specific malicious attacks (Ariadne, secure routing protocol (SRP)) [82]. More details on security protocols are presented in Chapter 8 and Chapter 3.
- *Economic incentives*, which aim at making selfish nodes forward for other, despite their selfish power policy, through different stimulating mechanisms [81].
- *Detection and reputation systems*, which enable nodes to adapt to changes in the network environment caused by misbehaving nodes [82].

Economic incentive employs payment and pricing schemes. [81] proposes a simple mechanism in order to stimulate the nodes for packet forwarding. A counter in each node that counts virtual currency (nuglet) forces the nodes to pay to forward their packets and to be paid when they forward some data for some other nodes. Similar approach is introduced in [83]. [84] applies *Broke Service* mechanisms, which allow node to enter a *broke* state whenever it runs out of virtual currency. [85] presents a pricing based joint user-and-network centric incentive mechanism that includes forwarding among selfish users by compensating the real and opportunity costs.

Detection and reputation systems monitor the behavior of network nodes. The reputation is opinion a node has of another nodes and it enables to make informed decisions about cooperation. *Second hand information* is reputation information obtained by others and it aims to prevent the misbehavior. *Trust* is another feature, reflecting the performance of a node in the policing protocol, which protects the base protocol [82].

Several models incorporate self-policing mechanisms to model cooperation in ad hoc networks: Marti's model [86], Market model [81], CONFIDANT [87], CORE [88], Context Aware Detection [89], GTFT [90], SORI [91]. [8, 82] provide more details and give a good overview of these models.

2.4.4 Summary

Self-organization can catch up the pecularities of ad hoc networks, providing mechanisms to design and foresee their behavior. The impact of self-organization reflects not only to ad hoc wireless structures, but can be applied to wired and hybrid structures as well, such as Internet (emergent phenomena), WWW [76], cellular networks [75], mesh and hybrid networks, sensor networks [92, 93], etc. Anderson and Willinger in [76] offer an inside look at self-organization phenomena as a modeling tool for next-generation communication networks (NGN) and all-IP solutions. Ad hoc networks play significant role in these future concepts.

2.5 Concluding Remarks

This chapter gives an overview of the new emerging areas in ad hoc networking: capacity and self-organization. It also covers the topology formation problems, as basis for understanding the structures and procedures in ad hoc network design. Intensive ongoing research highlights the importance of these new aspects, and shows up some parallel links to other disciplines such as sociology, biology, statistic physics, etc. Other basic concepts, not less important to ad hoc networking, are omitted from this chapter and will be discussed later. The following chapters of this book are dedicated to more pragmatic features concerning ad hoc wireless networks and their visionary role in future wireless communication.

References

[1] Melodia, T., Pompili, D., and Akyildiz, I. F., "On the Interdependence of Topology Control and Geographical Routing in Ad Hoc and Sensor Networks," *IEEE Journal on Selected Areas in Communications*, 23(3), March 2005, pp. 520–532.

[2] Zhang, X., Zhang, Q., Li, B., Zhu, W., and Yum, T. -S. P., "MultiServ: A Service-Oriented Framework for Multihop Wireless Network," *IEEE Journal on Selected Areas in Communications*, 23(6), June 2005, pp. 1146–1158.

[3] Ramanathan, R. and Rosales-Hain, R., "Topology Control of Multihop Wireless Networks using Transmit Power Adjustment," in *Proc. IEEE INFOCOM*, Tel-Aviv, Israel, March 2000, pp. 404–413.

[4] Takagi, H. and Kleinrock, L., "Optimal Transmission Ranges for Randomly Distributed Packet Radio Terminals," *IEEE Transactions on Communications*, 32(3), March 1984, pp. 246–257.

[5] Hou, T. C. and Li, V. O. K., "Transmission Range Control in Multihop Packet Radio Networks," *IEEE Transactions on Communications*, 34(1), January 1986, pp. 38–44.

[6] Finn, G. G., "Routing and Addressing Problems in Large Metropolitanscale Internetworks," *ISI Res. Rep. ISU/RR-87-180*, March 1987.

[7] Nelson, R. and Kleinrock, L., "The Spatial Capacity of a Slotted ALOHA Multihop Packet Radio Network with Capture," *IEEE Transactions on Communications*, 32(6), June 1984, pp. 684–694.

[8] Basagni, S., Conti, M., Giordano, S., and Stojmenovic, I., *Mobile Ad Hoc Networking*, IEEE Press, Wiley-Interscience, 2004.
[9] Gao, J., Guibas, L. J., Hershberger, J., Zhang, L., and Zhu, A., "Geometric Spanners for Routing in Mobile Networks," *IEEE Journal on Selected Areas in Communications*, 23(1), January 2005, pp. 174–185.
[10] Bao, L. and Garcia-Luna-Aceves, J. J., "Topology Management in Ad Hoc Networks," *MobiHoc'03*, Annapolis, Maryland, USA, June 1–3, 2003.
[11] Toussaint, G. T., "The Relative Neighborhood Graph of a Finite Planar Set," *Pattern Recognition*,12(4), 1980, pp. 261–268.
[12] Gabriel, K. R. and Sokal, R. R., "A New Statistical Approach to Geographic Variation Analysis," *Systematic Zoology*, 18, 1969, pp. 259–278.
[13] Lukovszki, T.,"New Results on Geometric Spanners and Their Applications," Ph.D. Thesis, University of Paderborn, 1999.
[14] Rajaraman, R., "Topology Control and Routing in Ad Hoc Networks: A survey," *SIGACT News*, 33, 2002, pp. 60–73.
[15] Li, X. -Y. and Stojmenovic, I., "Broadcasting and Topology Control in Wireless Ad Hoc Networks," *Handbook of Algorithms for Mobile and Wireless Networking and Computing* (A. Boukerche and I. Chlamtac, eds.), CRC Press, 2004.
[16] Xu, Y., Bien, S., Mori, Y., Heidemann, J., and Estrin, D., "Topology Control Protocols to Conserve Energy in Wireless Ad Hoc Networks," *Center for Embedded Networked Sensing Technical Report 6, UCLA*, USA, 2003.
[17] Guha, S. and Khuller, S., "Approximation Algorithms for Connected Dominating Sets," *Algorithmica*, 20(4), April 1998, pp. 374–387.
[18] Garey, M. R. and Johnson, D. S., *Computers and Intractability. A Guide to the Theory of NP-Completeness*, Freeman, Oxford, UK, 1979.
[19] Amis, A. D., Prakash, R., Vuong, T. H. P., and Huynh, D. T., "Max-Min D-Cluster Formation in Wireless Ad Hoc Networks," in *Proc. of IEEE INFOCOM 2000*, no. 1, March 2000, pp. 32–41.
[20] Sivakumar, R., Sinha, P., and Bharghavan, V., "CEDAR: A Core-Extraction Distributed Ad Hoc Routing Algorithm," *IEEE Journal on Selected Areas in Communications*, 17(8), August 1999, pp. 1454–1465.
[21] Chen, B., Jamieson, K., Balakrishnan, H., and Morris, R., "SPAN: An Energy-Efficient Coordination Algorithm for Topology Maintenance in Ad Hoc Wireless Networks," *ACM MOBICOM'01*, Rome, Italy, July 2001.
[22] Lin, C. R. and Gerla, M., "Adaptive Clustering for Mobile Wireless Networks," *IEEE Journal on Selected Areas in Communications*, 15(7), September 1997, pp. 1265–1275.
[23] Jia, L., Rajaraman, R., and Suel, T., "An Efficient Distributed Algorithm for Constructing Small Dominating Sets," *ACM Symposium on Principles of Distributed Computing PODC'01*, Newport, Rhode Island, August 2001.
[24] Basu, P., Khan, N., and Little, T. D. C., "A Mobility Based Metric for Clustering in Mobile Ad Hoc Networks," *International Workshop on Wireless Networks and Mobile Computing (WNMC2001)*, Scottsdale, Arizona, April 2001.
[25] Amis, A. D. and Prakash, R., "Load-balancing Clusters in Wireless Ad Hoc Networks," in *Proceedings of the 3rd IEEE Symposium on Application-Specific Systems and Software Engineering Technology*, Los Alamitos, CA, March 2000, pp. 25–32.

[26] Floreen, P., Kaski, P., Kohonen, J., and Orponen, P., "Lifetime Maximization for Multicasting in Energy-Constrained Wireless Networks," *IEEE Journal on Selected Areas in Communications*, 23(1), January 2005, pp. 117–126.

[27] Marks, R. J., Das, A. K., El-Sharkawi, M., Arabshahi, P., and Gray, A., "Minimum Power Broadcast Trees for Wireless Networks: Optimizing Using the Viability Lemma," in *Proceedings of IEEE International Symposium on Circuits and Systems*, 2002, pp. 245–248.

[28] Liang, W., "Constructing Minimum-Energy Broadcast Trees in Wireless Ad Hoc Networks," in *Proceedings of ACM MOBIHOC 2002*, pp. 112–122.

[29] Li, F. and Nikolaidis, I., "On Minimum-Energy Broadcasting in All-Wireless Networks," in *Proceedings of 26th Annual IEEE Conference on Local Computer Networks - LCN'01*, 2001.

[30] Marks, R. J., Das, A. K., El-Sharkawi, M., Arabshahi, P., and Gray, A., "Minimum Power Broadcast Trees for Wireless Networks: An Ant Colony System Approach," in *Proceedings of IEEE International Symposium on Circuits and Systems*, 2002.

[31] Wieselthier, J., Nguyen, G., and Ephremides, A., "On the Construction of Energy-Efficient Broadcast and Multicast Trees in Wireless Networks," in *Proceedings of IEEE INFOCOM 2000*, 2000, pp. 586–594.

[32] Pottie, G. J. and Kaiser, W. J., "Wireless Integrated Network Sensors," *Communications of the ACM*, 43(5), 2000, pp. 551–558.

[33] Cartigny, J., Simplot, D., and Stojmenovic, I., "Localized Energy Efficient Broadcast for Wireless Networks with Directional Antennas," in *Proceedings of IFIP Mediterranean Ad Hoc Networking Workshop (MED-HOC-NET 2002)*, Sardegna, Italy, 2002.

[34] Wu, J. and Dai, F., "Mobility-Sensitive Topology Control in Mobile Ad Hoc Networks," in *Proceedings of the 18th International Parallel and Distributed Processing Symposium (IPDPS'04)*, 1(1), 2004, p. 28a.

[35] Liu, Y., Hu, X., Lee, M. J., and Saadawi, T. N., "A Region-Based Routing Protocol for Wireless Mobile Ad Hoc Networks," *IEEE Network Magazine*, 18(4), July/August 2004, pp. 12–17.

[36] Frey, H., "Scalable Geographic Routing Algorithms for Wireless Ad Hoc Networks," *IEEE Network Magazine*, 18(4), July/August 2004, pp. 18–22.

[37] Xu, Y., Heidemann, J., and Estrin, D., "Geography-Informed Energy Conservation for Ad Hoc Networks," in *Proceedings of MOBICOM'01*, 2001.

[38] Blough, D. M. and Santi, P., "Investigating Upper Bounds on Network Lifetime Extension for Cell-Based Energy Conservation Techniques in Stationary Ad Hoc Networks," in *Proceedings of MOBICOM'02*, 2002.

[39] Li, L. and Halpern, J. Y., "Minimum Energy Mobile Wireless Networks Revisited," in *Proceedings of ICC'01*, June 2001, pp. 278–283.

[40] Rodoplu, V. and Meng, T. H., "Minimum Energy Mobile Wireless Networks," *IEEE Journal on Selected Areas in Communications*, 17(8), August 1999, pp. 1333–1344.

[41] Li, L., Halpern, J. Y., Bahl, V., Wang, Y. M., and Wattenhofer, R., "Analysis of a Cone-Based Distributed Topology Control Algorithm for Wireless Multi-Hop Networks," in *Proceedings of PODC'01*, August 2001, pp. 1702–1712.

[42] Wattenhofer, R., Li, L., Bahl, V., and Wang, Y. M., "Distributed Topology Control for Power Efficient Operation in Multihop Wireless Ad Hoc Networks," in *Proceedings of INFOCOM'01*, April 2001, pp. 1388–1397.

[43] Li, N., Hou, J. C., and Sha, L., "Design and Analysis of an MST-Based Topology Control Algorithm," in *Proceedings of INFOCOM'03*, March/April 2003, pp. 1702–1712.
[44] Wu, J. and Dai, F., "Mobility Control and Its Applications in Mobile Ad Hoc Networks," *IEEE Network Magazine*, 18(4), July/August 2004, pp. 30–35.
[45] Blough, D., Leoncini, M., Resta, G., and Santi, P., "The k-Neigh Protocol for Symmetric Topology Control in Ad Hoc Networks," in *Proceedings of MobiHoc'03*, June 2003, pp. 141–152.
[46] Liu, J. and Li, B., "MobileGrid: Capacity-Aware Topology Control in Mobile Ad Hoc Networks," in *Proceedings of ICCCN'02*, October 2002, pp. 570–574.
[47] Ramanathan, R. and Rosales-Hain, R., "Topology Control of Multihop Wireless Networks Using Transmit Power Adjustment," in *Proceedings of INFOCOM'00*, March 2000, pp. 404–413.
[48] Penrose, M., "The Longest Edge of the Random Minimal Spanning Tree," *Annals of Applied Probability*, 7, 1997, pp. 340–361.
[49] Penrose, M., "On k-Connectivity for a Geometric Random Graph," *Random Structures and Algorithms*, 15, 1999, pp. 145–164.
[50] Chen, H., Sivakumar, T. V. L. N., Huang, L., and Kashima, T., "Topology-Controllable Scatternet Formation Method and Its Implementation," *International Workshop on Wireless Ad-Hoc Networks (IWWAN) 2004*, Oulu, Finland, 2004.
[51] Salonidis, T., Bhagwat, P., Tassiulas, L., and LaMaire, R., "Distributed Topology Construction of Bluetooth Wireless Personal Area Networks," *IEEE Journal on Selected Areas in Communications,* 23(3), March 2005, pp. 633–643.
[52] Zhen, B., Park, J., and Kim, Y., "Scatternet Formation of Bluetooth Ad Hoc Networks," *36th Hawaii International Conference on System Sciences (HICSS'03)*, January 2003.
[53] Basagni, S. and Petrioli, C., "A Scatternet Formation Protocol for Ad hoc Networks of Bluetooth Devices," in *Proceedings of IEEE VTC Spring 2002*, 2002, pp. 424–428.
[54] Chiasserini, C. F., Marsan, M. A., Baralis, E., and Garza, P., "Towards Feasible Topology Formation Algorithms for Bluetooth-based WPANs," *36th Hawaii International Conference on System Sciences (HICSS'03)*, January 2003.
[55] Nesargi, S. and Prakash, R., "MANETconf: Configuration of Hosts in a Mobile Ad Hoc Network," *IEEE INFOCOM 2002*, New York, USA, June 2002.
[56] Meguerdichian, S., Koushanfar, F., Potkonjak, M., and Srivastava, M. B., "Coverage Problems in Wireless Ad-hoc Sensor Networks," in *Proceedings of IEEE INFOCOM 2001*, 3, April 2001, pp. 1380–1387.
[57] Gupta, P. and Kumar, P. R., "The Capacity of Wireless Networks," *IEEE Transactions on Information Theory*, 46, March 2000, pp. 388–404.
[58] Agarwal, A. and Kumar, P. R., "Capacity Bounds for Ad hoc and Hybrid Wireless Networks," *ACM SIGCOMM Computer Communications Review*, 34(3), July 2004, pp. 71–81.
[59] Xie, L. -L., and Kumar, P. R., "A Network Information Theory for Wireless Communication: Scaling Laws and Optimal Operation," *IEEE Transactions on Information Theory*, 50(5), May 2004, pp. 748–767.

[60] Liu, B., Liu, Z., and Towsley, D., "On the Capacity of Hybrid Wireless Networks," *IEEE INFOCOM 2003*, San Francisco, USA, 2003.

[61] Jindal, N., Mitra, U., and Goldsmith, A., "Capacity of Ad-Hoc Networks with Node Cooperation," *IEEE International Symposium on Information Theory*, Chicago, IL, June 2004.

[62] Yi, S., Pei, Y., Kalyanaraman, S., and Azimi-Sadjadi, B., "How is the Capacity of Ad Hoc Networks Improved with Directional Antennas," *Submitted to IEEE Journal on Selected Areas in Communications*, 2005.

[63] Grossglauser, M. and Tse, D. N. C., "Mobility Increases the Capacity of Ad Hoc Wireless Networks," *IEEE/ACM Transactions on Networking*, 10(4), August 2002.

[64] Perevalov, E. and Blum, R., "Delay Limited Capacity of Ad hoc Networks: Asymptotically Optimal Transmission and Relaying Strategy," *IEEE INFOCOM 2003*, San Francisco, USA, 2003.

[65] Sharma, G. and Mazumdar R. R., "Delay and Capacity Trade-off in Wireless Ad Hoc Networks with Random Mobility," *Available at*: www.ece.purdue.edu/~mazum/MONET2004.pdf

[66] El Gamal, A., Mammen, J., Prabhakar, B. and Shah, D., "Throughput-delay Trade-off in Wireless Networks," *IEEE INFOCOM 2004*, Hong Kong, March 2004.

[67] Spyropoulos, A. and Raghavendra, C. S., "Capacity Bounds for Ad-Hoc Networks Using Directional Antennas," *IEEE ICC 2003*, Seattle, USA, May 2003.

[68] Peraki, C. and Servetto, S., "On the Maximum Stable Throughput Problem in Random Wireless Networks with Directional Antennas," *ACM MobiHoc'03*, 2003.

[69] Negi, R. and Rajeswaran, A., "Capacity of Power Constrained Ad-Hoc Networks," *IEEE INFOCOM 2004*, Hong Kong, March 2004.

[70] Zhang, H. and Hou, J. C., "Capacity of Wireless Ad-hoc Networks under Ultra Wide Band with Power Constraint," *IEEE INFOCOM 2005*, Miami, USA, March 2005.

[71] Hubaux, J. -P., Gross, T., Le Boudec, J. -Y., and Vetterli M., "Toward Self-Organized Mobile Ad Hoc Networks: The Terminodes Project," *IEEE Communications Magazine*, January 2001, pp. 118–124.

[72] Robertazzi, T. G. and Sarachik, P. E., "Self-Organizing Communication Networks," *IEEE Communications Magazine*, January 1986, pp. 28–33.

[73] Milgram, S., "The Small World Problem," *Psychology Today*, May 1967, pp. 60–67.

[74] Barabasi, A. -L., LINKED: The New Science of Networks, Perseus Books Group, 2002.

[75] Dixit, S., Yanmaz, E., and Tonguz, O. K., "On the Design of Self-Organized Cellular Wireless Networks," *IEEE Communications Magazine*, July 2005, pp. 86–93.

[76] Alderson, D. and Wilinger, W., "A Contrasting Look at Self-Organization in the Internet and Next-Generation Communication Networks," *IEEE Communications Magazine*, July 2005, pp. 94–100.

[77] Musolesi, M., Hailes, S., and Mascolo, C., "Social Networks Based Ad Hoc Mobility Models," *3rd UK-Ubinet Workshop*, Bath, United Kingdom, February 2005.

[78] MacKenzie, A. B. and Wicker, S. B., "Game Theory and the Design of Self-Configuring, Adaptive Wireless Networks," *IEEE Communications Magazine*, November 2001, pp. 126–131.

[79] Prefoher, C. and Bettstetter, C., "Self-Organization in Communication Networks: Principles and Design Paradigms," *IEEE Communications Magazine*, July 2005, pp. 78–85.

[80] Van Dyke Parunak, H., and Brueckner, S. A., "Stigmergic Learning for Self-Organizing Mobile Ad-Hoc Networks (MANET's)," *3rd International Conference on Autonomous Agents and Multi-Agent Systems (AAMAS'04)*, Columbia University, New York City, USA, July 2004.

[81] Buttyan, L. and Hubaux, J. -P., "Stimulating Cooperation in Self-Organizing Mobile Ad Hoc Networks," *Mobile Networks and Applications*, 8(5), October 2003, pp. 579–592.

[82] Buchegger, S. and Le Boudec, J. -Y., "Self-Policing Mobile Ad Hoc Networks by Reputation Systems," *IEEE Communications Magazine*, July 2005, pp. 101–107.

[83] Blazevic, Lj., Buttyan, L., Capkun, S., Giordano, S., Hubaux, J. -P., and Le Boudec, J. -Y., "Self-Organization in Mobile Ad Hoc Networks: The Approach of Terminodes," *IEEE Communications Magazine*, June 2001, pp. 166–174.

[84] Shadpour, B., Valaee, S., and Li, B., "A Self-Organized Approach for Stimulating Cooperation in Mobile Ad Hoc Networks," *22nd Biennial Symposium on Communications*, Queen's University, Kingston, Ontario, Canada, May 31–June 3, 2004.

[85] Ileri, O., Mau, S. -C., and Mandayam, N. B., "Pricing for Enabling Forwarding in Self-Configuring Ad Hoc Networks," *IEEE Journal on Selected Areas in Communication*, 23(1), January 2005, pp. 151–162.

[86] Marti, S., Giuli T. J., Lai, K., and Baker, M., "Mitigating Routing Misbehavior in Mobile Ad Hoc Networks," *MOBICOM 2000*, 2000, pp. 255–265.

[87] Buchegger, S. and Le Boudec, J. -Y., "Performance Analysis of the CONFIDANT Protocol: Cooperation of Nodes – Fairness in Dynamic Ad-hoc Networks" *IEEE/ACM Symposium on Mobile Ad Hoc Networking and Computing*, Lausanne, Switzerland, June 2002.

[88] Michiardi, P. and Molva, R., "CORE: A Collaborative Reputation Mechanism to Enforce Node Cooperation in Mobile Ad Hoc Networks," *6th IFIP of Communications and Multimedia Security Conference*, Portoroz, Slovenia, 2002.

[89] Paul, K. and Westhoff, D., "Context Aware Detection of Selfish Nodes in DSR based Ad-hoc Networks," *IEEE GLOBECOM*, Taipeh, Taiwan, 2002.

[90] Srinivasan, V., Nuggehalli, P., Chiasserini, C. F., and Rao, R. R., "Cooperation in Wireless Ad Hoc Networks," *IEEE INFOCOM 2003*, San Francisco, USA, 2003.

[91] He, Q., Wu, D., and Khosla, P., "SORI: A Secure and Objective Reputation-Based Incentive Scheme for Ad Hoc Networks," *WCNC 2004*, Atlanta, USA, March 2004.

[92] Kochhal, M., Schwiebert, L., and Gupta, S., "Role-based Hierarchial Self Organization for Wireless Ad hoc Sensor Networks," *WSNA'03*, San Diego, USA, September 2003.

[93] Catterall, E., Van Laerhoven, K., and Strohbach, M., "Self-Organization in Ad Hoc Sensor Networks: An Empirical Study," in *Proceedings of Artificial Life VIII, The 8th International Conference on the Simulation and Synthesis of Living Systems*, Sydney, Australia, MIT Press, 2002, pp. 260–264.

3

Multiple Access and Routing

3.1 Introduction

Ad hoc networking includes a variety of networks comprising WLANs, WPANs, WBANs and WSNs that will have to provide a large range of capabilities in terms of coverage/bit rate and play an important role in the deployment and functioning of systems without any assistance from existing infrastructures. These networks cover a short range domain with restricted mobility and specific trade-offs between power-consumption/data-rates/ quality-of-service/spectrum-allocation, and are suited in different deployment scenarios. The functionalities that can fulfill these requirements are concentrated in the three lower layers of the protocol stack, i.e., physical, data and network layer. A lot of research has been completed towards achieving better system performances through improving functionalities concentrated in lower network-oriented layers. Adaptation of network solutions to wireless medium, available bandwidth and traffic load in heterogeneous network environment is foreseen as a necessity in future network design. Regulations about frequency allocation and underlying techniques are under jurisdiction of international standardization and regulative bodies (e.g., IEEE family of standards, Federal Communications Commission — FCC, ETSI). This chapter gives a brief overview of the most relevant physical layer issues, MAC approaches and routing protocols implemented and foreseen for today's and future ad hoc networks.

3.2 Physical Layer

Different ad hoc networking concepts (e.g., WPANs, WLANs, sensor networks) cope with physical layer problems in specific ways trying to provide better coverage, higher spectrum efficiency and higher data rates. Current wireless networking solutions use infrared waves, radio frequency signals and microwaves for transmission at the physical layer. Recently, the UWB radio raised as a new emerging technology for ad hoc networking. The variety of available devices (from sensors to satellite receivers) and the shared wireless medium with its limitations pose specific problems in the physical layer design. Major issues in physical layer, related to ad hoc networking, are mentioned in the following paragraphs.

3.2.1 Frequency Ranges

The definition of *frequency range* specifies the spectrum usage and allocation. Several frequency ranges are applicable for ad hoc networking including the freely available Industrial - Scientific-Medical (ISM) bands. Conventional spectrum allocation covers 2.4/5 GHz bands, which are already used by IEEE 802.15.1 (only in 2.4 GHz) [1], IEEE 802.11x [2], ETSI HIPERLAN/2 [3] and ARIB MMAC standards [4], and considers newly created IEEE 802.11n TASK Group for High Throughput WLAN [5]. The clear need for wireless systems moving towards the upper frequency ranges in order to offer additional capacity and higher spectral efficiency addresses new spectrum range around 17 GHz, which is being considered by CEPT (17.1–17.3 GHz) in [6] and ITU (17.3–17.7 GHz) in [7]. This band does not provide interference to other wireless systems and is expected to be globally approved. The 60 GHz band is allocated worldwide for broadband wireless communications. It is already standardized in Japan (by ARIB) and addressed in several European projects, e.g., Broad Way [8].

The limited bandwidth resources, overcrowded with different systems and technologies, require careful sharing in order to achieve higher *spectrum efficiency*. The technical areas that can influence and improve the overall spectrum efficiency are: air interface technologies, antenna technologies, advanced signal processing, new network topologies and cooperation of networks.

3.2.2 Characteristics and Limitations of the Physical Medium

Wireless physical medium, as unstable medium with possible unpredictable disturbances, poses serious problems in the definition and modeling of physical channels. These channels are affected with fading, frequency offset, phase jitter, impulse and thermal noise, interference, frequency and time spreading effects. The *multipath propagation* of transmitted signal can cause undesirable effects deteriorating the system performances. The transmitted signal experiences reflection, scattering and diffraction during multipath transmission, resulting in *small-scale fading* effects. *Large-scale* fading reflects the propagation losses. The presence of *fading* originates several problems such as: rapid changes of signal strength, frequency spreading (due to Doppler shifts on different multipath signals) and time spreading (due to multipath propagation delays). Spreading causes frequency or time expansion of transmitted signal respectively [9]. A lot of work was dedicated to resolving these problems which resulted in the development of different multicarrier modulation techniques, equalization, statistical fading models, etc.

Interference, either *inter-carrier (ICI)* or *inter-symbol (ISI)*, can significantly reduce the system efficiency. It can be caused by different technologies coexisting in the same frequency band, thermal noise, or large number of same technology devices (self-interference) [10]. Interference mitigation and cancellation techniques should be developed to increase the spectrum efficiency. Different methods for interference avoidance propose schemes for adaptive frequency hopping, frequency reuse (not applicable for OFDM-based systems), enhanced version of frequency rolling (proposed for Bluetooth) [11], exploitation of different activity modes, transmit power control (in IEEE 802.11h), etc.

The interference imposes the important problem of *coexistence* of different technologies in the same frequency band (e.g., ISM band). The problem of efficient spectrum sharing and coexistence was addressed by some interest groups [12, 13]. New emerging technologies, such as Wi-Fi, that are lately getting popularity, are affecting the already crowded bandwidth boldering the problem of efficient spectrum sharing [14]. This implies need for solutions for spectrum scenarios and arrangements, radio interface design, dynamic frequency selection and network topologies.

3.2.3 Relevant Procedures and Parameters

There are several physical layer procedures that can be exploited for improving the ad hoc networking performances. Often addressed by academia and industry, they try to offer the best possible system/device adaptation to available medium and its imperfections. Powerful coding and channel-state aware decoding schemes are combined with adaptive modulations to show superior performance and capacity. New spreading codes and interleaving techniques are proposed in order to achieve better spatial capacity and robustness. Convolutional or block turbo codes and Viterbi coding are also included as powerful options. Many studies of optimization and system parameter analysis were performed trying to offer solution that can adapt on particular requirements in the dynamic and constantly changing wireless environment.

Parameters that precisely define the quality characteristics of the physical medium are Signal-to-Noise Ratio (SNR), Bit Error Rate (BER), Packet Error Rate (PER), etc. Physical layer parameters can influence the system behavior making *adaptation* to different channel conditions an important issue. Physical layer may need to adapt to rapid SNR changes in wireless links caused by unstable medium and mobility. Different techniques include: power control, multi-user detection, new multiple-access schemes (e.g., spatial division multiple access-SDMA), directional/smart antennas, virtual antenna arrays (VTT), advanced receiving algorithms, Multiple-Input-Multiple-Output (MIMO) and other approaches exploiting micro diversity, Complementary-Code-Keying CCK (for Wi-Fi) and software radio. Some of these problems are in close connections to MAC related functionalities.

3.2.4 Proprietary Techniques and Systems

The realization of different systems that are used to build up today's ad hoc networks is based on different techniques. Each of them has its own specifics that accommodate certain system requirements (coverage, data rate, etc.) in particular networking environments.

Infrared LANs [15, 16] use the wavelength band, which is between the visible spectrum and the microwaves (780–950 nm, i.e., in the light spectrum). Signals can be transmitted omnidirectionally (nondirect IR LANs) or directionally (direct IR LANs), providing different coverage. Three data rates were identified by the IrDA (Infrared Data Association): 115 kbps, 1.15 Mbps and 4 Mbps. Infrared LANs are not bandwidth limited and are more resistible to electromagnetic noise than the spread-spectrum systems. However, they require the unobstructed Line-Of-Sight (LOS) and can not penetrate opaque objects.

A *microwave* is a short radio-wave that varies from 1 mm to 30 cm in length that, unlike the longer radio waves, can pass through the ionosphere. This makes microwave technology more suitable for long-range applications. Microwaves use narrowband transmission with single-frequency modulations and are set up mostly for 5.8 GHz band [17]. They request license to operate in an appropriate frequency band that makes them interference free, but also the most expensive WLANs on the market. They are mostly used for applications such as communications with satellite.

Spread-spectrum techniques, usually used for WLAN and some WPAN operation (e.g., IEEE 802.11b, IEEE 802.15.1.0 and Bluetooth, HomeRF), are spreading the signal power over a wide band of frequencies, making it less sensitive to noise and interference. They can use frequency-hopping (FH) spread spectrum or direct-sequence (DS) spread-spectrum approach. In FH approach, the data signal is modulated with carrier signal that hops from frequency to frequency as a function of time over a wide, predefined, range of frequencies. In DS approach, the data signal is combined with a higher data-rate bit sequence, thus spreading the signal over the whole frequency band [18].

Orthogonal Frequency Division Multiplexing (OFDM) is foreseen for applications in the 5 GHz-band wireless LANs, such as IEEE 802.11a/g, Hiperlan2 and MMAC, and new standards such as 802.11h and 802.16.a [19, 20]. OFDM is a parallel data transmission scheme and presents a special case of multi-carrier transmission over a number of lower-rate sub-carriers. The implemented coding schemes provide robustness in frequency selective fading channels and make complex equalizers unnecessary. The drawbacks are synchronization sensitivity and a large peak-to-average-power (PAP) ratio [21].

The new emerging *Ultra Wide Band (UWB)* technology introduces a new way to use and share the spectrum. The UWB communication is based on transmission of signals widely spread in the frequency domain. These signals are using the same spectrum as a variety of signals from conventional narrow-band systems, however the UWB signals are transmitted with lower power (so they do not cause undesired interference). The 802.15.3a is adopted to cover the UWB physical layer [22]. More details can be found in [23]. UWB technology is defined as any wireless transmission scheme that occupies a bandwidth of more than 25% of a center frequency higher than 1.5 GHz. Even though UWB originates from radar-based technology, it recently became increasingly attractive for low-cost consumer applications. Implementation of UWB technology will open new horizons in future WLANs and WPANs [24]. Within the IEEE, the recently established Task Group 3a (TG3a) is defining an alternative physical-layer (Alt-PHy) that is to be based on UWB radio technology, complementary to the existing 802.15.3 PHY standard. The emerging Alt-PHy standard will support high data rate consumer applications for multimedia distributions, typically in a home [25]. Also, recently established IEEE 802.15.4a Task Group (TG4a) is using Alt-PHy approach to enrich WPAN capabilities. All in all, UWB techniques show great spatial capacity, potential compliance with global unlicensed operation and implementation advantages.

For future WB-PANs, modulation techniques such as OFDM/DS-CDMA or A/SHF-CDMA are foreseen to work in the 60 GHz band at possible data rate of 1 Gbps for very low mobility cases [26]. More flexible spectrum use and improved spectrum efficiency can be achieved with solutions offered by software defined radio (SDR) [27] and cognitive radio technology [28].

3.2.5 Summary

Functionality and application domains of wireless ad hoc networks are redesigned with new emerging technologies. Sustainable requirement for higher throughput, range and reliability, problem of coexistence in unlicensed frequency bands and power consumption, adaptation to medium constrains and networking in heterogeneous environment are all the driven forces in physical layer development, design and standardization.

The most important technical features of existing short range technologies are listed in Table 3.1.

The physical layer developments in ad hoc networking domain are oriented towards new solutions for variety of applications in emerging sensor networks, ultra wideband technologies, development of adaptive approach and support for mobility and higher bit rates for wireless broadband multimedia communication. They are followed by intensive standardization efforts.

Table 3.1. Comparison of different wireless systems

Standard	Physical Layer: OFDM	FHSS	DSSS	Freq. band	Type	Data rates	Coexistence	Range	Adv.	Disadv.
Bluetooth / 802.15		√1600 hops/s		2.4 GHz ISM	PAN	Variable (max 732.2 Kbps)	Interference with other systems in ISM band	10 m (up to 100 m with increased power)	First real "ad hoc" PAN technology; Low cost	Interference; Low data rate
802.11a (Wi-Fi 5.2 GHz)	√			5 GHz UNII	LAN	54 Mbps	Interference with other systems in ISM band	∼20 m		High cost; Not scalable to personal devices
802.11		√	√	2.4 GHz ISM	LAN	1 and 2 Mbps	Interference with other systems in ISM band	50–100 m	High data rates; Wide deployment; Conditional advantage of larger range (except 802.11a)	Low data rate
802.11b (Wi-Fi 2.4 GHz)			√	2.4 GHz ISM	LAN	5.5 and 11 Mbps	Interference with other systems in ISM band	50–100 m		No backward compatibility
IrDA	Optical infrared			780–990 nm (optical wavelength)	Short range, PAN	9.6–115 Kbps; extensions 1.4 and 16 Mbps	No interference	<2 m	No interference; Low cost; Low power consumption	Requires LOS; Very short range for certain purposes
HomeRF		√50 hops/s		2.4 GHz ISM	LAN	0.8 and 1.6 Mbps	Interference with other systems in ISM band	50 m	Lower power than 802.11; Higher data rate than Bluetooth	Not an open standard
UWB	Multiband OFDM, Pulse Position Modulation, Wideband DSSS or frequency hopping			3.1–10.6 GHz (short pulse 0.1–2 ns)	PAN, sensor networks, piconets	100 Mbps–1 Gbps	No interference	1–10 m	Short range High Data Rate (∼1000 Kbps/m^2); Low power consumption in Low Data Rate; No interference	Synchronization; Sensitivity to NLOS; Ongoing regulatory procedure

3.3 MAC Layer

The main responsibility of the Medium Access Control (MAC) protocol in ad hoc wireless networks is to control and manage the access and packet transmission through the shared channel in a distributed manner, with minimum possible overhead involved. The scarce radio spectrum and the limited available bandwidth should be accessed and used fairly and utilized efficiently by all nodes in the network. Also, ad hoc wireless networks have to adapt to node mobility and corresponding dynamic topology.

The characteristics of the wireless medium and the ad hoc nature of the environment significantly influence MAC layer design which focuses on different problems such as: distributed operation, synchronization, hidden/exposed terminal, fairness, throughput and delay, capability of power control, adaptive rate control, resource reservation, real-time traffic support, use of directional antennas, scalability, mobility of nodes, etc. Design of the MAC protocol can be focused on different characteristics depending on what the target network parameters are and what type of enhancement and optimizations are being addressed.

This subchapter gives a general overview of different approaches in medium access control techniques, with no aim of detailed description. The appropriate references are notified. It also presents the classification of existing MAC protocols and relevant problems to broadcasting wireless medium. Some of the intriguing techniques that incorporate important mechanisms of power control, QoS and security will be covered in separate chapters of this book.

3.3.1 The Hidden/Exposed Terminal Problem

The radio interface of each node uses broadcasting and copes with limited wireless transmission range. Collisions and higher probability of packet losses due to transmission errors are also specifics of the wireless medium. They may severely reduce the channel utilization and the throughput. The problems of *hidden terminal* and *exposed terminal* are among the most remarkable problems posed in front of the CSMA/CA based MAC layer design.

The hidden and exposed terminal problems are unique to wireless communications [29]. The *hidden terminal problem* occurs when two or more terminals (e.g., node A and node C), which have disjoint transmission ranges start transmitting towards the same receiver (e.g., node B). This results in packet collision, as shown in Fig. 3.1.

The *exposed terminal problem* occurs when a node (e.g., node C) that falls into the transmission range of an active node (e.g., node B transmits to node A) wants to start a transmission to another inactive node within its range, but outside the range of the active node (e.g., node C wants to transmit to

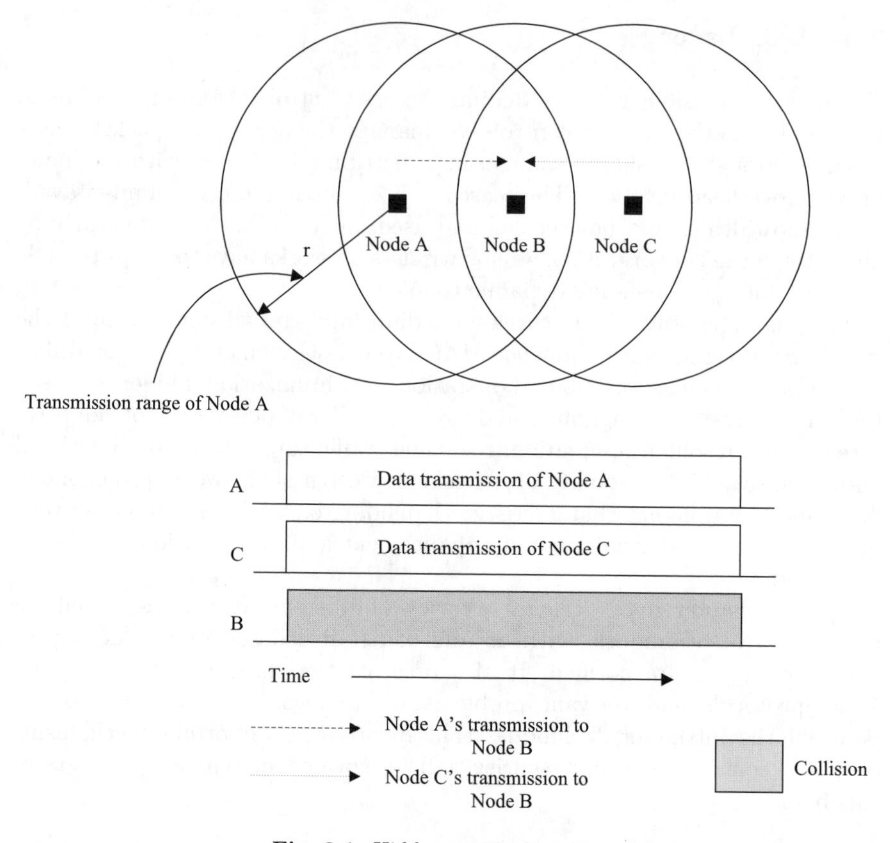

Fig. 3.1. Hidden terminal problem

node D), as presented in Fig. 3.2. According to the CSMA scheme, that node has to postpone the transmission, which results in throughput decrease.

The problem of hidden and exposed terminals can significantly reduce the throughput, especially in high load conditions. Therefore, this problem yields careful considerations in MAC layer functionalities design. A possible solution to cope with hidden terminal problem is to introduce the *Request-To-Send/Clear-To-Send (RTS/CTS) mechanism.* This results in signaling overhead and may not work well in multi-hop networks. The exposed terminal problem can be reduced with the use of a scheduler in order to schedule the transmitter/receiver pairs and to achieve both QoS and fairness.

3.3.2 Classification

Medium access approaches can be *contention-based* or *contention-free.* The most commonly considered contention-based approach is CSMA/CA (Carrier

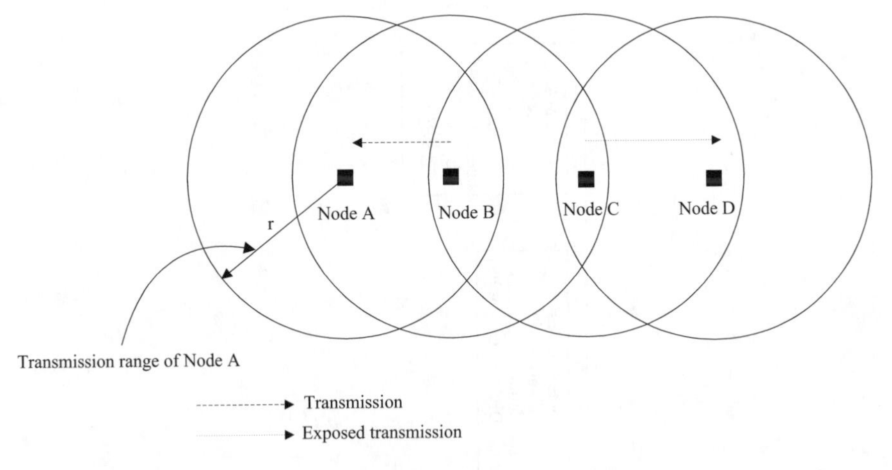

Fig. 3.2. Exposed terminal problem

Sense Multiple Access with Collision Avoidance) and the most commonly considered contention-free approach is TDMA (Time Division Multiple Access). Collision avoidance is an important issue in contention-based medium access control. Karn was the first to suggest the collision avoidance for ad hoc networks [30] in MACA protocol. Many improvements followed this proposal [31–33].

Ever since it was accepted as the basis for the IEEE 802.11 standard, the CSMA/CA received a lot of attention and enhancements through different variations [34, 35]. The IEEE 802.11 standard offers two types of services: contention-based provided by the Distributed Coordination Function (DCF), which is CSMA/CA based, and contention-free, provided by the Point Coordination Function (PCF), which is polling based.

The TDMA approaches present fixed, contention-free methods, where time is divided into synchronized repeating frames and time slots, though the contention usually occurs during the resource (bandwidth) reservation phase.

Contention based ad hoc wireless networking protocols can be classified into three basic categories:

- *Contention-based protocols*
- *Contention-based protocols with reservation mechanisms*
- *Contention-based protocols with scheduling mechanisms.*

Fig. 3.3 provides a detailed classification tree [33]. The [36, 37] give an extensive overview of existing MAC protocols. This subchapter presents only the distinction between the major categories.

Contention-based protocols do not have any bandwidth reservation mechanisms and follow a contention-base channel access policy. All nodes contend

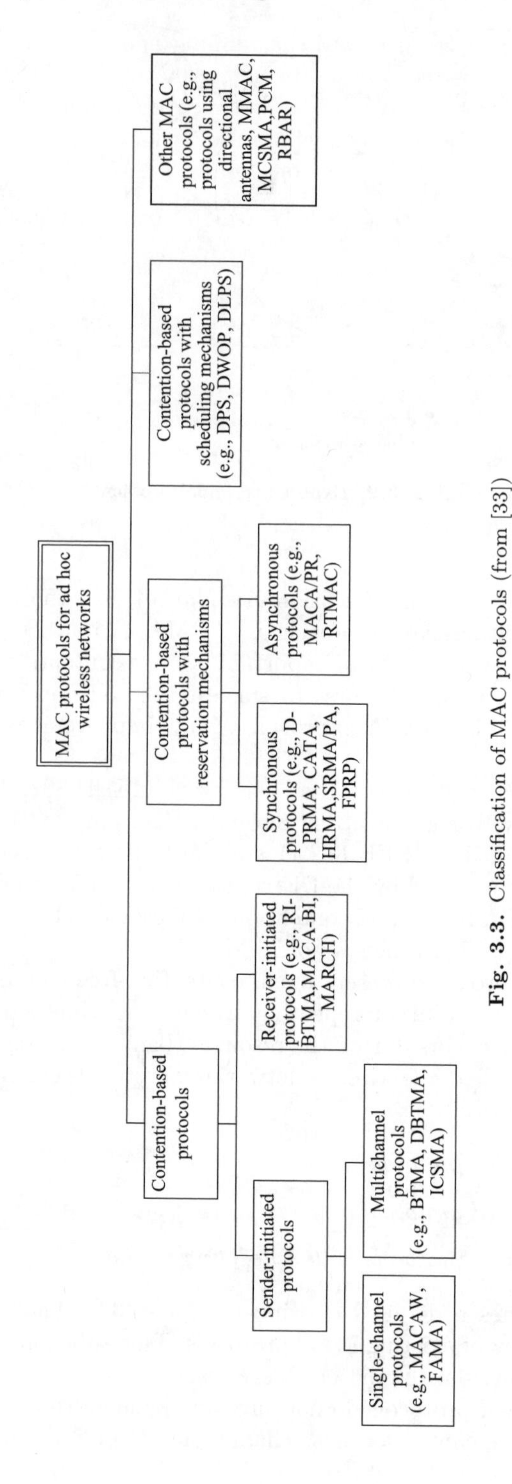

Fig. 3.3. Classification of MAC protocols (from [33])

for the channel simultaneously and the winning node gains the access to the channel. The contention based protocols can have a *carrier sensing operation* and/or *collision avoidance dialog* between the sender and the intended receiver. There is no bandwidth guarantee, which makes these protocols inappropriate for transmitting real-time traffic and enabling QoS guaranties. They can be sender (sender initiate packet transmission) and receiver (receiver initiate contention resolution protocol) initiated. Sender-initiated protocols can be single channel and multi channel depending on how much of the available bandwidth is actually used; either the entire bandwidth is used by one user or is divided into several simultaneous channels used by several nodes.

In wireless ad hoc networks that rely upon a carrier sensing random access approach (e.g., 802.11 DCF), appropriate solution to complex medium phenomena such as hidden- and exposed-terminal problem must be found. According to the DCF, the station must sense the channel before starting its own transmission. If the medium is found idle for an interval longer than *Distribute InterFrame Space (DIFS)*, the station starts the transmission, otherwise the station must continue monitoring the medium until it is found idle for more than a DIFS period. Then, the station waits for a random *back-off interval.* After the random back-off interval expires, the station starts the transmission, but only if the medium is sensed free. If the medium is sensed busy, then the back-off procedure is invoked again. The back-off algorithm is implemented in many media access protocols.

The hidden-terminal problem can be alleviated by extending the DCF basic mechanism through a *virtual carrier sensing.* The transmission phase is preceded with an exchange of two control frames (RTS and CTS frames, as in MACA [30], FAMA [32]) to announce the upcoming frame transmission to the receiver, and, to indicate receiver's readiness to receive the data frames. This handshaking dialog that can include more different frames in different protocols (e.g., acknowledgement — ACK, data sending — DS , request-to-request-to-send RRTS [MACAW]; ready-to-send RTS [MACA-BI], etc.) generally reduces the collisions and improves the system performances, but introduces some overhead and delay. If acknowledgement mechanism is implemented, then the receiving station sends the *acknowledgement frame (ACK)* after successful reception of data frame and waiting for a *Short InterFrame Space (SIFS)* interval. Further modification of this spaces and procedures leads to system improvements, such as enhanced QoS (as in IEEE 802.11e). Examples of contention-based MAC protocol are MACA [30], MACAW [31], FAMA [32], BTMA [38], MACA-BI [39] and others (see Fig. 3.3).

Contention-based protocols with reservation mechanisms are mostly based on TDMA philosophy and implement mechanisms for reserving bandwidth in order to support real-time traffic and provide QoS guarantees. They can require synchronization among all nodes in the network (synchronous protocols) or use relative time information for effecting reservations (asynchronous protocols).

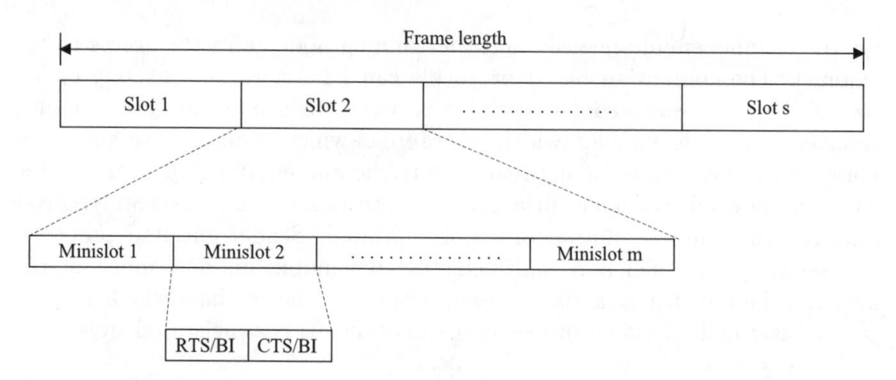

Fig. 3.4. Frame structure protocol with reservation mechanism (D-PRMA [40])

An example of frame structure implementing these mechanisms, introduced in the D-PRMA (distributed packet reservation multiple access) protocol is shown in Fig. 3.4 [40]. The slots in the frame are divided into minislots. RTS/BI (BI-bussy indication) and CTS/BI fields are used to place the requirements and avoid the hidden terminal problem.

Different protocols are based on different frame structures that reflect the reservation mechanism and incorporated functionalities (see [33]). Different implementations can be found in CATA[41], HRMA [42], SRMA/PA [43], FPRP [44], MACA/PR [45], RTMAC [46] and other protocols. Contention mechanisms are generally used for reservation of time slot.

Contention-based protocols with scheduling mechanisms focus on nodes' channel access transmission scheduling, providing fair treatment of all nodes, and on packet scheduling enforcing priorities between different packet flows. These mechanisms can take into consideration various factors that can influence network performances such as: remaining battery power at nodes, packet delay, traffic load, type of packet flow, etc.

Most of the protocols are based on two basic schemes: *distributed priority scheduling (DPS)* and *multi-hop coordination* [47]. The DPS is based on the IEEE 802.11 distributed coordination function mechanisms combined with piggybacking the priority information on RTS/CTS/DATA/ACK packets, as shown in Fig. 3.5. The transmitting packets update the scheduling table with priority tags, marking each source-destination pair. Each node's scheduling table gives the rank of the node with respect to other nodes in the neighborhood.

Multi-hop coordination schemes balance the *priority index* between up-stream and down-stream packets achieving better end-to-end delay performance. These schemes can utilize the time sensitive traffic on ad hoc wireless networks.

Variations of these mechanisms are implemented in different scheduling MAC protocols, such as DWOP [48] and DLPS [49]. Adaptive approach

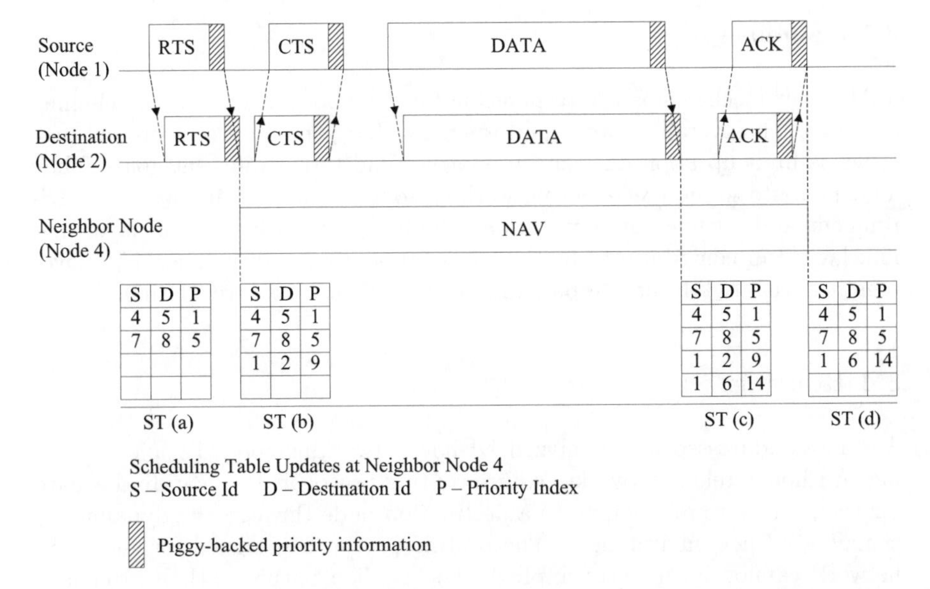

Fig. 3.5. Piggybacking and scheduling table update mechanisms in DPS [47]

and combination with other techniques (e.g., ARQ, feedback mechanisms, directional antennas) can further enhance the possibilities for corrections in end-to-end delay and QoS achievements.

3.3.3 Other MAC Protocols

There are a number of MAC protocols that cannot be strictly classified into the previously mentioned categories. They combine some of the previous approaches, require radical modifications or introduce new mechanisms. The MAC protocols that use directional antennas [50–53] to improve the throughput and directional busy tone-base MAC protocol that improves the spectral reuse and increase the channel capacity [54] can be classified within this group. Another important group of MAC protocols are power control protocols, such as PCM [55–57]. More details are presented in Chapter 4. Interesting adaptation principle (rate adaptation) is implemented in RBAR [58] that improves overall system throughput. The efficient overcoming of exposed terminal problem in ad hoc networks is proposed in ICSMA [33], improving throughput and channel access delay. There are protocols that deal with multihop connections [59] and resolve the use of multiple channels for data transmission (MMAC) [60], and (MCSMA) [61]. These do not cover the variety of recently proposed schemes. Ongoing research on this topic investigates any possibility to achieve better system performances and design the most appropriate access method.

3.3.4 Summary

MAC layer mechanisms are responsible for the nodes' access to the channel and affect the overall network performances. Large number of proposed protocols is built up upon different approaches and target different goals. They all contribute to more efficient networking solutions for specific wireless environment and ad hoc communications. Joint design with physical layer and link layer can contribute to higher efficiency and reliability. The MAC layer together with link layer can play important role in cross-layer design.

3.4 Routing

The most addressed and analyzed ad hoc networking related topic is routing. Ad hoc wireless networks require routing protocols that can find a path for the packets from a source to a destination node through the dynamically changing ad hoc environment. The routing protocols designed for the wired network cannot be directly applied. The ad hoc nature with its dynamic topology, mobility of nodes, absence of established infrastructure and sensitive transmission media makes these solutions more demanding. Routing in ad hoc wireless networks has to cope with specifics such as limited bandwidth, high error rates, dynamic topology, resource poor devices, power constraints and hidden and exposed terminal problems. The routing protocols for wireless ad hoc networks must be fully distributed, adaptive and effective in sense of number of involved nodes and route computational time. They also have to minimize the circulating routing information, quickly converge to optimal routes avoiding loops, optimally use the resources, provide certain level of QoS and support time sensitive traffic. The complexity of the ad hoc routing protocols can vary from simple modifications of Internet protocols to complex multilevel hierarchical schemes. Most of them were developed under set of assumptions that all nodes have homogeneous resources and transmission range, bi-directional links, and support scalability (operation in large networks). The issue of routing in ad hoc networks was often addressed, which resulted in the design of a huge number of different protocols. Each of them aims to solve some challenges, introduce improved approach to previous solutions and compromise between the requirements and the constraints. Along with the previously mentioned requirements, this poses an uneasy task. The reader should see [62–68] for more details. Some of the ad hoc routing protocols are under consideration of Mobile Ad Hoc Networks (MANET) Working Group [69].

The scope of this chapter is to give a short overview of major routing protocols and their characteristics, since the subject of routing protocols is already thoroughly analyzed in many published books and papers. The following paragraph will address some of the wireless ad hoc networks features and present classification approach.

3.4.1 Classification of Routing Protocols

Routing protocols for ad hoc networks can be classified based on different criteria. Most common categories rely on: routing information *update mechanisms*; use of *temporal information* for routing; *routing topology*; utilization of *specific resources*. Differentiations between the categories are not always sharp, and some protocols can be included in more categories, depending on relevant approaches and/or their combinations. Fig. 3.6 presents classifications of routing protocols with most significant representatives, based on [33]. Other approaches can be found in [70–75].

3.4.1.1 Routing Protocols Based on Routing Information Update Mechanisms

Updated and reliable information about nodes and routes is crucial for successful and efficient routing in dynamic ad hoc environments. This information is essential for route discovery and maintenance. Fast and efficient updates, not overloading the network, are desirable features in protocol design. Depending on routing information *update mechanisms*, ad hoc wireless protocols can be classified into three major categories:

- Table driven (proactive) routing protocols
- On-demand (reactive) routing protocols
- Hybrid routing protocols.

Information update mechanisms can be incorporated within a protocol as a separate solution or successfully combined to adjust the particular protocol architecture and desired network parameters. The following paragraph presents global characteristics of these major categories through several most significant representatives and discusses the different possibilities for their improvements and combinations.

Table Driven Routing Protocols *Table driven routing protocols* require that every node maintain the network topology information in form of updated routing tables. Route creation and maintenance is accomplished through some combination of *periodic* and *event-triggered routing updates*. Periodic updates occur at particular time intervals, regardless of the mobility and traffic characteristics. However, time intervals can be adjusted to particular zones (as in Fisheye Routing Protocol). Event driven updates mostly depend on mobility of nodes than can result in link addition or removal. Table driven protocols are derived from the traditional distance vector (DV) protocols [76], link state (LS) protocols [77] and their combinations, originally developed for wire-line Internet.

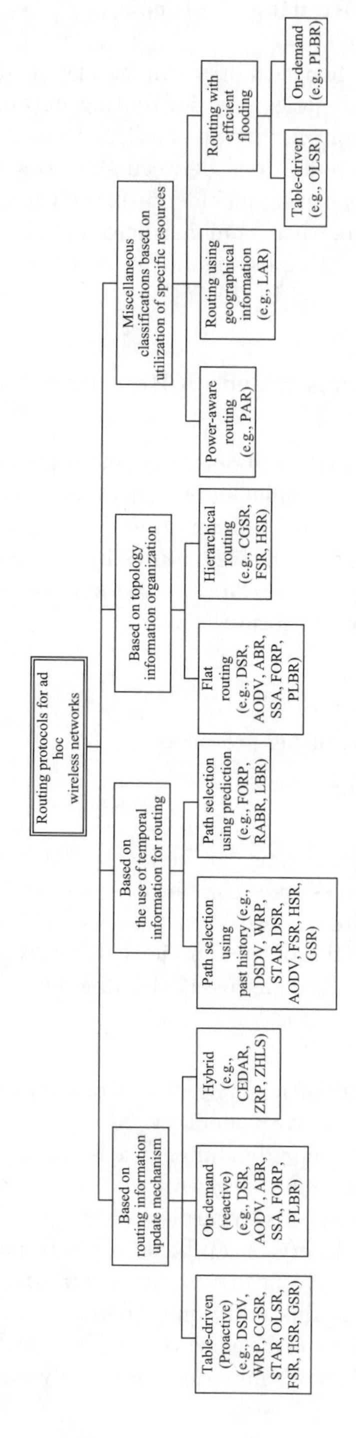

Fig. 3.6. Classification of routing protocols.

The proactive protocols aim to increase the amount of topology information stored at each node in order to avoid loops and speed up protocol convergence. They optimize the flooding information, combine DV and LS feature and dynamically adjust the size of route updates and update frequency.

The well known representatives of this group of routing protocols are using:

- various *distance vector approaches*, such as distance-vector routing protocol (DSDV) [78], and its derivatives such as gateway switch routing protocol (CGSR) [79];
- various *link-state routing approaches*, such as wireless routing (WRP) protocol [80], source-tree adaptive routing (STAR) protocol [81], Optimized Link State Routing (OLSR) protocol [82], Topology Dissemination Based on reverse-Path-Forwarding (TBRPF) protocol [83];
- *merging distance vector* and *link state behavior*, such as Fisheye State (FSR) protocol [84], and
- *combination of different approaches* (hierarchical, clustering, location awareness, on-demand etc.).

Table driven protocols have the advantage of route availability at any time, which can be useful in interactive applications. Their disadvantage is that they produce significant overheads in case of large networks and high mobility. Some of these protocols that are referred as starting concepts for continuous protocol enhancements are presented in the following paragraphs.

Destination Sequenced Distance-Vector Routing Protocol (DSDV) [78], is one of the first routing protocols proposed for the ad hoc wireless networks. As a table-driven protocol, it updates the *routing table* at every node, keeping the latest information about network topology status.

Tables consist of information for all possible destinations from each node, presenting next hop node and the distance to the destination node calculated according to some metrics. Entries in the classical routing tables are enriched with a sequence number, which are increased with every update (see Fig. 3.7). This approach prevents loops, copes with count-to-infinity problem and provides faster convergence. The updated routing tables eliminate the delay in route set-up process. However, the updates due to broken links and node mobility result in heavy control overhead, which makes this protocol inappropriate for ad hoc wireless environment with limited bandwidth and dynamic topology.

Wireless Routing Protocol (WRP) [80], is similar to DSDV, but it updates and maintains *multiple tables*: distance table (DT), routing table (RT), link cost table (LCT) and a message retransmission list (MRL). Those tables contain information that can lead to faster convergence and involve fewer table updates. This protocol requires higher processing power to maintain the multiple tables. At high mobility cases, it involves significant control overhead, which makes it inappropriate for large and highly mobile ad hoc wireless networks.

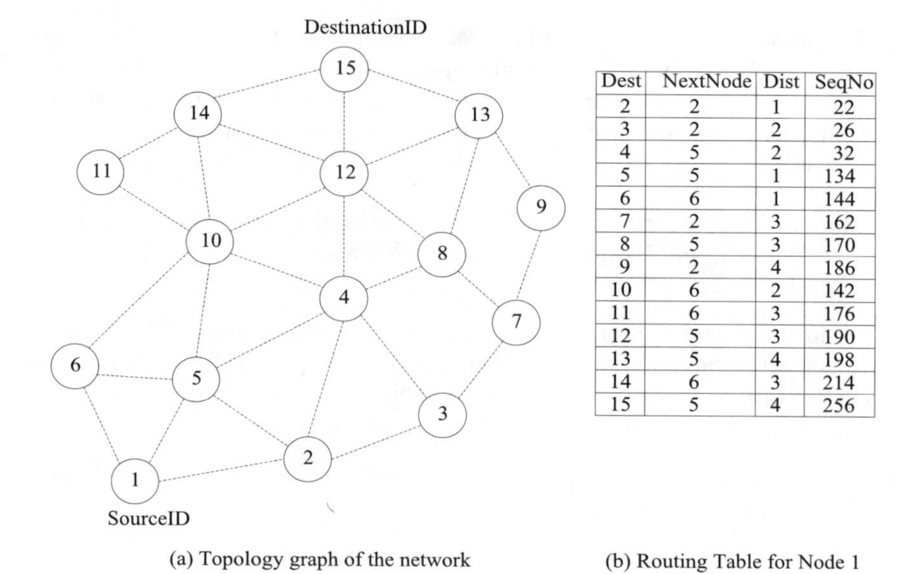

Dest	NextNode	Dist	SeqNo
2	2	1	22
3	2	2	26
4	5	2	32
5	5	1	134
6	6	1	144
7	2	3	162
8	5	3	170
9	2	4	186
10	6	2	142
11	6	3	176
12	5	3	190
13	5	4	198
14	6	3	214
15	5	4	256

(a) Topology graph of the network

(b) Routing Table for Node 1

Fig. 3.7. Route establishment with DSDV

Optimized Link-State Routing (OLSR) Protocol [82] is an optimization of a classical link state protocol adapted for operation in mobile wireless networks. It introduces the *multiple-relay (MPR) nodes* to reduce duplicate broadcast packets and achieve efficient flooding. Each node k selects a minimal set of multipoint relay nodes, *MPR (k)*, among its one-hop neighbors. They provide connectivity to all reachable two-hop nodes, desirably in a symmetric manner (see Fig. 3.8). When node k wants to send the message it is flooding it to all of its multiple-relay nodes. The node retransmits the message only if it has not received the message before. To reach the desired destination the message continues through the chain of follow up multiple-relay nodes. Each node defines also a set of neighbor nodes, for which it acts as their MPR, as its *multipoint relay selector* set of nodes. MPR node is responsible for maintaining the routing information within its selector set of nodes. Periodic *topology control (TC)* messages flow through each MPR's selector set updating the routing tables and reflecting the network dynamism.

Topology Dissemination Based on Reverse-Path-Forwarding (TBRPF) Protocol [83] is a link state routing protocol which applies different techniques for reducing overhead. Each TBRPF node *computes a shortest path tree* to all reachable network nodes and propagates only a part of this tree, so-called *reportable structure (RT)*, to its neighbors. Two types of updates (periodic and differential) are used to maintain updated routing information. They report on topology structure and changes in neighbor status, including only minimum-hop paths from each neighbor to every other neighbor.

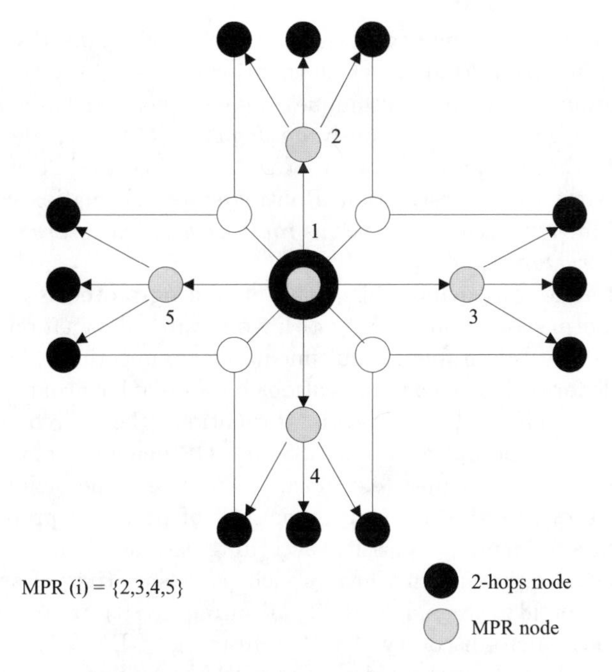

Fig. 3.8. Flooding in OLSR protocol

To reduce the control information, the updates can be combined with Hello messages.

Source-Tree Adaptive Routing (STAR) Protocol [81], is a variation of table-driven protocols, which implements a concept for *least overhead routing* (LORA) rather than optimum routing (ORA). LORA approach does not guarantee the optimum path with respect to the defined metrics, but attempts to provide feasible paths with least overhead. Every node updates and broadcasts the *source-tree* information that consists of wireless links used by the nodes in its preferred path to the destinations and build up a topology graph. STAR has very low communication overhead compared with other table-driven protocols.

On Demand Routing Protocols *On-demand routing protocols* do not maintain the network topology information continuously. They perform a path discovery process and exchange routing information only when the source node demands it. In order to start the packet transmission, the source node checks if the requested route toward the destination node exists. If such route does not exist, it performs a path discovery procedure. Two nodes that have not communicated before do not maintain a route between each other. The route discovery often consists of flooding of request messages through the network.

A reactive protocol manages the path through *path discovery*, *path maintenance* and *path deletion* (which is optional). Data forwarding is accomplished according to two main techniques: *source routing* and *hop-by-hop*. Path discovery is triggered asynchronously *on demand*. Network nodes update the routing state through path discovery process, storing the information about discovered paths to the destination. Routing state information can be maintained with different techniques such as *route caches, temporary routing tables*, and *logical structures*.

Since initiated on demand, this group of routing protocols does not offer any prior information about quality of links towards destination nodes (e.g., bandwidth, delay), desirable in multimedia communications. They may not be applicable for real-time communications because of long route set up delay.

In order to reduce the overhead information, the search area can be redesigned with some optimizations [85–87]. On demand approach is often combined with other features (e.g., location update, zone definition, hierarchical structure) resulting in many variations of proposed protocols, which are sometimes difficult to classify according to the dominating criteria. Many protocols represent this group such as Associativity Based Routing (ABR) protocol [88], Signal Stability Routing (SSR) protocol [89], Relative Distance Micro-discovery Ad Hoc Routing (RDMAR) protocol [90], Temporally-Ordered Routing Algorithm (TORA) routing protocol [91], Adaptive Distance Vector (ADV) protocol [92], etc. The benefit of the on demand approach is that signaling overhead is significantly reduced compared to the table driven solutions. The drawback is the introduction of route acquisition latency, as a result of a route discovery process.

Ad Hoc On-Demand Distance Vector Routing (AODV) Protocol [93] is widely accepted routing protocol for ad hoc mobile networks based on *hop-by-hop* routing model. It provides reactive routing based on a route discovery on demand cycle combined with a broadcast network search and unicast reply containing discovered path.

AODV nodes maintain a route table in which next-hope routing information for destination nodes is stored. The route discovery process is triggered if the originating node do not have information about the destination node. Route discovery procedure starts with creation of a route request (RREQ) packet, which contains destination node's IP address, the last known destination sequence number and the source IP address and source sequence number. The sequence numbers (as in DSDV) are time marks for freshness of the route. RREQ packet also contains a hop counter, started from zero and a RREQ ID, which is increasing each time the node initiates a new route request. The source node broadcast RREQ packet towards the destination node. As a node forwards RREQ, it sets up a reverse path to the originating node. A destination sends route reply (RREP) messages to the source. The selected route is composed based on the most fresh route information and smallest number of hops. An example of propagation and reverse path set-up in AODV is presented in Fig. 3.9.

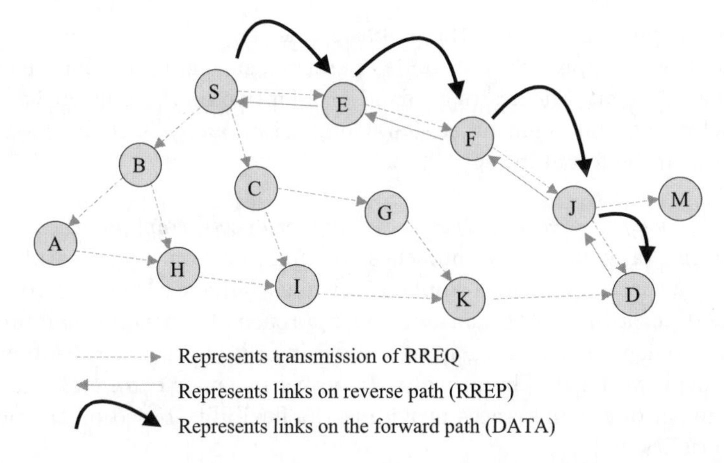

Fig. 3.9. Propagation and path set-up in AODV

As a result of the discovery phase and a request-reply packet transmission, a new routing state is created at the nodes. The state consists of the *routing tables* entries that record the next-hope node to the destination of the forward path. Route maintenance is based on periodic beaconing messages. AODV deals with route breaks creating route error messages.

To improve the protocol performance and reduce overhead, AODV can implement specific optimization techniques called *expanding ring search* [94], where the incrementally larger areas of the network are searched until a route to the destination is discovered. It can also implement a number of optional features (e.g., gratuitous RREP, RREP acknowledgements) to improve operation in a wide range of scenarios.

Dynamic Source Routing (DSR) protocol [95] is another reactive routing protocol, based on the *source routing* techniques. During the route discovery cycle, data packets contain strict source routes that specify each node along the path to the destination. Each node maintains a *route cache*, which contains records for multiple paths. The DSR protocol is supported by two mechanisms: Route Discovery and Route Maintenance. The route discovery is triggered by the source node, if it does not have path information into its cache. The discovery is based on flooding of route requests (RREQ) packets. Intermediate nodes react to RREQ control packet with sending reply message (RREP) to the source node (concatenating the information about the path from the source and the path towards the destination from the cache), discarding the packet if it was already received or appending its own identification (ID) into the broadcast packet to its neighbors. Route maintenance procedure reacts on a broken link to the next-hop removing it from the cache and sending backward to the source node Route Error message. Then, the source node triggers a new route discovery procedure.

Many improvements to these basic mechanisms enhance the node functionality (e.g., promiscuous node operation, salvaging, random delays) or cache management schemes (e.g., fixed or adaptive lifetime, negative cache). More details can be found in [96–98]. Comparison analysis of the on-demand protocols can be found in [64].

Hybrid Routing Protocols *Hybrid routing protocols* combine the best features of the previous two approaches, defining certain *zones* within which the *table-driven* approach is applied, whereas outside the zones routing is performed according to the *on-demand* approach. Each node maintains the network topology information up to k nodes in order to reduce the latency and the protocol overhead. The proactive/reactive behavior can be also adjusted to certain set of circumstances providing the flexibility based on the network characteristics [99].

There are several protocols within this group: Zone Routing Protocol (ZRP) [99], DDR [100], CEDAR [101], ZHLS [102], etc. ZRP effectively combines the best characteristics of both proactive and reactive protocols. CEDAR integrates routing and QoS support in efficient way while ZHLS uses the geographical information of the nodes to form non-overlapping zones, significantly reducing the storage requirements and the communication overhead in dynamic situations. The major characteristics of the hybrid routing approach will be represented through the ZRP protocol.

Zone Routing Protocol (ZRP) divides the network into several *zones* [99]. A zone $Z(k,n)$ for a node n with radius k, is defined as a set of nodes at a distance not greater than k hops:

$$Z(k,n) = \{i|H(n,i) \leq k\}$$

where $H(i,j)$ is the distance in number of hops between node i and j The node n is called a *central node*, while the node b is called a *peripheral node* if $H(n,b) = k$, see Fig. 3.10.

Each node performs *proactive routing* (check its routing table) within the zone and *reactive routing* outside the zone. If the destination node is not found within the zone (S's zone), the peripheral nodes start the search outside the zone, within the newly defined zone regions (e.g., B's zone). For intrazone routing ZRP defines the Intrazone Routing Protocol (IARP), while for interzone routing ZRP utilizes the Interzone Routing Protocol (IERP).When the destination node is found, the reply is generated and sent back towards the source node. The ZRP protocol maintains proactive routing information locally (within the routing zone), while reactively acquiring routes to destinations beyond the routing zone. Within the zone, the proactive routing protocol provides updated and detailed information about each node surrounding topology. *Border-casting* is used in place of traditional broadcasting to improve the efficiency of global reactive routing protocol [71]. Since there is no coordination among nodes, the zones can overlap which

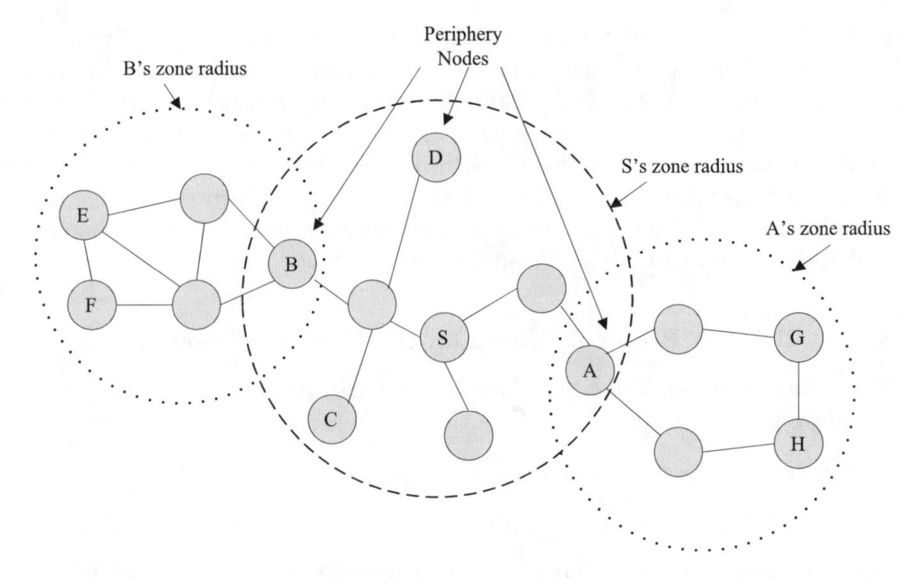

Fig. 3.10. A ZRP routing zone with 2-hop radius

can significantly decrease the performance. An improvement of ZRP with a definition of non-overlapping zones and construction of a routing *forest* is proposed in Dynamic Distributed Routing (DDR) [100]. New version of ZRP, ZRPv2 was introduced in [103], where the border-casting is performed on hop-by-hop basis, eliminating the need for maintaining an extended routing zone.

3.4.1.2 Routing Protocols Based on the Use of Temporal Information for Routing

The routing protocols within this group are using temporal information (e.g., lifetime of the wireless links, lifetime of the selected paths) to make routing decision. The protocols within this group can be classified as:

- Routing protocols using past temporal information;
- Routing protocols that use future temporal information.

They can take into consideration different metrics collected from the network history (i.e., remaining battery power) or rely on prediction of future behavior (prediction of link availability). Fig. 3.6 shows that some of these protocols already belong to previously mentioned classes, using proactive/reactive method for necessary updates (e.g., DSDV, GSR, AODV, DSR).

Flow-Oriented Routing Protocol (FORP), [104] employs a prediction-based *multi-hop-handoff* mechanism to cope with effects of path breaks on real-time packet flows. It is foreseen for IPv6 based ad hoc networks with QoS requirements. FORP uses prediction-based mechanisms that utilize the mobility and location information of nodes to estimate the link expiration time (LET). The minimum of these values on a route is marked as route expire time (RET). FORP requires the availability of GPS in order to identify the location of the nodes, to predict their direction of movement and transmission range and to estimate LET. The adaptive proactive route reconfiguration mechanism performs well in highly dynamic ad hoc environment.

3.4.1.3 Routing Protocols Based on the Routing Topology

Based on the network topology, ad hoc wireless routing protocols can be classified as:

- Flat topology routing protocols
- Hierarchical topology routing protocols.

Advantages and disadvantages of both categories are described in the following paragraphs.

Flat Topology Routing Protocols Flat architectures do not assume grouping of nodes and are more appropriate for smaller networks. Connections are established between nodes that are in close proximity. Routing between any two nodes is constrained only by the connectivity conditions. These protocols have *equal nodes* responsibility without grouping nodes in any manner. This results in lower processing power, better wireless spectrum usage and capacity. However, they produce more overhead in high mobility case and larger networks. Many important reactive protocols belong to this category (see Table 3.2).

Hierarchical Topology Routing Protocols Unlike the flat topology routing protocols, the hierarchical topology routing protocols can consider either one or multiple topology levels. The ad hoc nodes are grouped into *clusters*. The grouping is performed based on some criteria: location, particular functionality, etc. The clusters can be overlapping or completely disjoint. Once the network has been initialized and the clusters have been established, *cluster leader selection* and *revocation algorithms* take place. There are different approaches to these algorithms (e.g., weighted based, lowest ID). The control within the cluster is usually performed on a centralized manner by cluster leader or completely distributed with no cluster leader. The nodes that are usually located within the boundaries of multiple clusters, or serve as routers between clusters perform the *gateway* functionalities.

In order to improve the network scalability, clusters can be organized in multilevel hierarchy. The hierarchy includes logical and physical topologies and the logical associate addressing schemes. In this approach, each cluster becomes a node at the next highest cluster level. Routes can be organized as hierarchical routing tree.

To create the clusters and to maintain current information about cluster members and gateway availability, many cluster-based protocols require periodic overhead. To overcome this disadvantage and to reduce the fraction of overhead, some protocols introduce on-demand approach and create clusters only when needed [105]. Other disadvantages include the centralization of routes through cluster leaders, which can affect their battery lifetime. This can be overcomed if the cluster leaders are selected among nodes with higher resources (e.g., in military communications). The hierarchical approach better suits Internet, where the number of nodes is larger and there is no high sensitivity to overhead information.

Hierarchical routing and hierarchical addressing schemes are beneficial to the network robustness and increase the routing flexibility and route lifetime. They can enhance the resource allocation and management, reduce the size of the routing tables and provide better network scalability.

Hierarchical State Routing (HSR) Protocol is a distributed multi-level hierarchical routing protocol that performs clustering in different levels [33]. The first level of physical clustering is performed among nodes that are reachable in a single wireless hop (see Fig. 3.11). The second level of clustering is done between the nodes that are elected as *cluster leaders* of each of these first-level clusters. The cluster hierarchy can continue to be build up upon physical (geographical positions) or some logical clustering scheme (logical relations). Cluster leaders (cluster heads) are responsible for resource allocation and membership management at every level of clustering. They exchange the link state and topology information among their peers in the neighborhood clusters. This exchange is done over a set of hops consisting of gateway nodes (which belong to multiple clusters) and cluster heads. The path between two cluster heads is called *virtual link (tunnel)*. The cluster heads flood the information to the lower levels providing all nodes with hierarchy topology information.

Though this method can reduce the routing table size and offers some benefits through resource efficient resource management performed by cluster leaders (e.g., allocation of spreading codes), overhead involved in the leader election process and multiple level hierarchy information do not suit perfectly the ad hoc wireless concept. It is more appropriate in Internet or some military applications, where cluster heads can be chosen between devices with more significance.

Fisheye State Routing (FSR) Protocol [84], aims at reducing the routing overhead by introducing mechanisms for different frequency updates between close and distant nodes. It is proactive protocol based on the so-called "*fisheye technique*" and combines the distance vector and link-state behavior. It maintains a topology table and a route table at each node, transmitting the link-state packets to the neighbors instead of flooding them (method borrowed from the Global State Routing protocol [106]).

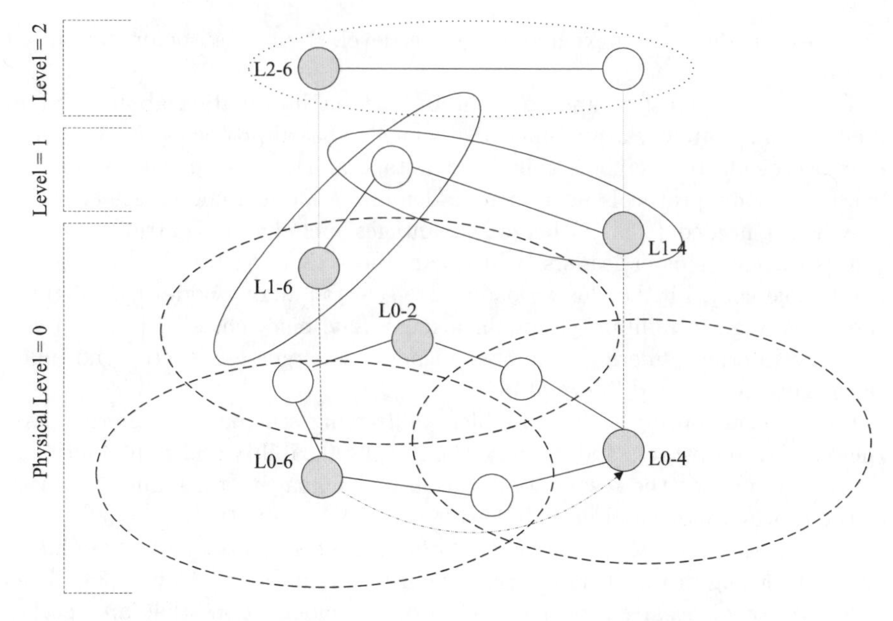

Fig. 3.11. Example of HSR multi-level clustering

FSR introduces the notion of *scope* to define regions of networks with different accuracy in routing information (see Fig. 3.12). This keeps the immediate neighborhood topology information maintained at a node more precise compared to the information about nodes farther away form it.

FSR is based on an optimally route-route cost trade off. FSR is suitable for large and highly mobile ad hoc networks. However, different mobility patterns can significantly influence its benefit.

3.4.1.4 Routing Protocols Based on Utilization of Specific Resources

The specifics of ad hoc networking, which has to be performed under unstable and variable conditions and imposed limitations, invoke necessity of particular optimizations towards some networking resources. Protocols within this category are not homogeneous in sense of utilization of specific resources. This category includes:

- Power aware routing
- Geographical approaches
- Routing with efficient flooding
- Security Aware Protocols.

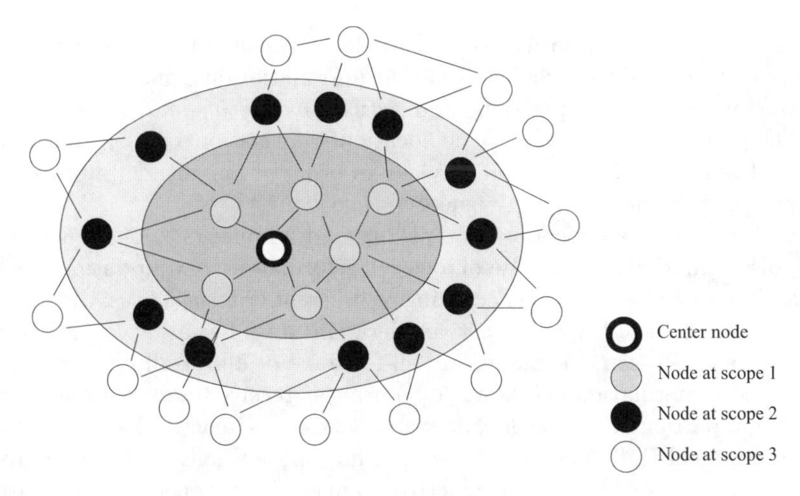

Fig. 3.12. An example of scopes in the FSR protocol

The protocols within this category can be partly overlapping with some previously mentioned classes. Next paragraphs give a short description of the major features of each category. Some of them are elaborated in more details in the following chapters of this book.

Power Aware Routing Power constraints are of special interest for ad hoc wireless networks. Ad hoc nodes have certain requirements towards portability, weight, size and battery life, which, if carefully taken into consideration, can improve the efficient utilization of energy and increase the lifetime of the network. The sensitive nature of the wireless links affects the energy consumption and efficiency of the ad hoc network. On the other hand, the energy efficiency of a routing protocol is influenced by different aspects of the physical/MAC layer, architecture, transport layer solutions and applications. Design of power aware protocols [107, 108] should consider appropriate routing metrics such as: minimal energy consumption per packet, minimum variance in node power level, minimum cost per packet, maximized network connectivity, minimized maximum node cost. These protocols take a variety of approaches towards energy efficiency performing powering down unutilized nodes, load balancing, transmission power adjustments, localization awareness, etc. This important issue will be addressed in more details in Chapter 4.

Geographical Approaches The knowledge about the geographical position of the nodes can shorten and facilitate the routing process eliminating the necessity for route discovery and maintenance. In position-based approaches, the nodes must be able to determine their own position, applying some type of positioning mechanisms [86, 109]. The node forwards packets to its one-hop neighbors, based on their physical position and the physical position of the destination node. The position-based (or location-aware) routing consists of *location service*, which provides location information and maps the ID of a node to its geographic position, and actual *routing of data packets*, based on the previous information. The location information can be allied into different stages of routing and can yield an order of magnitude improvement in performance and bandwidth/energy usage.

Location Services depend on complexity and number of nodes, query process, dissemination and storage of location information, and updating mechanisms. They can be classified according to different criteria [71] as: distributed vs. centralized, reactive vs. proactive, hierarchical vs. flat or, according to the type, as: some-to-some, some-for-all, all-for-some or all-for-all.

Location information necessary for routing is received and maintained differently in particular protocols. For example, a simple reactive location service (RLS) protocol [110, 111] stores location information only at the node itself and retrieves it on query in a reactive manner. Homezone [112] and Home Agent [113] location services introduce the concept of a virtual home region of a node and maintain up-to-date position information for each node in the defined region through hashing the identifier of a node to its position. The concept of quorum (used in distributed database design) and some optimization criteria are implemented in *quorum-based scheme* [114] to provide and update location information.

The knowledge of a node location is an important issue in wireless networks with and without infrastructure. Many different ad hoc techniques for *location discovery* were developed stemming from the highly dynamic nature of mobile wireless devices and difficulties in providing infrastructure support and utilizing distributed approach in managing location information. Location discovery is fundamentally based on two phases: a *measurement phase* (measures distance, angular/optical set of anchor points) and a *combining phase* that combines the measurements to produce the final estimates.

These techniques have to cope with randomly deployed nodes, rapid infrastructure installation with *self-calibration*, localization in the presence of obstacles with non-line-of-sight additional measurements, distributed operation and constraints according to embedded processing power within the nodes. The localization process also requires some system integration effort, since nodes need to coordinate and cooperate on multiple levels. Interconnections to physical layer (signal strength, radio connectivity, measurement techniques, accuracy validation) attract attention in latest proposed techniques, [115–118]. The context-awareness becomes very important providing framework for location reference (relative or global) [119]. New solutions are offered for integration of ad hoc sensors localization system [120, 121] in a novel concept of emergency networks [122]. More details on this topic can be found in [123, 124].

The actual *routing of data packets* based on position information can be separated into three distinct categories:

- *Greedy routing* that works by forwarding the packets towards the destination and is implemented in the Most-Forward within Radius

(MFR) policy [125], the Nearest with Forward Progress (NFP) scheme [126], Geographic Distance Routing (GEDIR) [127] and *compass routing* scheme [128];

- *Direct flooding* where the packets are broadcasted in general direction to the destination as in DREAM [129] and LAR [86, 130]; and
- *Hierarchical routing* which combines position-based and non-position-based routing algorithms in different geographical zones. Example can be found in GRID [131] and Terminode [132] routing protocols where routing is done by means of a position-independent protocol at the local level and a greedy variant at the long-distance level. Interesting multi-zone approaches can be found in [133, 134].

There will be increase in location aware applications in the future. Combinations with context-awareness and service discovery procedures are also foreseen as beneficial solutions in highly dynamic ad hoc environments.

Routing with Efficient Flooding The route discovery process usually implements some query dissemination mechanism. Many of the existing on-demand routing protocols perform *flooding* of control packets (e.g., RouteRequest) in order to find a feasible path with the required constraints. Flooding of control packets produces significant traffic load, wastage of bandwidth and increase the number of collisions. In case of high mobility and intensive topology changes flooding can provoke *broadcasting storm.* Flooding can be combined with different routing techniques and optimizations.

Existing routing protocols that employ efficient flooding mechanisms include preferred link-based routing (PLBR) protocol [135] and optimized link state routing (OLSR) protocol [82]. They combine the flooding with on-demand and table-driven techniques.

In distance routing effect algorithm for mobility (DREAM) [129] each node maintains a position database that stores location information for every other node. An entry in the database includes a host's ID, location and time when the entry was created. Each host regularly floods packets to update its database. The *temporal resolution* and *spatial resolution* mechanisms are proposed to control the accuracy of location information.

Security Aware Protocols Ad hoc networks are more vulnerable to security attacks compared to wired or infrastructured networks. The security requirements include confidentiality, integrity, availability and non-reputations, similar with other networks. Different attacks can be classified as internal and external and can require security actions on different protocol layers. Different techniques to overcome the attacks include cryptography and key management approach. A number of security protocols SAR [136], SEAD [137], ARAN [138] are trying to offer solutions and improve the network security. Security features concerning ad hoc networking will be analized in Chapter 8.

3.4.1.5 Other Techniques

The completely distributed nature and the ability to operate without any infrastructure makes ad hoc wireless networks an important candidate for the future wireless systems. Continuous topology changes, strong requirements in minimizing battery consumption and limited network resources make the problem of routing in wireless ad hoc networks a real challenge. Specific routing protocols emerge in order to fulfill the new requirements. Multicast is one of the areas in mobile ad hoc networks which is to play a key role in future wireless systems [33, 139]. It follows the mobility aware mesh-based approaches. Other issues are scalability [140] and QoS guaranties. Inter layer and cross layer optimizations address routing as extremely important feature in optimization process. Self organization also has to be followed with particular routing procedures [141]. In future wireless ad hoc networks adaptive routing techniques will contribute to efficient communications in unstable and changing ad hoc environments.

3.4.2 Summary

The routing in ad hoc networking is still an open issue. Combinations of different features and approaches already exist, while there is a constant emerge of new ideas day by day. The most common protocols, up to date, based on basic proactive and reactive approaches and their main characteristics are summarized in Tables 3.2 and 3.3.

Variety of different protocols aim to achieve better routing efficiency and overall system performance. It is evident that while many interesting ideas have been proposed, still other efforts are required to fully understand how all the needs can be satisfied in a cost efficient way.

Demanding research areas will continue to search the possibility for best solution for loop avoidance, fast convergence and localized reaction upon topology changes, possibility of multiple routes information, unidirectional link support, QoS support, support for network partition, scalability, etc.

Together with physical layer and especially with MAC layer design, routing is a leading technique that offers capabilities for optimization and network enhancements. Some of these procedures will be analyzed in the following chapters.

3.5 Concluding Remarks

This chapter provides an overview of intriguing aspects and specifics of physical and MAC layers relevant to ad hoc wireless networking. It also gives a classification of routing protocols with briefing of their main characteristics.

Table 3.2. Comparison of on-demand protocols [71]

Protocol		AODV	DSR	LAR	TORA	ZRP
Route discovery	Query	Flooding	Flooding	Selective flooding	Flooding	Selective flooding
	Storing mechanism	Hop-by-hop	Accumulation	No	Hop-by-hop	Hop-by-hop/ accumulation
	Reply transmission	Unicast	Unicast	Flooding	Flooding	Unicast
	Intermediate node reply	Yes	Yes	No	No	Yes
	Multiple routes	No	Yes	No	Yes	Yes
Route maintenance	Proactive/ reactive	Proactive (hello)	Reactive	Reactive	Proactive (hello)	Proactive (IARP)
	Local partial Repair	No	Yes	No	Yes	Yes
	Route deletion	Soft	Soft	Soft	Hard	Soft
Route state		Route table	Route cache	Cache at source	DAG	Route tables

Table 3.3. Comparison of proactive protocols [71]

		DSDV	FISHEYE	OLSR	WRP
Classification		DV	Hybrid DV and LS	LS	PFA
Route update	Main content	<D,H,NH,SN>	<D,NL,SN>*	MPR set	<D,H,NH, PR,SN>*
	Target nodes	Neighbors	Neighbors	All	Neighbors
	Transmission	Periodic/on event	Periodic*	Periodic	Periodic/on event
Maintain information		Forwarding table	Topology table	Topology table	Distance table
		Setting table time	Neighbor list	Neighbor list	Routing table
			Routing table	Routing table	Link cost table
					Mess. retransm. List

Notes:
D — destination, NH — next hop, H — number of hops, NL — neighbor list, SN — sequence number, PR — predecessor node, LS — link state, PFA — path finding algorithm.

The chapter aims to give a broad understanding of this important part of ad hoc network design through which a variety of system characteristics and performances can be created.

References

[1] Bluetooth Core Specification Version 1.2, November 2003. http://www.bluetooth.org

[2] IEEE 802.11 Local and Metropolitan Area Networks: Wireless LAN Medium Access Control (MAC) and Physical Layer Specifications, ISO/IEC 8802-11:1999, at http://grouper.ieee.org/groups/802/11

[3] CEPT T/R 22-06, "*Harmonized Radio Frequency Bands for High Performance Radio Local Area Networks (HIPERLAN) in the 5 and 17 GHz frequency range.*"

[4] Multimedia Mobile Access Communications Systems (MMAC), standardized by the Association of Radio Industries and Businesses (ARIB), http://www.arib.or.jp

[5] IEEE P802.11 Standardization Committee, "*Minutes of the High Throughput Study Group Channel Model Special Committee Teleconference on 19 June 2003*," IEEE 802.11-03/460r0, June 2003.

[6] CEPT T/R 22-06 "*Harmonized Radio Frequency Bands for Higher Performance Radio Local Area Networks (HIPERLAN) in the 5 GHz and 17 GHz Frequency Range*".

[7] ITU Study Groups, Documents 8A-9B/58-E, "*Spectrum Aspects of Fixed Wireless Access*," 1998.

[8] "*BroadWay, the Way to Broadband Access at 60 GHz*," IST-2001-32686. http://www.ist-broadway.org

[9] Rappaport, T. S., *Wireless Communications Principles and Practice*, Prentice Hall, 1996.

[10] Popovski, P., Yomo, H., Guarracino, S., and Prasad, R., "Adaptive Mitigation of Self-Interference in Bluetooth Scatternets," *Proceedings on the 7th Conference on Wireless Personal Multimedia Communications (WPMC)*, September 2004.

[11] Popovski, P., Yomo, H., Aprili, S., and Prasad, R., "Frequency Rolling: A Cooperative Frequency Hopping for Mutually Interfering WPANs," *The 5th ACM International Symposium on Mobile Ad Hoc Networking and Computing (MOBIHOC)*, May 2004.

[12] Landsford, J., "MEHTA: A Method for Coexistence between Co-Located 802.11b and Bluetooth Systems," IEEE 802.15-00/36r0, November 2000.

[13] Chiasserini, C. F. and Rao, R. R., "A Comparison between Collaborative and Non-collaborative Coexistence Mechanisms for Interference Mitigation in ISM Bands," *VTC 2001 Spring, IEEE 53rd VTC*, 3, 2001, pp. 2187–2191.

[14] Ferro, E. and Potori F., "Bluetooth and Wi-Fi Wireless Protocols: A Survey and a Comparison," *IEEE Wireless Communications*, 12(1), 2005, pp. 12–26.

[15] Infrared Radio Association (IrDA), at http://www.irda.org.

[16] Williams, S., "IrDA: Past, Present and Future," *IEEE Personal Communications*, 7(1), February 2000, pp. 11–19.

[17] Lee, T. H., Samavati, H., and Rategh, H. R., "5-GHz CMOS Wireless LANs," *IEEE Transactions on Microwave Theory and Techniques*, 50(1), January 2002, pp. 268–280.

[18] Viterbi, A. J., *CDMA Principles of Spread Spectrum Communication*, Addison — Wesley, 1998.

[19] Morikura, M. and Matsue, H., "Trends for IEEE 802.11 Based Wireless LAN," *IEICE Trans.*, J84-B(11), November 2001, pp. 1918–1927.

[20] Prasad, R., "*OFDM for Wireless Multimedia Communications Systems*," Artech House, 2004.

[21] Hara, S. and Prasad, R., "*Multicarrier Techniques for 4G Mobile Communications*," Artech House, 2003.

[22] Hirt, W. and Porcino, D., "*Pervasive Ultra-wideband Low Spectral Energy Radio Systems (PULSERS),*" white paper, WWRF/WG4/UWB-subgroup, contr. to WWRF7 meeting, Eindhoven, December 2–4, 2002.

[23] Tafazolli, R., "*Technologies for the Wireless Future*," John Willey and Sons, 2005.

[24] Forester, J., Green, E., Somayazulu, S., and Leeper, D., "Ultra-Wideband Technology for Short- or Medium-Range Wireless Communications," *Intel Technology Journal*, 2nd quarter, 2001. http://developer.intel.com/technology/itj/q22001/articles/art_4.htm

[25] Gandolfo, P., "XtremeSpectrum — SG3a CFA response," IEEE P802.15 ALT PHY Study Group, Document 02031r0P802-15_SGAP3-CFAResponseAltPhy.ppt, January 2002.

[26] Prasad, R., "60 GHz Systems and Applications," *Second Annual Workshop on 60-GHz WLAN System and Technologies*, Kungsbacka, Sweden, May 15–16, 2001.

[27] Burracchini, E., "The Software Radio Concept," *IEEE Communications Magazine*, September 2000, pp. 138–143.

[28] Costlow, T., "Cognitive Radios will adapt to Users", *IEEE Intelligent Systems*, May/June 2003, pp. 7.

[29] Tobagi, F. A. and Kleinrock, L., "Packet Switching in Radio Channels: Part II — The Hidden Terminal Problem in Carrier Sense Multiple Access Modes and the Busy Tone Solution," *IEEE Transactions on Communications*, 23(12), 1975, pp. 1417–1433.

[30] Karn, P., "MACA — A New Channel Access Method for Packet Radio," in *Proceedings of ARRL/CCRL Amateur Radio 9th Computer Networking Conference*, September 1990.

[31] Bhargavan, V., Demeers, A., Shenker, S., and Zhang, L., "MACAW — A Media Access Protocol for Wireless LAN's," in *Proceedings of ACM SIGCOMM '94*, September 1994.

[32] Fullmer, C. L. and Garcia-Luna-Aceves, J. J., "Floor Acquisition Multiple Access (FAMA) for Packet Radio Networks," in *Proceedings of ACM SIGCOMM*, 1995, pp. 212–225.

[33] Siva Ram Murthy, C. and Manoj, B. S., *Ad Hoc Wireless Networks: Architectures and Protocols*, Prentice Hall Communications Engineering and Emerging Technologies Series, 2004.

[34] Official Homepage of the IEEE 802.11 Working Group, http://grouper.ieee.org/groups/802/11.

[35] IEEE Standard 802.11, "Wireless LAN Medium Access Control (MAC) and Physical Layer (PHY) Specifications," August 1999.

[36] Bertsekas, D. and Gallager, R., *Data Communications*, 2nd edition, Prentice Hall, 1992.
[37] Ilyas, M., *The Handbook of Ad Hoc Wireless Networks*, CRC Press, 2003.
[38] Deng, J. and Haas, Z. J., "Dual Busy Tone Multiple Access (DBTMA): A New Medium Access Control for Packet Radio Networks," *Proceedings of IEEE ICUPC 1998*, 1, pp. 973–977, October 1998.
[39] Talucci, F. and Gerla, M., "MACA-BI (MACA by Invitation): A Wireless MAC Protocol for High Speed Ad Hoc Networking," *Proceedings of IEEE ICUPC 1997*, 2, pp. 913–917, October 1997.
[40] Jiang, S., Rao, J., He, D., and Ko, C. C., "A Simple Distributed PRMA for MANETs," *IEEE Transactions on Vehicular Technology*, 51(2), pp. 293–305, March 2002.
[41] Tang, Z. and Garcia-Luna-Aceves, J. J., "A Protocol for Topology-Dependent Transmission Scheduling in Wireless Networks," *Proceedings of IEEE WCNC 1999*, 3(1), pp. 1333–1337, September 1999.
[42] Tang, Z. and Garcia-Luna-Aceves, J. J., "Hop-Reservation Multiple Access (HRMA) for Ad Hoc Networks," *Proceedings of IEEE INFOCOM 1999*, 1, pp. 194–201, March 1999.
[43] Ahn, C. W., Kang, C. G., and Cho, Y. Z., "Soft Reservation Multiple Access with Priority Assignment (SRMA/PA): A Novel MAC Protocol for QoS-Guaranteed Integrated Services in Mobile Ad Hoc Networks," *Proceedings of IEEE Fall VTC 2000*, 2, pp. 942–947, September 2000.
[44] Zhu, C. and Corson, M. S., "A Five-Phase Reservation Protocol (FPRP) for Mobile Ad Hoc Networks," *ACM/Baltzer Journal of Wireless Networks*, 7(4), pp. 371–384, July 2001.
[45] Lin, C. R. and Gerla, M., "Real-Time Support in Multi-Hop Wireless Networks," *ACM/Baltzer Journal of Wireless Networks*, 5(2), pp. 125–135, March 1999.
[46] Manoj, B. S. and Siva Ram Murthy, C., "Real-Time Traffic Support for Ad Hoc Wireless Networks," *Proceedings of IEEE ICON 2002*, pp. 335–340, August 2002.
[47] Kanodia, V., Li, C., Sabharwal, A., Sadeghi, B., and Knightly, E., "Distributed Priority Scheduling and Medium Access in Ad Hoc Networks," *ACM/Baltzer Journal of Wireless Networks*, 8(5), pp. 455–466, September 2002.
[48] Kanodia, V., Li, C., Sabharwal, A., Sadeghi, B., and Knightly, E., "Ordered Packet Scheduling in Wireless Ad Hoc Networks: Mechanisms and Performance Analysis," *Proceedings of ACM MOBIHOC 2002*, pp. 58–70, June 2002.
[49] Karthigeyan, I., Manoj, B. S., and Siva Ram Murthy, C., "A Distributed Laxity-Based Priority Scheduling Scheme for Time-Sensitive Traffic in Mobile Ad Hoc Networks," to appear in *Ad Hoc Networks Journal*.
[50] Nasipuri, A., Ye, S., You, J., and Hiromoto, R. E., "A MAC Protocol for Mobile Ad Hoc Networks Using Directional Antennas," *Proceedings of IEEE WCNC 2000*, 1, pp. 1214–1219, September 2000.
[51] Ko, Y. B., Shankarkumar, V., and Vaidya, N. H., "Medium Access Control Protocols Using Directional Antennas in Ad Hoc Networks," *Proceedings of IEEE INFOCOM 2000*, 1, pp. 13–21, March 2000.

[52] Dyberg, K., Farman, L., Eklof, F., Gronkvist, J., Sterner, U., and Rantakokko, J., "On the Performance of Antenna Arrays in Spatial Reuse TDMA Ad Hoc Networks," in *Proceedings of IEEE MILCOM*, Anaheim, California, October 2002.

[53] Bao, L. and Garcia-Luna-Aceves, J. J., "Transmission Scheduling in Ad Hoc Networks with Directional Antennas," in *Proceedings of ACM MOBICOM*, Atlanta, Georgia, September 2002.

[54] Huang, Z., Shen, C. C., Srisathapornphat, C., and Jaikaeo, C., "A Busy Tone-Based Directional MAC Protocol for Ad Hoc Networks," *Proceedings of IEEE MILCOM 2002*, October 2002.

[55] Monks, J. P., Bhargavan, V., and Hwu, W. W., "A Power Controlled Multiple Access Protocol for Wireless Packet Networks," in *Proceedings of IEEE INFOCOM*, Anchorage, Alaska, April 2001.

[56] ElBatt, T. and Ephremides, A., "Joint Scheduling and Power Control for Wireless Ad-Hoc Networks," in *Proceedings of IEEE INFOCOM*, New York, June 2002.

[57] Fahmy, N. S., Todd, T. D., and Kezys, V., "Ad Hoc Networks with Smart Antennas Using IEEE 802.11-Based Protocols," in *Proceedings of IEEE ICC*, 2002.

[58] Holland, G., Vaidya, N., and Bahl, P., "A Rate-Adaptive MAC Protocol for Multi-Hop Wireless Networks," *Proceedings of ACM MOBICOM 2001*, pp. 236–251, July 2001.

[59] Ramanathan, R. and Hain, R., "Topology Control of Multihop Radio Networks using Transmit Power Adjustment," in *Proceedings of IEEE INFOCOM*, Tel Aviv, Israel, 2000.

[60] So, J. and Vaidya, N. H., "A Multi-Channel MAC Protocol for Ad Hoc Wireless Networks," *Technical Report*: http://www.crhc.uiuc.edu/~nhv/papers/jungmin-tech.ps, January 2003.

[61] Nasipuri, A., Zhuang, J., and Das, S. R., "A Multi-Channel CSMA MAC Protocol for Multi-Hop Wireless Networks," *Proceedings of IEEE WCNC 1999*, 1, pp. 1402–1406, September 1999.

[62] Broch, J., Maltz, D. A., Johnson, D., Hu, Y. -C., and Jetcheva, J., "A Performance Comparison of Multi-Hop Wireless Ad Hoc Network Routing Protocols," in *Proceedings of the 4th Annual ACM/IEEE International Conference on Mobile Computing and Networking (MobiCom)*, Dallas, Texas, October 1998, pp. 85–97.

[63] Das, S. R., Castaneda, R., and Yan, J., "Comparative Performance Evaluation of Routing Protocols for Mobile, Ad Hoc Networks," in *Proceedings of the 7th International Conference on Computer Communications and Networks*, Lafayette, LA, October 1998, pp. 153–161.

[64] Das, S. R., Perkins, C. E., and Royer, E. M., "Performance Comparison of Two On demand Routing Protocols for Ad Hoc Networks," in *Proceedings of the IEEE Conference on Computer Communications (INFOCOM)*, Tel Aviv, Israel, March 2000, pp. 3–12.

[65] Johansson, P., Larsson, T., Hedman, N., Mielczarek, B., and Degermark, M., "Scenario-based Performance Analysis of Routing Protocols for Mobile Ad-Hoc Networks," in *Proceedings of the 5th ACM/IEEE International Conference on Mobile Computing and Networking (MobiCom)*, Seattle, WA, August 1999, pp. 195–206.

[66] Lee, S. -J., Toh, C. -K., and Gerla, M., "A Simulation Study of Table-Driven and On-Demand Routing Protocols for Mobile Ad-Hoc Networks," *IEEE Netvork*, 13(4), July/August 1999, pp. 48–54.
[67] Royer, E. M. and Toh, C. -K, "A Review of Current Routing Protocols for Ad-Hoc Mobile Networks," *IEEE Personal Communications*, 6(2), April 1999, pp. 46–55.
[68] Lang, D., "A Comprehensive Overview About Selected Ad Hoc Networking Routing Protocols," *Technical Report*, Department of Computer Science, Technische Universität München, 2003.
[69] Macker, J. and Corson, M. S., Internet Engineering Task Force (IETF) Mobile Ad Hoc Networks (MANET) Working Group Charter, http://www.ietf.org/html.charters/manet-charter.html
[70] Perkins, C. E., *Ad Hoc Networking*, Addison-Wesley, 2000.
[71] Ilyas, M., *The Handbook of Ad Hoc Wireless Networks*, CRC Press, 2003.
[72] Toh, C. -K., *Ad Hoc Wireless Networks: Protocols And Systems*, Pearson Education, Inc., 2001.
[73] Aggelou, G., *Mobile Ad Hoc Networks: From Wireless LANs to 4G Networks*, McGraw Hill, 2005.
[74] Basagni, S., Conti, M., Giordano, S., and Stojmenovic, I., *Mobile Ad Hoc Networking*, IEEE Press, Wiley-Interscience, 2004.
[75] Iwata, A., Chiang, C. C., Pei, G., Gerla, M., and Chen, T. W., "Scalable Routing Strategies for Ad Hoc Wireless Networks," *IEEE Journal on Selected Areas in Communications*, 17(8), August 1999, pp. 1369–1379.
[76] Malkin, G. S. and Steenstrup, M. E., "Distance-Vector Routing," in M. Steenstrup (ed.). *Routing in Communications Networks*, Prentice-Hall, 1995, pp. 83–98.
[77] Tanenbaum, A. S., *Computer Networks*, Pearson Education, Inc., 2003.
[78] Perkins, C. E. and Bhagwat, P., "Highly Dynamic Destination-Sequenced Distance-Vector Routing (DSDV) for Mobile Computers," *SIGCOMM '94: Computer Communications Review*, 24(4), October 1994, pp. 234–244.
[79] Chiang, C. C., Wu, H. K., Liu, W., and Gerla, M., "Routing in Clustered Multi-Hop Mobile Wireless Networks with Fading Channel," *Proceedings of IEEE SICON 1997*, April 1997, pp. 197–211.
[80] Murthy, S. and Garcia-Luna-Aceves, J. J., "An Efficient Routing Protocol for Wireless Networks," *ACM Mobile Networks and Applications Journal, Special Issue on Routing in Mobile Communication Networks*, 1(2), October 1996, pp. 183–197.
[81] Garcia-Luna-Aceves, J. J. and Spohn, M., "Source-Tree Routing in Wireless Networks," *Proceedings of IEEE ICNP 1999*, October 1999, pp. 273–282.
[82] Clausen, T. H., Hansen, G., Christensen, L., and Behrmann, G., "The Optimized Link State Routing Protocol, Evaluation Through Experiments and Simulation," *Proceedings of IEEE Symposium on Wireless Personal Mobile Communications 2001*, September 2001.
[83] Bellur, B., Ogier, R. G., and Templin, F. L., "Topology Broadcast Based on Reverse-Path Forwarding (TBRPF)," *IETF Internet Draft, draft-ietf-manet-tbrpf- Ql.txt* (work in progress), March 2001.
[84] Gerla, M., Hong, X., and Pei, G., Fisheye State Routing Protocol (FSR) for Ad Hoc Networks, draft-ietf-MANET-fsr-02.txt, IETF MANET Working Group — Internet Draft, December 2001.

[85] Castaneda, R. and Das, S. R., "Query Localization Techniques for On-demand Routing Protocols in Ad Hoc Networks," *in Proceedings of the 5th Annual ACM/IEEE International Conference on Mobile Computing and Networking (MobiCom)*, Seattle, August 1999, pp. 186–194.

[86] Ko, Y. -B. and Vaidya, N. H., "Location-Aided Routing (LAR) in Mobile Ad Hoc Networks," in *Proceedings of the 4th ACM/IEEE International Conference on Mobile Computing and Networking (MobiCom)*, Dallas, Texas, October 1998, pp. 66–75.

[87] Lee, S. -J., Royer, E. M., and Perkins, C. E., "Ad Hoc Routing Protocol Scalability," *International Journal on Network Management*, 2002, pp. 97–114.

[88] Toh, C. -K., "Associativity-Based Routing for Ad Hoc Mobile Networks," *Wireless Personal Communications*, 4(2), pp. 1–36. March 1997.

[89] Dube, R., Rais, C. D., Wang, K. Y., and Tripathi, S. K., "Signal Stability-Based Adaptive Routing for Ad Hoc Mobile Networks," *IEEE Personal Communications Magazine*, February 1997, pp. 36–45.

[90] Aggelou, G. and Tafazolli, R., "RDMAR: a bandwidth-efficient routing protocol for mobile ad hoc networks," *Proceedings of the 2nd ACM international workshop on Wireless mobile multimedia*, Seattle, Washington, USA, 1999, pp. 26–33.

[91] Park, V. D. and Corson, M. S., *Temporally-Ordered Routing Algorithm* (TORA) version 1: Functional Specification. Internet-Draft, draft-ietf-manettora-spec01.txt, August 1998.

[92] Boppanam, R. V. and Konduru, S. P., "An Adaptive Distance Vector Routing Algorithm for Mobile Ad Hoc Networks," *Proceedings of IEEE INFOCOM 2001*, Anchorage, AK, April 22–26, 3, 2001, pp. 1753–1762.

[93] Perkins, C. E. and Royer, E. M., "The Ad Hoc On-Demand Distance Vector Protocol," in C. E. Perkins (ed.), *Ad Hoc Networking*, pp. 173–219. Addison-Wesley, 2000.

[94] Lee, S. -J., Belding-Royer, E. M., and Perkins, C. E., "Ad hoc on-demand distance vector routing scalability," *ACM SIGMOBILE Mobile Computing and Communications*, 6(3), July 2002, pp. 94–95.

[95] Johnson, D. B. and Maltz, D. A., "Dynamic Source Routing in Ad Hoc Wireless Networks," in T. Imielinski and H. Korth (eds), *Mobile Computing*, pp. 153–181. Kluwer Academic Publishers, 1996.

[96] Hu, Y. -C. and Johnson, D. B., "Caching Strategies in On-Demand Routing Protocols for Wireless ad Hoc Networks," in *Proceedings of the Sixth Annual IEEE/ACM International Conference on Mobile Computing and Networking (MOBICOM 2000)*, Boston, MA, August 2000, pp. 231–242.

[97] Hu, Y. -C. and Johnson, D. B., "Implicit Source Routing in On-Demand Ad Hoc Network Routing," in *Proceedings of the Second Symposium on Mobile Ad Hoc Networking and Computing* (MobiHoc 2001), October 2001, pp. 1–10.

[98] Lee, S. -J. and Gerla, M., "AODV-BR: Backup Routing in Ad Hoc Networks," in *Proceedings of the Wireless Communications and Networking Conference (WCNC)*, Chicago, IL, September 2000.

[99] Pearlman, M. R. and Haas, Z. J., "Determining the Optimal Configuration for the Zone Routing Protocol," *IEEE Journal on Selected Areas in Communications*, 77(8), August 1999, pp. 1395–1414.

[100] Nikaein, N., Labiod, H., and Bonnet, C., "Distributed Dynamic Routing Algorithm for Mobile Ad Hoc Networks," *Proceedings of ACM MobiHoc 2000*, Boston, MA, 2000, pp. 19–27.

[101] Sinha, P., Sivakumar, R., and Bharghavan, V., "CEDAR: A Core Extraction Distributed Ad Hoc Routing Algorithm," *IEEE Journal on Selected Areas in Communications*, 17(8), August 1999, pp. 1454–1466.

[102] Ng, M. J. and Lu, I. T., "A Peer-to-Peer Zone-Based Two-Level Link State Routing for Mobile Ad Hoc Networks," *IEEE Journal on Selected Areas in Communications*, 17(8), August 1999, pp. 1415–1425.

[103] Samar, P., Pearlman, M. R., and Haas, Z. J., "Hybrid Routing: The Pursuit of an Adaptable and Scalable Routing Framework for Ad Hoc Networks," in M. Ilyas (ed.), *Handbook of Ad Hoc Wireless Networks*, Chapter 14, CRC Press, 2002.

[104] Su, W. and Gerla, M., "IPv6 Flow Handoff in Ad Hoc Wireless Networks Using Mobility Prediction," *Proceedings of IEEE GLOBECOM 1999*, December 1999, pp. 271–275.

[105] Gerla, M., Kwon, T., and Pei, G., "On Demand Routing in Large Ad Hoc Wireless Networks With Passive Clustering," in *Proceedings of the IEEE Wireless Communications and Networking Conference (WCNC)*, September 2000.

[106] Chen, T. -W. and Gerla, M., "Global State Routing: A New Routing Scheme for Ad-Hoc Wireless Networks," *Proceedings of IEEE ICC*, Atlanta, GA, June 8–11, 1998, pp. 171–175.

[107] Jones, C. E., Sivalingam, K. M., Agrawal, P., and Chen, J. -C, "A Survey of Energy Efficient Network Protocols for Wireless Networks," *Wireless Networks*, 7(4), 2001, pp. 343–358.

[108] Singh, S., Woo, M., and Raghavendra, C. S., "Power-Aware Routing in Mobile Ad Hoc Networks," *Proceedings of ACM MOBICOM 1998*, October 1998, pp. 181–190.

[109] Bettstetter, C. and Krausser, R., "Scenario-Based Stability Analysis of the Distributed Mobility-Adaptive Clustering (DMAC) Algorithm," in *Proceedings of the 2nd Annual Symposium on Mobile Ad hoc Networking and Computing*, Long Beach, California, October 2001.

[110] Fusler, H., Mauve, M., Kasemann, M., Vollmer, D., and Hartenstein, H., "A Comparison of Routing Strategies for Vehicular Networks," *Technical Report TR-03-2002*, Department of Computer Science and Mathematics, University of Mannheim, Germany, 2002.

[111] Hsiao, P., "Geographic Region Summary Service for Geographical Routing," *ACM Mobile Computing and Communications Review*, October 2001, pp. 25–39.

[112] Giordano, S. and Hamdi, M., "Mobility Management: The Virtual Home Region," *Technical Report SSC/1999/037*, October 1999.

[113] Stojmenovic, I., "Home Agent Based Location Update and Destination Search Schemes in Ad Hoc Wireless Networks," *Technical Report TR-99-10*, Computer Science, SITE, University of Ottawa, Canada, September 1999.

[114] Haas, Z. J. and Liang, B., "Ad Hoc Mobility Management with Uniform Quorum System," *IEEE/ACM Transactions on Networking*, 7, pp. 228–240, 1999.

[115] Capkun, S., Hamdi, M., and Hubaux, J. P., "GPS — free Positioning in Mobile Ad Hoc Networks," in *Proceedings of Hawaii International Conference on System Sciences*, HICCSS-34, January 2001.
[116] Fontana, R. and Gunderson, S., "Ultra Wideband Precision Asset Location System," in *Proceedings of IEEE Conference on Ultra Wideband Systems and Technologies*, May 2002.
[117] Bahl, P. and Padmanabhan, V., "An In — Building RF — based User Location and Tracking System," in *Proceedings of INFOCOM 2000*, Tel Aviv, Israel, 2, March 2000, pp. 775–784.
[118] Girod, L. and Estrin, D., "Robust Range Estimation Using Acoustic Multimodal Sensing," in *Proceedings of the IEEE/RSJ International Conference on Intelligent Robots and Systems (IROS)*, Maui, Hawaii, October 2001.
[119] Dixit, S. and Wu, T., *Content Networking in the Mobile Internet*, John Willey and Sons, 2004.
[120] Savarese, C., Rabay, J., and Langendoen, K., "Robust Positioning Algorithms for Distributed Ad Hoc Wireless Sensor Networks," in *Proceedings of USENIX Technical Annual Conference*, June 2002.
[121] Savvides, A., Han, C. C., and Srivastava, M. B., "Dynamic Fine — Grained Localization in Ad Hoc Networks of Sensors," in *Proceedings of Fifth Annual International Conference on Mobile Computing and Networking, Mobicom*, Rome, Italy, July 2001, pp. 166–179.
[122] Widens Project, IST-FP6, http://www.widens.org.
[123] Nicolescu, D. and Nath, B., "Ad Hoc Positioning System," in *Proceedings of IEEE GlobeCom*, November 2001.
[124] WhereNet, http://www.wherenet.com.
[125] Takagi, H. and Kleinrock, L., "Optimal Transmission Ranges for Randomly Distributed Packet Radio Terminals," *IEEE Transactions on Communications*, 32(3), 1984, pp. 246–257.
[126] Hou, T. C. and Li, V. O. K., "Transmission Range Control in Multihop Packet Radio Networks," *IEEE Transactions on Communications*, 34(1), 1986, pp. 38–44.
[127] Stojmenovic, I. and Lin, X., "Loop — Free Hybrid Single — Path/Flooding Routing Algorithms with Guaranteed Delivery for Wireless Networks," *IEEE Transactions on Parallel and Distributed Systems*, 12, 2001, pp. 1023–1032.
[128] Kranakis, E., Singh, H., and Urrutia, J., "Compass Routing on Geometric Networks," in *Proceedings of 11th Canadian Conference on Computational Geometry*, Vancouver, August 1999.
[129] Basagni, S., Chlamtac, I., Syrotiuk, V., and Woodward, B., "A Distance Routing Effect Algorithm for Mobility (DREAM)," in *Proceedings of the 4th Annual ACM/IEEE International Conference on Mobile Computing and Networking (MobiCom)*, Dallas, TX, October 1998, pp. 76–84.
[130] Ko, Y. B. and Vaidya, N. H., "Location — Aided Routing (LAR) in Mobile Ad Hoc Networks," in *Proceedings of MOBICOM 1998*, pp. 66–75; *Wireless Networks*, 6(4), July 2000, pp. 307–321.
[131] Liao, W. H., Tseng, Y. C., and Sheu, J. P., "GRID: A Fully Location — Aware Routing Protocol for Mobile Ad Hoc Networks," in *Proceedings of IEEE HICSS*, January 2000.
[132] Blazevic, L., Buttyan, L., Capkun, S., Giordano, S., Hubaux, J. -P., and Le Boudec, J. -Y, "Self-Organization in Mobile Ad Hoc Networks: The

Approach of Terminodes," *IEEE Communications Magazine*, June 2001, pp. 166–175.

[133] Amouris, K. N., Papavassiliou, S., and Li, M., "A Position Based Multi — Zone Routing Protocol for Wide Area Mobile Ad Hoc Networks,"in *Proceedings of the 49th IEEE Vehicular Technology Conference*, Amsterdam, 1999, pp. 1365–1369.

[134] Li, J., Jannoti, J., De Couto, D. S. J., Karger, D. R., and Morris, R., "A Scalable Location Service for Geographic Ad Hoc Routing," in *Proceedings of ACM MobiCom*, Boston, MA, 2000, pp. 120–130.

[135] Sisodia, R. S., Manoj, B. S., and Siva Ram Murthy, C., "A Preferred Link-Based Routing Protocol for Ad Hoc Wireless Networks," *Journal of Communications and Networks*, 4(1), pp. 14–21, March 2002.

[136] Yi, S., Naldurg, P., and Kravets, R., "Security — Aware Ad Hoc Routing for Wireless Networks," in *Proceedings of ACM MOBIHOC 2001*, October 2001, pp. 299–302.

[137] Hu, Y., Johnson, D. B., and Perrig, A., "SEAD: Secure Efficient Distance Vector Routing for Mobile Wireless Ad Hoc Networks," in *Proceedings of IEEE WMCSA 2002*, June 2002, pp. 3–13.

[138] Sanzgiri, K., Dahill, B., Levine, B. N., Shields, C., and Royer, E. M. B., "A Secure Routing Protocol for Ad Hoc Networks," in *Proceedings of IEEE ICNP 2002*, November 2002, pp. 78–87.

[139] Ruiz, P. M. and Gomez-Skarmeta, A. F., "Mobility — Aware Mesh Construction Algorithm for Low Data — Overhead Multicast Ad Hoc Routing," *Journal of Communications nad Networks, Special Issue on Mobile Ad Hoc Wireless Networks*, 6(4), December 2004, pp. 331–342.

[140] Adjih, C., Baccelli, E., Clausen, T. H., Jacquet, P., and Rodolakis, G., "Fish Eye OLSR Scaling Properties," *Journal of Communications nad Networks, Special Issue on Mobile Ad Hoc Wireless Networks*, 6(4), December 2004, pp. 343–351.

[141] Blazevic, L., Giordano, S., and Le Boudec, J. -Y., "Self — Organizing Routing," *Cluster Computing Journal*, 5(2), April 2002.

4

Cross-Layer Optimization

4.1 Introduction

The central building paradigm in network designs, which unifies variety of solutions and devices and allows their mutual communication is the layering concept. This aged philosophy is inbuilt in many network protocols which today gained global popularity, such as OSI and TCP/IP based networks. The layered concept was primarily created for wired networks and naturally follows their architectural design. Designing wireless networks with strict layering principle did not fulfill the expectation raised in wireline network design. The ad hoc mobile networks oppose strict layered protocol design, because of their dynamic nature, infrastructureless architecture, limited resources, mobility of nodes and time varying unstable links and topology.

The concept of cross-layer design is based on architecture where the layers can exchange information in order to improve the overall network performances. Promising results achieved by cross-layer optimizations initiated significant research activity in this area. A search for answers to conceiving challenges has opened new horizons towards new way of understanding of network functionality.

The following chapter aims to present the major ideas in cross-layer paradigm. The introduction of layering concept is followed by brief overview of cross-layering principles, presenting examples of evolutionary and revolutionary cross-layer approaches. The generic cross-layer design develops towards the referent 4G architectures and will probably be inbuilt (at least partially through power awareness) in the future ad hoc network. The problem of cross-layering is being elaborated in many ongoing and recent projects.

4.2 Layered Approach

The layering principle of network design has served for about 30 years as a framework of modular building concepts for complex networking systems. It

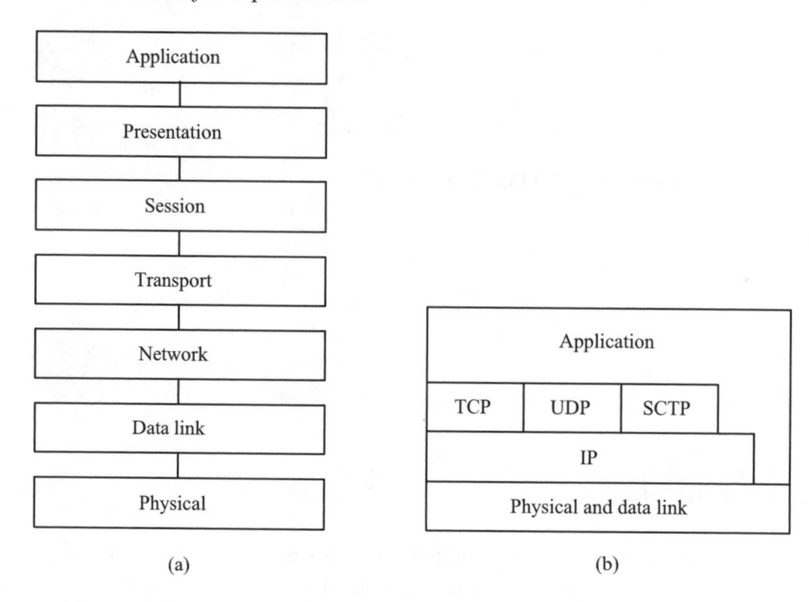

Fig. 4.1. (a) OSI layered architecture (b) Current Internet architecture

is based on isolating the functions of different conceptual layers into closed structures with minimum interaction through specified interfaces between the layers. This strict layered approach designs layers' functionality in a black-box manner. The algorithmic design of each layer is isolated from other layers and takes into account the pre-defined interfaces towards neighboring layers. The header added to transported data at each layer reflects the layer's functionality and interactions.

ISO (International Organization for Standardization) proposed the *open system interconnection (OSI)* model, which was primarily designed to enable multi-vendor computers to interact and communicate. The layered OSI architecture [1] presents a general framework for building modular systems (see Fig. 4.1(a)). It divides the network functionalities, which are involved in provisioning end-to-end data transmission, into hierarchical layers containing sub-tasks (sub-functions). OSI defines seven layers with hierarchy that goes from physical to application layers. The physical layer defines a virtual bit pipe between any pair of nodes connected through a physical communication channel. The data link control (DLC) layer provides reliable two-way transmission of packets to the network layer. The network layer provides the transport layer with a route (transmission links between the source and the destination). The transport layer is responsible for end-to-end communication. The upper layers are application oriented and provide functionalities concerning application, semantics of data presentation and dialog control. Today, OSI is still a *reference model*, often used to describe and outline the different levels of networking protocols and their relationships with each other.

Another important layered structure is the current Internet architecture, which relies on the *IP-based protocol stack* (see Fig. 4.1(b)). The focal point in this architecture is the IP protocol. In order to introduce various features and adapt it to the different and new services offered by the Internet, the IP protocol can be combined with several control protocols: *UDP* (User Datagram Protocol — protocol with light requirements), *TCP* (Transport Control Protocol — protocol for reliable point-to-point data transport) and *SCTP* (Session Control Transport Protocol — protocol for reliable transport of message-oriented applications with support for multi-streaming and multi-homing). Lately, *RTP* (Real-time Transport Protocol) is designed to work over UDP and support real-time applications. IP can be also enhanced in order to meet the QoS requirements introducing *Diff-Serv* (Differentiated services) and *IntServ* (Integrated Services) QoS features (see Chapter 5).

The layering concept was primarily designed for wired networks. One of the most successful implementations of the OSI principles is the IP-based protocol stack, although it inherits potential flows and weakness of the layering approach. Major inconvenience of the layered design is that it is highly rigid and strict and does not show any flexibility in dynamic environments. Even though the layering approach was serving the networking designers for almost three decades, with principles widely adopted through various implementations and applications, it could not follow the growth of more demanding applications and explosion in penetration of wireless technologies.

Wireless systems work in scarce conditions concerning limited frequencies and available channels. The dynamic wireless systems are affected by: *small-scale channel variations* and *large scale channel variations*. The former are due to fading, scattering and multi-path and can change the channel state within a few milliseconds, while the latter are due to user location and interference from the surrounding environment and are much slower. The large scale variations can not be covered by strictly layered protocols in a best manner.

In wireless environment (e.g., wireless LANs), where the users communicate over the shared medium in order to capture the communication over multi-access channel, the layering model is upgraded with a *MAC sub-layer*. The MAC design slightly redefines the functionality of network layer. It should reflect the communication over a single channel, as well as multi-access/multi-hop communication scenarios in mobile ad hoc networks.

The new wireless networks, such as 3G and beyond [2], are expected to be all-IP using the IP-based protocol stack (often called *TCP/IP protocol stack*) in order to ensure interoperability. Ad hoc networks will be an integral part of this scenario.

The layering approach should provide end-to-end communication design independent of type of information source, network topology or devices. However, the generalization of this approach leads to *rate distortion effect of the universal architecture* resulting in some performance losses.

4.3 Cross-Layer Approach

4.3.1 Definition and General Remarks

The traditional layered approach does not fit the wireless communication system design perfectly due to multiple users accessing scarce and changeable transmission media. The performances of such systems can be optimized by considering some vertical coupling between the layers. These inter- or cross-layer interactions provide useful information allowing improvement of network performance.

Cross-layer optimization defines a general concept of communication between layers, considering certain smart interactions between them, and resulting in network performance improvements. It aims in coupling the functionality of network layers with the goal of boosting system-wide performance [3, 4].

Traditional approach concerning OSI layered model [1, 3] can recognize a subset of possible cross-layer interactions depicted in Fig. 4.2.

In ad hoc network design, cross-layer interactions can boost the performance behavior by effective adaptation to the dynamic environment. The traditional settings within the closed layers do not follow in a best possible manner the unstable conditions of dynamic nodes and wireless links, making obvious the need for more flexible approach. Vertical communications provide a shortcut in relevant information delivery resulting in proper settings on particular layer's activities. The key to practical cross-layer optimization is to find an appropriate abstraction of each layer [3] and adequate coupling mechanisms.

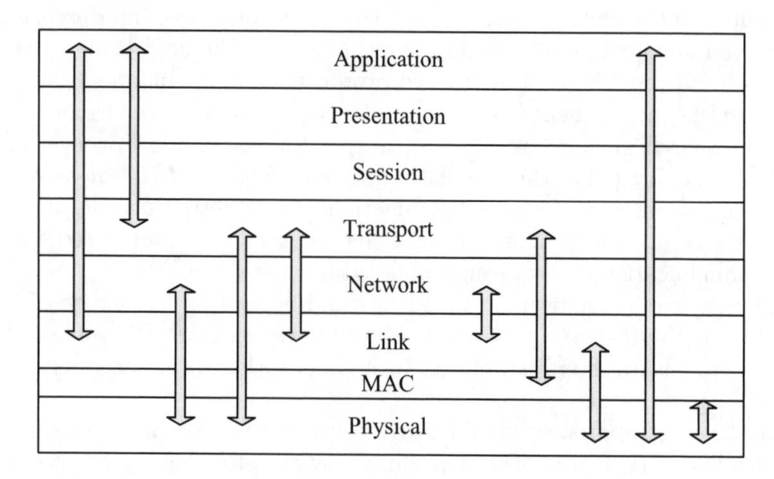

Fig. 4.2. Possible cross-layer communications

Ad hoc networking is a multilayer problem [5]. The physical layer must adapt to rapid changes in link characteristics. The multiple access control layer needs to minimize collisions and allow fair access and semi-reliable transport of data over the shared wireless links in the presence of rapid changes and hidden or exposed terminals (see Chapter 3). The network layer is responsible to determine and distribute information used to calculate paths and to maintain efficiency with changeable links and variable bandwidth. Each layer in an ad hoc network depends on other layers, either directly or not. So, the cross-layer design that supports *adaptability* and *optimization* is needed.

Benefits of cross-layer information exchanges increase the cost of the system in sense of additional *signaling* needed to extract relevant parameters from one layer and offering it to another layer/layers [6, 7]. Also, more complex control plane and corresponding transmission occupied resources, increased computation complexity of the involved protocols and overhead information has to be considered. It means that whenever cross-layer approach is proposed, it must be carefully analyzed in each particular network scenario combined with carefully chosen parameters involved in optimizations.

4.3.2 Relevant Parameters

Relevant cross-layer parameters, which could be exchanged between layers, depend on functionalities considered for cross-layer interaction, and on specific constraints and objectives of the end user applications. Potentially useful cross-layer information differentiates as [3]:

- *Channel state information (CSI)*: channel response estimation (in time and in frequency domains), location information, terminal speed, signal strength, interference level, interference modeling and condition number;
- *QoS related parameters*: delay, throughput, bit error rate (BER) and packet error rate (PER) (especially for end-to-end requirements);
- *Resource information* (concerning the nodes): multi-user reception capabilities, number and type of antennas and battery depletion level;
- *Traffic pattern*: data traffic information, knowledge of data rate (constant or variable), data burstiness, data fragmentation, packet size and information about queue sizes.

Adequate cross-layer information exchanges and proper mechanisms and algorithms have to be designed in order to optimize the gain relevant to the overall system performance. Table 4.1 gives an example of functions relevant to different layers, which participate in cross-layer exchanges in ad hoc networking [7].

Table 4.1. Relevant functions involved in cross-layer design

Layer	Information
Application layer	Topology control algorithm, Server location, Network Map
Transport layer	Congestion window, Timeout clock, Packet losses rate
Network layer	Routing affinity, Routing lifetime, Multiple routing
MAC/Link layer	Link bandwidth, Link quality, MAC packet delay
Physical layer	Node's location, Movement pattern, Radio transmission range, SNR information

4.4 Categorization of Cross-Layer Approaches

Cross-layer approach in network design introduces a paradigm shift in understanding of protocols, layers, incorporated functionalities in the protocol stack and the overall system construction. The cross-layer design involves collaboration between three major fields: *wireless networking, signal processing* and *information theory* [8]. Each of them focuses on solving different features. However, their major goal is to create a network which can be highly adaptive and QoS efficient by sharing information between different processes or modules in the system. A lot of research has been done mainly developed into two directions, following evolutionary or revolutionary approach in cross-layer design [8].

The *evolutionary approach* is more concentrated on *compatibility* than on performance and provides optimization and enhancements by some vertical shortcut through the layers, keeping in background the existing protocol stack. Usually two or three layers are involved in cross-layer optimizations, even though all layers of the protocol stack may participate. The evolutionary approach to cross-layer design extends the existing layered structure in order to maintain compatibility. It can be done through simple, effective (i.e., *evolutionary basic*) solutions, which extends parts of the strict layering structure or through *system-wide cross-layer* solutions, which design the stack-wide layer interdependencies to optimize the overall network performance [9]. Most cases implement simple cross-layer solutions including two or three layers which are sharing information. Evolutionary approach is also referenced as *layer-related* approach, and will be elaborated in the next subsection.

The *revolutionary approach* (or *alternative* approach), which is free of existing layered concept, provides new and more philosophical definitions of functional entities and their interactions, i.e., cooperativeness. It is concentrated more on the *performances* and does not compromise to maintain compatibility, which weaknesses its universal dimension. Therefore, the revolutionary cross-layer design may result in problems with compatibility and cost. That is the major reason why the majority of existing cross-layer solutions are evolutionary ones. However, the revolutionary approach opens new horizons in understanding network functionality based on metrics originated from information theory, network capacity, topology, sensor network peculiarities, context based information, etc. (see [8, 9]).

4.5 Evolutionary Layer-Related Approaches

Depending on directions of information flow between the layers in the protocol stack, and for the sake of convenience, the cross-layer feedback is often categorized as follows [3]:

- *Upper to lower layers.* Information from upper layer influences the decision (metrics) on lower layer. For example, application layer (delay or loss constraints of the application) communicates with the link layer (adapt the error correction mechanisms), or application layer (user defined application priority) communicates with the transport layer (TCP adapts the receiver window).
- *Lower to upper layers.* Information from the lower layer influences one or more upper layers. Following examples present such feedback: transport layer (TCP packet loss information) communicates with application layer (application adapts its sending rate), physical layer (transmit power and bit-error rate) communicates with link/MAC layer (adapt error correction mechanisms), etc.

In ad hoc wireless environment, with power constraints and limited transmission area, mobile nodes, link variations on per-hop and time basis, interference, and absence of any infrastructure [10], the cross-layer design should create protocols within an integrated and hierarchical framework that takes advantage of the interdependence between the layers. In order to cope with the different time scales of these variations, inter-layer adaptations as well as intra-layer adaptations can be performed either locally or with respect to the global system constraints. This leads to *adaptive* cross-layer design, which identifies the information exchanged across the protocol layers and the way it is adopted. Each layer protocol and metrics needs to be designed with global system constraints.

Some protocols positioned within certain layer exchange information with lower and upper layers simultaneously. Several parts of the cross-layer system can be correlated initiating the *cyclic dependence* (e.g., dependence between physical/MAC/physical layer), which makes it difficult to differentiate if the dominant cross-layer information flow is in down or up direction. The highly complex, large and dynamic ad hoc and sensor networks provide a new multiple-time-level-space-scale dependence behavior [7]. A good understanding of these dependencies can lead to more revolutionary approach in cross-layer design.

Table 4.2. List of layer-related studies

Physical layer	Channel state is used to adapt the throughput [15–18]
MAC layer	The retransmission number at the MAC level may indicate the quality of the link, timer [27]
Network layer	Using quality of transmission information in routing algorithms [45–48]
Transport layer	The MAC layer could adapt the error control scheme among TCP retransmission timers and update the information on the delay for better QoS [52, 53]

4.5.1 Layer-Related Designs

Layer-related classification of cross-layer optimizations gives insight in relevant parameters and functions/protocols involved in cross-layer design and originated from the particular layer. Good overviews of cross-layer optimization can be found in [2, 3, 7–11]. Table 4.2 presents a list of recent cross-layer studies related to different layers, which are mentioned in this chapter.

The presented table does not cover all existing research efforts. This field is continuously open to new emerging concepts. Closer look into each layer-related design specifics, with some general remarks, follows in the next subsections.

4.5.2 Physical Layer Related Cross-Layer Optimization

4.5.2.1 Relevant Information on Physical Layer

The physical layer is responsible for transmission of bits, aiming to achieve minimum bit error rate. The most relevant metrics that describe the physical layer features are the bit error rate (BER) and SNR (signal-to-noise ratio). Also, SNIR (signal-to-noise-interference ratio) captures the interference effect from the environment. Another important parameter is the transmitting power (depends on the operation mode) and battery status. The battery aware physical layer declares a battery status to which it may adapt the operation mode, i.e., coding and modulation schemes. Optimal transmitting power is important in power-based optimizations. An order of magnitude reduction in transmit power is possible by properly exploiting the macrodiversity inherent in dense ad hoc networks and the gains can be achieved by simply decentralized and cross-layer protocols [12, 13].

Particular physical layer design and information gained from physical layer conditions can beneficiary influence the upper layer solutions. On the other hand, the user/application requirements can require particular physical operation mode.

4.5.2.2 Interaction with Upper Layers

Examples of influence of physical layer metrics and mutual interactions with other layers are numerous. The application layer can tune the physical layer parameters to improve the application layer throughput. The BER can influence network layer solution by selecting an appropriate interface and route.

Harmonization of link/MAC layer protocols with the aid of the information forwarded from physical layer (BER, SNR, SINR, transmit power, etc.) was often addressed in the literature. Authors in [14] investigate the effect of BER and transmit power on packet length in order to achieve minimum consumed energy. They showed that *optimal* consumed power for particular BER is achieved for packets of certain size. Similar investigation of the effect of SNR is reported in [15–18] (also see Section 4.5.1.2).

A set of physical layer/MAC mechanisms based on a cross-layer dialogue are proposed in [19–21]. In [19] system efficiency is improved by means of automatic rate adaptation mechanism in packet switched CDMA access networks. It reduces power consumption and inter-cell interference. In [21] the information from the physical layer (Rayleigh fading environment) obtained through measurements is used in the design of a reservation based MAC scheme through a slotted ALOHA procedure.

Song and Li in [22, 23] investigate the possibility of cross-layer optimizations for OFDM-based wireless networks, trying to build the bridge between the physical and the MAC layer and to balance the efficiency and fairness of wireless resource allocation. They approach the utility-based cross-layer design through finding the global optimum of an objective function and demonstrated the significant performance gain.

The cooperation between physical and MAC layer is observed in [7] where the use of antenna arrays in presence of fading and interference and exploiting the macrodiversity (and multihop diversity) improves the energy and/or bandwidth efficiency.

4.5.3 Link/MAC Layer Related Cross-Layer Optimizations

4.5.3.1 Relevant Information on Link/MAC Layer

The function of link/MAC layer is to ensure reliable packet transmission with minimal overhead through different mechanisms such as forward error correction (FEC), automatic repeat request (ARQ) procedures, avoiding/reducing collisions, data fragmentation into packets (frames), scheduling, etc. The relevant information which can be offered to other layers is: current FEC scheme, number of retransmitted frames, frame (packet) length, time moment when transmission can happen, etc.

The link establishment and access schemes are crucial features of wireless communication, which promote the link/MAC related cross-layer optimizations as extremely important and elaborated area.

4.5.3.2 Boosting the Link-Layer Related Information Exchange

The actual realization of the cross-layer optimizations at the link layer is achieved with embedded *protocol boosters* [24] which are transparent enhancements of the protocol communication between the end entities (see Fig. 4.3). Protocol boosters can perform adaptation procedures towards upper layers [25] and lower layers [15–18], assisting the process of information exchanges between various layers without a need for actual protocol format changes, leading to improved system performances.

4.5.3.3 Interactions with Upper Layers

Link parameters can influence the user decisions. Depending on the link conditions (e.g., link throughput), the user may decide which application to run.

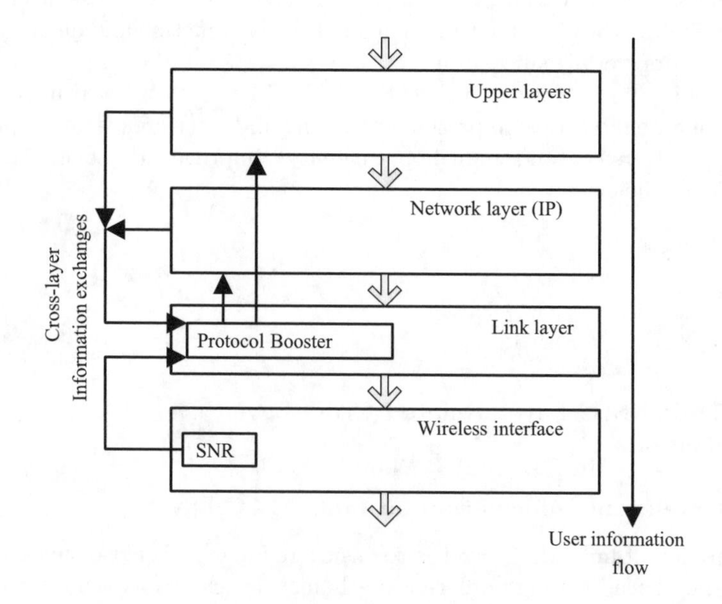

Fig. 4.3. Implementation of a link-layer protocol booster

Application. Multi-service link layer [26] may provide the *application* with certain QoS requirements (i.e., delay, jitter) by adapting the link layer FEC/ARQ mechanisms according to particular service classes. That may result in increase in processing overhead and power consumption.

Transport. The low SNR and poor channel conditions result in retransmissions and delays on link layer. To avoid large number of retransmissions on *TCP layer* (which will reduce the throughput), the cooperation between the link and the TCP layer can result in increased number of link layer retransmission and decreased power consumption [27].

Network. Example of interactions between the link and the *network layer* can be found in Mobile-IP [28], where the link layer hand-off information (signal strength monitored continually at the link layer) can be used to reduce the hand-off latency. Similar approach is used in Cellular IP [29].

A new cross-layer design is proposed in [30] which improves the performance of ad hoc networks under Rayleigh-fading channel (see *interactions with physical layer*). Cross-layer design with involvement of MAC layer is investigated in [21, 31].

4.5.3.4 Interactions with Physical Layer

The interactions with physical layer can affect different features on link-layer. Various aspects of these inter-layer dependencies are elaborated in [32]. Using signal processing tools, authors investigate and analyze throughput-delay benefits of recently introduced class of xNDMA (network diversity multiple access) and BNDMA (blind-NDMA) protocols. They employ rotational invariance and factor analysis techniques to achieve better system utilization and system capacity, while being insensitive to multipath effects and synchronization errors [33]. Joint physical/MAC layer design is offered in [34]. The authors propose the broadcast protocol for multicast routing. Investigating the capacity to exchange information between the nodes, they exploit the signal separation principles (through training sequences which are selected at random from a reduced set). Results are throughput gain and more fair distribution of channel capacity among nodes with a very different number of neighbors.

Examples of cross-layer design that combines physical and link/MAC layer interactions are presented in the following text.

4.5.3.5 Effects on Optimal Packet Length

Wireless links exhibit time-dependent characteristics which lead to channel instability. In addition, the BER in wireless environments takes on variable and high values. A key issue in future pervasive communication is the adaptation to such conditions.

This section elaborates the effects of choosing the optimal frame length, under various network scenarios (different numbers of sending stations) and various channel conditions (different BER values), on the throughput performance of contending IEEE 802.11a systems in Rayleigh fading channels. An optimal frame length is the value for the length of the frame that delivers the maximum possible throughput at the MAC layer. The discussion is based on an analytical model derived in [15].

The analysis of the maximum IEEE 802.11a throughput achievable under a certain SNR, transmission rate and number of stations contending for the medium shows that the throughput depends on the length of the MAC Service Data Unit (MSDU). Fig. 4.4 and 4.5 depict this case. NoS is the Number-of-Sending stations contending for the medium.

It is obvious from Fig. 4.4 and 4.5 that there is an optimal MSDU size that delivers the maximum throughput under certain SNR and/or NoS values. Namely, in deteriorated channel conditions (low SNR values) it is desirable to use smaller MSDUs since they provide higher values for the throughput. However, good channel conditions (high SNR values) yield the use of maximum possible MSDU size. Maximum assumed value for the MSDU size is 1500 bytes. Fig. 4.5 depicts the dependence of the maximally achievable throughput on the MSDU size under channel conditions that can be regarded as modest (SNR = 19 dB) for the 36 Mbps mode of operation. It is clear that every value for the NoS imposes an optimal MSDU size for delivering the highest possible throughput.

Introduction of an adaptive MSDU length selection guarantees maximum throughput under different channel conditions. The curve for variable MSDU

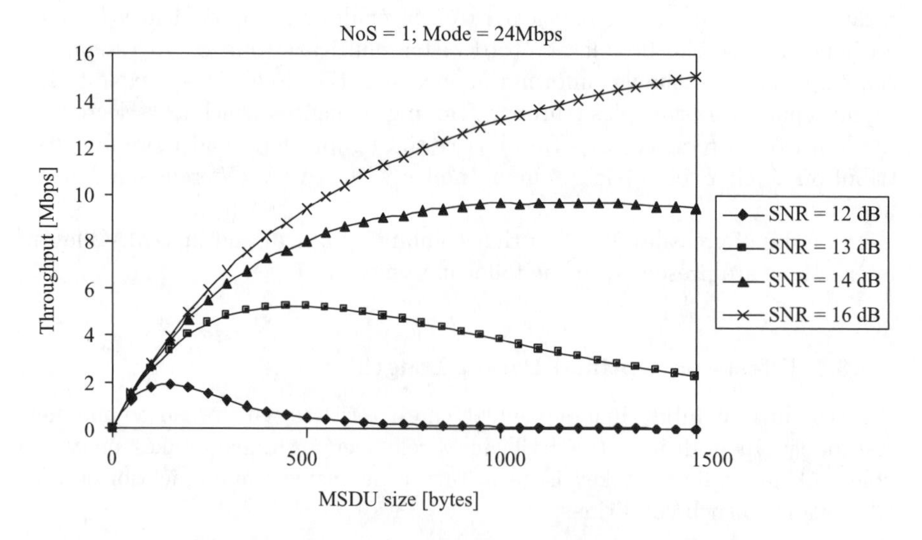

Fig. 4.4. Throughput vs. MSDU size for various SNRs

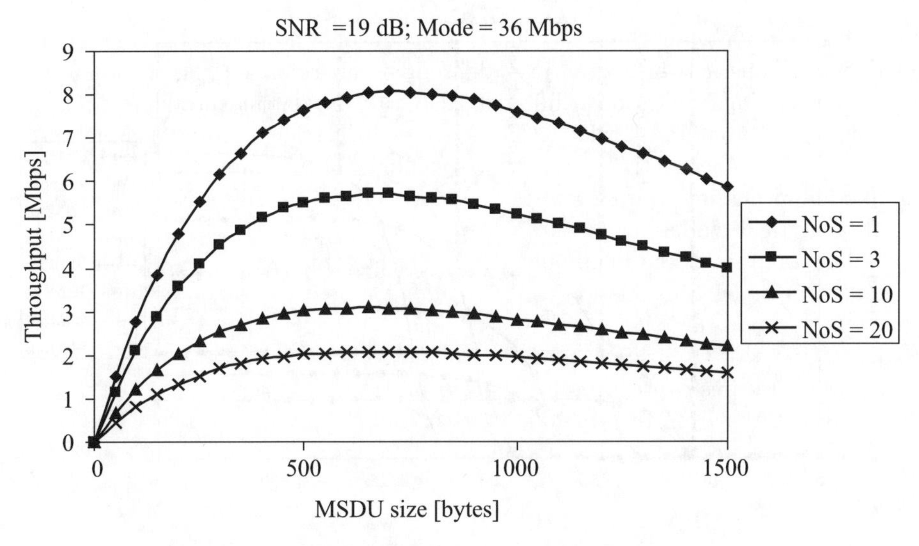

Fig. 4.5. Throughput vs. MSDU size for various NoSs

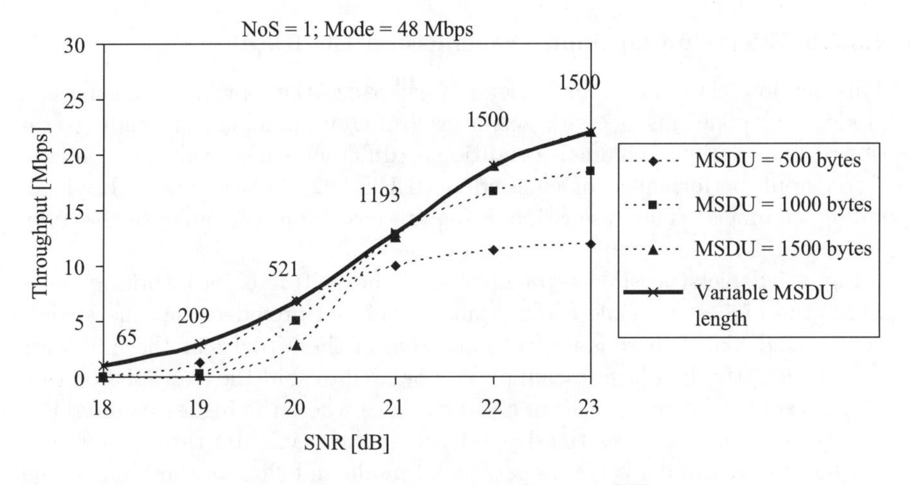

Fig. 4.6. Maximum throughput vs. SNR

size in Fig. 4.6 shows that the throughput in those circumstances always outperforms the throughput achievable with the use of fixed length MSDUs. The numbers by the points in the variable MSDU length curve represent the current MSDU size.

More details can be found in [15].

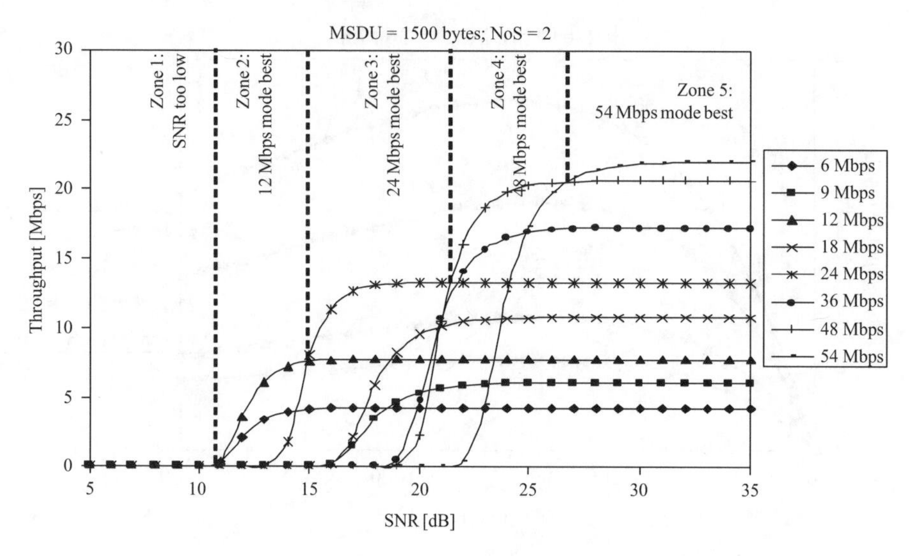

Fig. 4.7. Throughput vs. SNR, operational mode fixed

4.5.3.6 Effects on Optimal Transmission Bit Rate

This section elaborates the effects of choosing the optimal transmission mode, under various network scenarios (different numbers of sending stations) and various channel conditions (different BER values), on the throughput performance of contending IEEE 802.11a systems in Rayleigh fading channels. The discussion is again based on the analytical model derived in [15].

Fig. 4.7 depicts possible segmentation of the SNR axis into working zones where the highest possible throughput is achieved provided that the MSDU size is fixed. So, if there is no fragmentation of the MSDUs at the link layer then an adaptive link layer technique can be used to shift the transmission rate of the communicating entities in a certain zone where the highest throughput is obtained. Fig. 4.8 gives the dependence of the maximum throughput that can be achieved under certain operational mode and NoS on the SNR. Fixed MSDU values of 1500 bytes are assumed.

Fig. 4.9 shows MSDU vs. SNR dependency for three operational modes. Numbers next to the curves represent the throughput achievable with the corresponding MSDU size under the current channel state (SNR). It is clear that when the MSDU size is fixed, a similar value for the throughput can be obtained with all operational modes, but under different channel conditions. In this case, an adaptive choise of the transmission rate for various SNRs will provide the highest possible throughput.

More details can be found in [15].

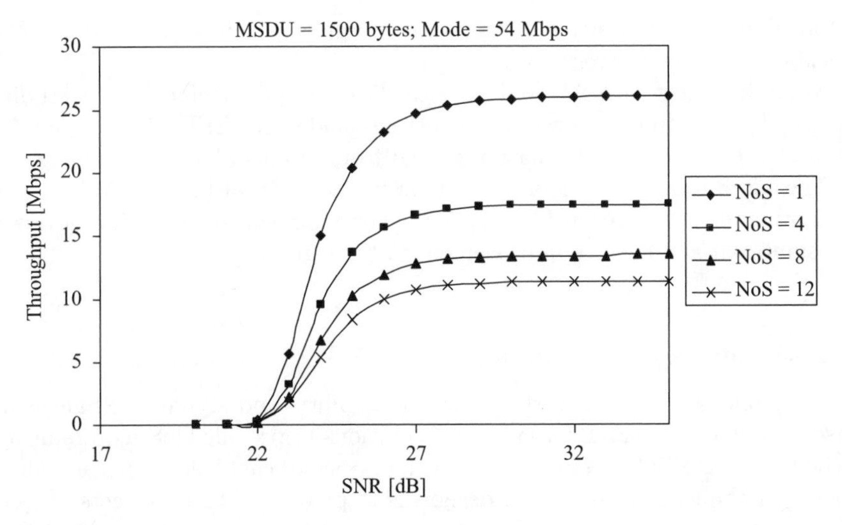

Fig. 4.8. Throughput vs. SNR, NoS fixed

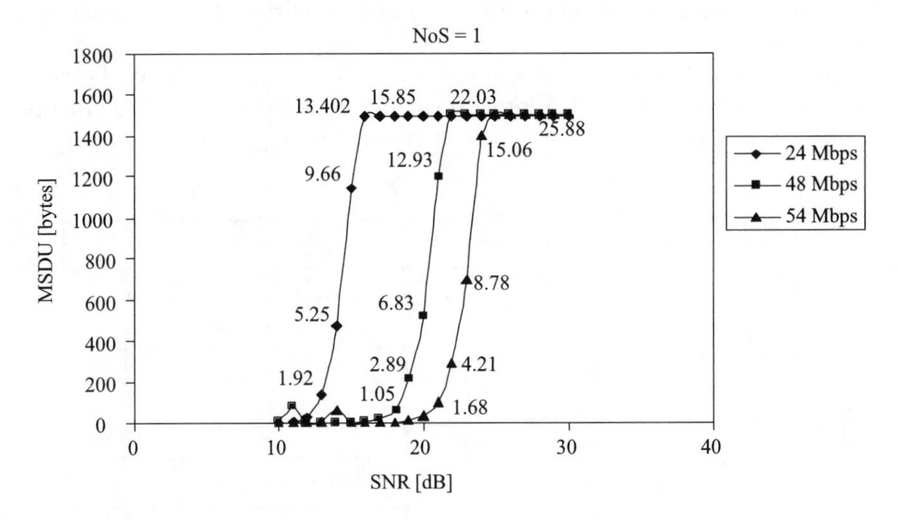

Fig. 4.9. MSDU vs. SNR for three operational modes

4.5.3.7 Effects on Power Control

Authors in [10, 35] show the importance of definition of power control in ad hoc networks. They show that the optimal transmit power can control the connectivity property of a network and through that the capacity and throughput. COMPOW protocol presented in [35] provides power aware routes and reduces the MAC layer contentions. Goldsmith and Wicker in [10] present a detailed analysis of design challenges for energy constraints in ad hoc wireless networks relevant to cross-layer design.

The joint effect of MAC and physical layer on power efficiency is studied in [20] by selecting a power transmitting mode for IEEE 802.11a WLAN orchestrated with the information gained from physical layer.

The adaptation of the error control mechanisms [18] at the link/MAC layer, and transmit power control at physical layer can result in reduction of power consumption and achieve improvement in throughput [3].

4.5.3.8 Effects on Scheduling

The mobile ad hoc networks, due to mobility and dynamic changes in resources, have to search for enhanced solutions to provide QoS requirements. Scheduling, as a link layer technique, can play significant role in such enhancements. A challenge to scheduler designs is to predict all three aspects of QoS: throughput, loss and delay. The scheduler aims to guarantee prescribed QoS with efficient resource utilization over wireless fading links [36, 37]. Many authors dedicated their work to investigate and define related metrics and mechanisms which can influence the scheduler activities. The following text presents some of them.

Authors in [38] propose a new architecture based on integration between the cross-layer concept and priority queueing depicted in Fig. 4.10. In this architecture, a node is aware of the state of the other components of the

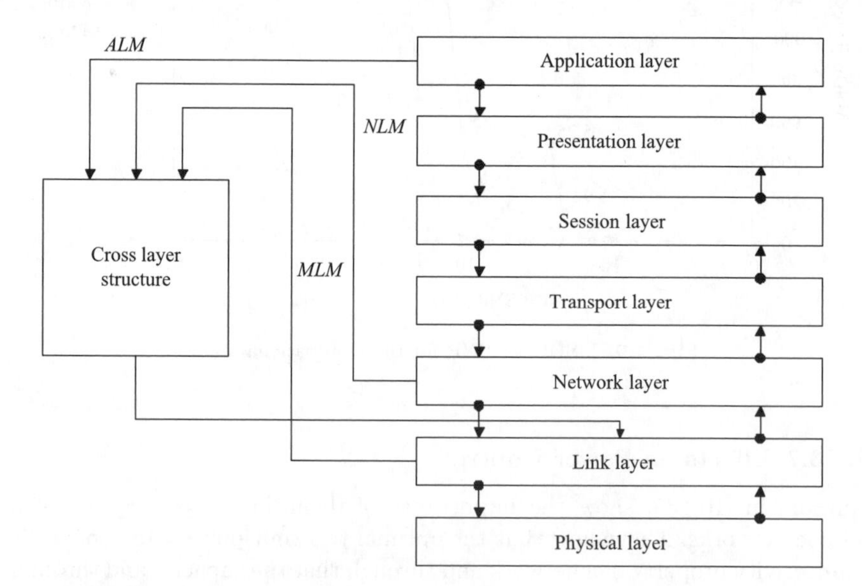

Fig. 4.10. Cross-layer structure — information introduced from the structure to the MAC layer (NLM, MLM and ALM) defines the scheduling

network. The information shared between the nodes includes the MAC layer metrics (MLM), which contains MAC layer buffer size, and network layer metrics (NLM), which contains network topology, path gains and transmission power (see also Chapter 5).

The proposed cross-layer concept is based on a class traffic differentiation and different queueing services to each class. The MAC layer is designed with *spatial reuse time division multiple access (STDMA)* techniques [39], which provides transmission slots in a deterministic way. The *priority queueing* is used to differentiate each service class. If the queue of the certain link is loaded with *high priority traffic*, this queue will need *more slots* with *strict gaps* between them to control the jitter. The *lower priority traffic* needs *less slots* with *variable gaps* between the slots. In order to eliminate the unwilling delays of queueing in FIFO (first-in-first-out) queues, the FPQ (fixed priority queueing) and WFPQ (Weighted Fair Priority queueing) protocols are introduced [40], where the packets with higher priority wait less. The proposed method, which is based on two metrics, reduces the global delay and satisfactorily meets the different traffic class requirements. Other metrics, such as antenna type and transmission power can further improve the scheduling effects.

Another cross-layering for multi-user scheduling is proposed in [36]. The proposed scheduler design forms a building block for end-to-end QoS support in multihop networks. It accounts for both the channel variations and the status of user queues, and includes admission control and scheduling polices for multiple users with different QoS requirements (QoS guaranteed and best-effort classes). The scheduler allocates certain number of time slots to each QoS-guaranteed user. The number of actually scheduled time slots depends on data availability in the queue and the transport capacity. The unused time slots are occupied by best-effort users. According to the Little's theorem, the average delay per packet through the wireless link can be calculated as:

$$T = \frac{Nwl}{\lambda \cdot (1 - P_d)} \tag{4.1}$$

where Nwl is the average number of packets in the queue, λ is data arrival rate, P_d is the packet dropping (or overflow) probability and $(1 - P_d)$ is the probability of a correctly received packet. Through combination with admission policy, the delay can be reasonably bounded. The proposed cross-layer scheduler guarantees prescribed QoS, achieves efficient bandwidth utilization, enjoys low-complexity implementation, isolates QoS guaranteed service and provides scalability. Efficient bandwidth utilization is accomplished with adaptive modulation and coding schemes.

In [41] authors propose *adaptive scheduling scheme,* differentiating between different types of traffic in *downlink* and *uplink* transmission. They adapt the power and bandwidth assignment according to the application's QoS requirements and wireless channel conditions. At MAC layer they define power-adaptive MAC states (High-QoS, Media-QoS and Low-QoS), which are scheduled according to resource availability and overall QoS. The physical and MAC layer design is based on RNS-OFDM (Residue Number System frequency hopping OFDM). MAC scheduler influences appropriate transmission format (coding schemes and modulation) separately for uplink (more classes) and downlink (one class) adaptation. For instance, in bad channel conditions the user can transmit through a larger number of tones for a longer period of time. Three classes of users are differentiated (gold user, silver user and bronze user). In limited bandwidth conditions, the user's class can be degraded.

Example of cross-layer design with adaptive scheduling is reported in [42–44]. Authors in [42] propose priority adaptive scheduler placed in cross-layer design which can significantly improve Internet performance over wireless links.

4.5.4 Network Related Cross-Layer Optimizations

Network layer lies in the middle of the protocol stack and is strongly affected by cross-layer optimizations with lower and upper layers. It captures the networking philosophy, so it is the first layer to reflect the new networking paradigms.

4.5.4.1 Relevant Information on Network Layer

Network layer functions are routing, addressing, selecting the network interface and maintaining the IP connectivity in foreign networks (see Chapter 3). Single hop and multi hop routing raise the neighbor discovery and connectivity aspects, which is one of the first steps in initialization of randomly distributed nodes in ad hoc mobile networks. Power constraint is an important issue in neighbor discovery, as well as in connectivity design. The connectivity is heavily influenced by the ability to adopt various parameters at the link layer such as rate, power and coding, since the connectivity can be performed even on links with low SNIR if the parameters are well adopted [10]. Scalability of routing protocols has mainly focused on self-organization issues (see Chapter 2). Overlaying structures which solve different discovery aspects on application layer (e.g., service discovery, context awareness) influence the choice of routing protocol in order to improve the overall system performances.

4.5.4.2 Interaction with Lower Layers

Interaction of networking layer with physical and MAC layer, resulting in rate adaptation scheme, can enlarge the performance margins. Authors in [45] investigate the interactions of routing protocols (DSR) with different rate adaptation schemes (IARA — interference aware routing adaptation, RA — rate adaptation and DSR — with no rate adaptation) on MAC-physical layer. Time varying routing metrics can further influence the results. Power aware routing schemes are often addressed in the literature. Handoff in Mobile IP, i.e., initiation/completion events, and currently used interfaces are also affected by lower layer status [28]. In [46] authors propose a cross-layer routing aware of physical layer. The network layer is able to set the parameters at the physical layer (i.e., a transmission power on a per hope basis to achieve a global power optimization), and achieve the throughput improvements (by local parameters control). The resulting routing protocol can control the interface settings and get the right feedback from the physical layer.

4.5.4.3 Interactions with Upper Layers

Network layer interacts with transport (i.e., fast retransmit in Mobile IP) and application layers (exchanging location information, harmonize service discovery and enable context awareness). A good illustration of this mechanisms is provided in [47]. Authors in [48] propose a CrossROAD protocol, showing that the overlaying cross-layer P2P system can optimize the MANET performances (drastically reducing the network overhead). They propose a cross-layer interactions exchanging topology information to build the routing tree in combination with proactive routing protocols.

Examples of cross-layer design, which combines routing with service discovery processes, are presented in the following subsection.

4.5.4.4 Efficient Service Discovery

This section deals with optimizations of the service discovery procedure in wireless ad hoc networks. Two methods for efficient service discovery are presented: *Routing Assisted Service Discovery (RASD)* [25] and *Clustering Based Service Discovery (CBSD)* [49]. These methods take into consideration the possible interaction between the service layer and the networking layer by associating a certain routing protocol to the service discovery procedure. Also, the CBSD method includes various service distribution schemes on application layer.

Routing Assisted Service Discovery With RASD, service discovery information is being *piggybacked* on routing information. The actual realization of this approach is achieved by a link-layer protocol booster [24] which is a transparent enhancement of end-to-end protocol communication. Namely, the service and the network layer pass information about services and routes to the booster which is performing the piggybacking of the service related information on the routing information. Boosters at sending and receiving stations are responsible for assembling and disassembling packets containing both service and routing relevant information. In this manner, cross-layer system design

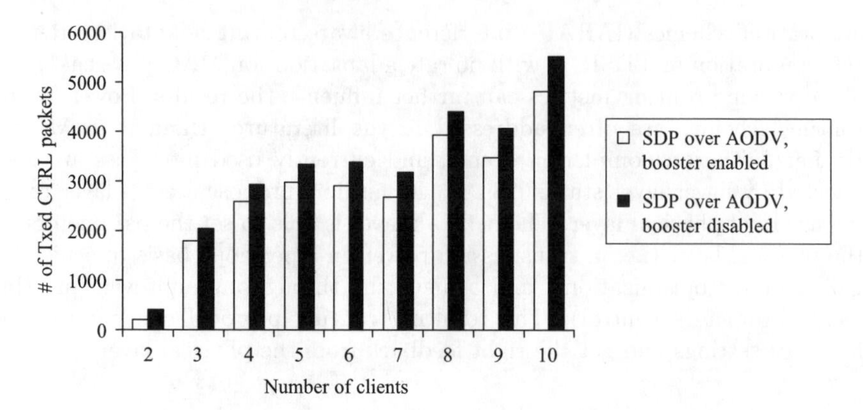

Fig. 4.11. Service discovery over AODV

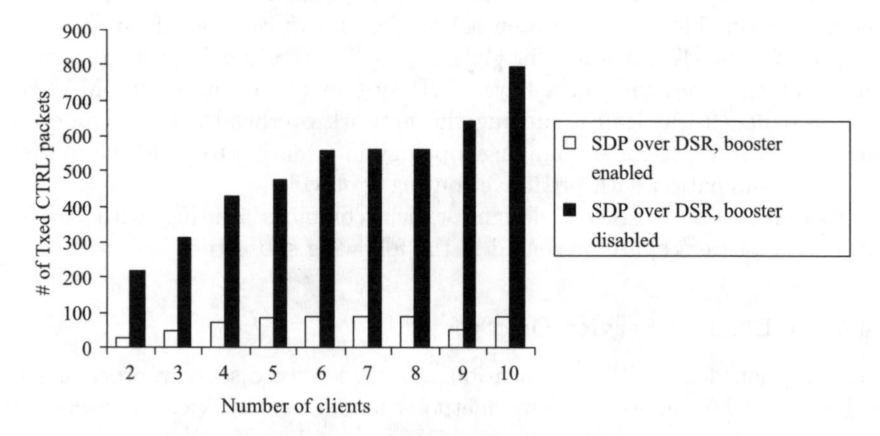

Fig. 4.12. Service discovery over DSR

where the service and the network layer interact with the link layer, while at the same time there is no need to actually impose changes in the service and the network layer protocol formats, is introduced.

The benefits of using the RASD approach are given in Figs 4.11 and 4.12. Results clearly show *low number of generated control messages* (low control overhead) if the process of service discovery relies on the routing procedure in the network (cross-layer interactions) compared to the case when there are no information exchanges across the protocol stack (no cross-layer interactions). Bars for booster enabled/disabled present the cases when the cross-layer interactions are enabled/disabled respectively. It is evident that the approach where DSR provides the network layer support outperforms AODV network layer support in terms of generated control overhead in the network.

Extensive details on the RASD approach can be found in [25] and [49].

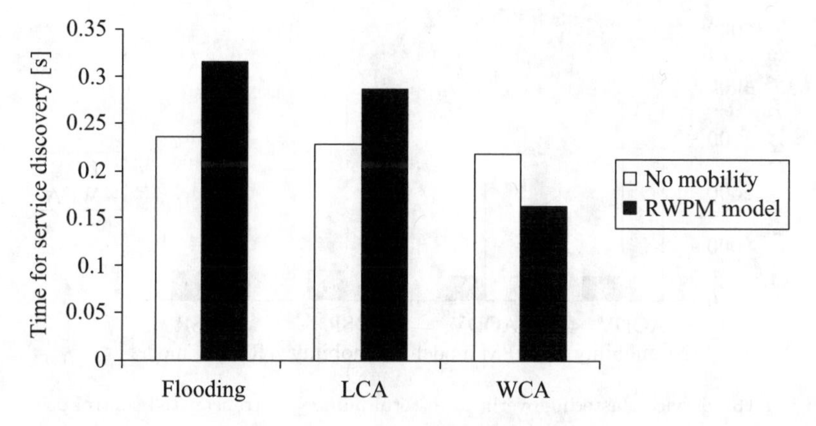

Fig. 4.13. Average time for service discovery with AODV routing support

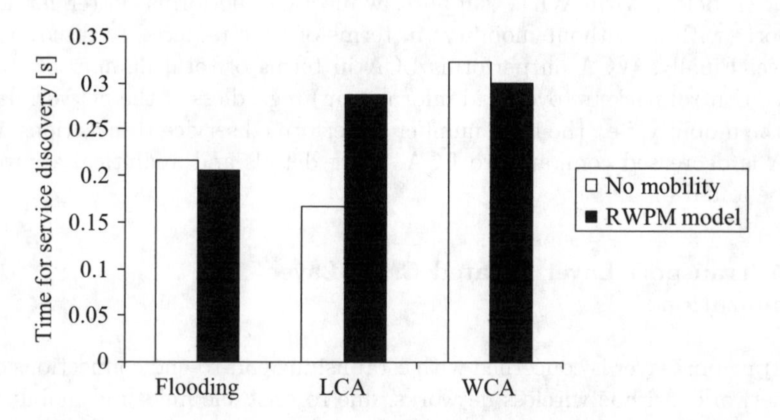

Fig. 4.14. Average time for service discovery with DSR routing support

Clustering Based Service Discovery in Ad Hoc Networks The CBSD method involves a mechanism for service dissemination on application layer whose choice is based on inter-layer interactions with the networking layer. There are three possibilities for service dissemination on application layer: flooding, *Linked-Cluster Algorithm (LCA)* [50] and *Weighted Clustering Algorithm (WCA)* [51].

The benefits of using the CBSD approach are summarized in Figs 4.13, 4.14 and 4.15.

Figs 4.13, 4.14 and 4.15 show that the analyzed service distribution mechanisms perform differently for AODV and DSR. Namely, when AODV is used then the *time for service discovery* is minimized with WCA service distribution, especially with mobility present. However, if DSR is used then *minimum service discovery time* is achieved with flooding (if mobility is present) or LCA

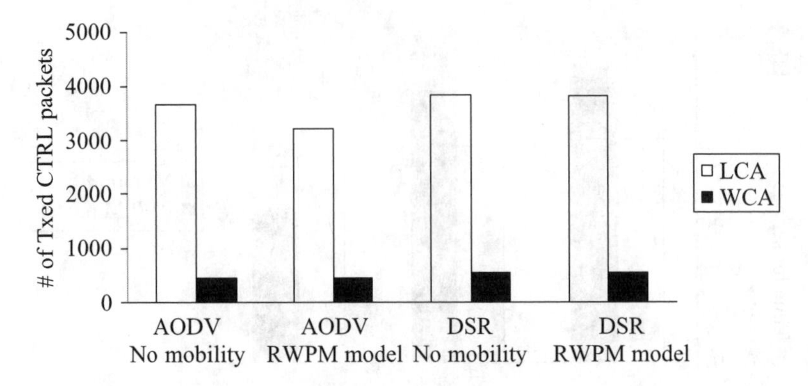

Fig. 4.15. Service clustering overhead in total number of transmitted control packets

(if there is no mobility). If we compare LCA and WCA then we see that WCA performs better with AODV support, while LCA performs better for DSR support (with or without mobility) in terms of time required to discover the service. Finally, WCA outperforms LCA in terms of total number of transmitted control packets (overhead information) regardless of the network layer and the mobility, i.e., the total number of performed service transactions with WCA is decreased compared to LCA. More details and analytical approach can be found in [49].

4.5.5 Transport Layer Related Cross-Layer Optimization

The transport layer is concerned with establishing end-to-end connections over the network. Ad hoc wireless networks, due to unstable links and mobility of nodes, are characterized by large delays, packet losses and high bit error rates. Transport protocol (e.g., TCP) interprets this as congestion loss, which reduces its throughput. Transport layer parameters, which can be used in cross-layer information exchanges can be seen in Table 4.1. Relevance of each of them depends on particular transport protocol and application.

TCP traffic can benefit from MAC/link layer retransmission policy through joint efforts with lower and upper links [52]. The authors in [53] incorporate user feedback into the protocol stack. They use link layer connection/disconnection information to improve the TCP performance. Joint error control using BER and handoff notification is also considered in link transport layer information exchange resulting in power savings and improved throughput [3]. TCP layer may provide packet loss and goodput information to the applications in joint effort to meet QoS requirements. A systematic approach to jointly design TCP congestion control and MAC algorithms proposed in [52] not only improves the performance, but makes the interactions between MAC/TCP layer more transparent using optimization techniques.

4.5.6 Application Layer-Related Cross-Layer Optimization

The application layer in cross-layer design can play a significant role in network efficiency. It is the interface to the user and captures its requirements communicating them to the lower layers [52]. For ad hoc wireless networks, any QoS guarantees are questionable if the application does not adapt to the QoS parameters offered by the network [10]. The application layer can communicate to other layers the application's QoS needs, i.e., the delay tolerance, acceptable delay variations and packet loss rate, required throughput. The required QoS through lower layer adaptations can result in good overall performance, even under poor network conditions [10]. The benefits of lower layer adaptation to application requirements are mostly obvious in video and multimedia demanding applications. Examples can be found in [54–57].

The cross-layer design schemes for QoS delivery, with information exchange among application, MAC and physical layer are proposed in [36, 38, 41].

4.6 Revolutionary (Alternative) Cross-Layering Approaches

Evolutionary cross-layer designs are based on traditional layering paradigm. Deeper understanding of ad hoc nodes' behavior, new ways of link abstraction and cooperation between the nodes (cooperative diversity), routing awareness, capacity limits, future wireless structure design factor initiated alternative approaches to cross-layer design [58–61]. Instead of following the line, revolutionary approaches introduce new level of abstraction in definition of functional modules (i.e., conceptual levels), and defining interactions between them [8, 42]. Examples can be found in WSN (wireless sensor networks). Application-specific communication architectures, WSNs, which can act as stand-alone architectures, do not need to implement strict layering principles. WSNs design cross-layering concerning power-efficiency [62]. Another example is Shannon Mappings [63–65], which provides direct analogue source symbol to channel symbol conversion.

4.7 Generic Cross-Layer Design

The generic cross-layer design for mobile ad hoc networks is proposed in [7] and [66] presenting architecture capable to cope with interlayer interactions. The core component of this architecture is the *network status*, which is a repository for information that the network protocol collects through the stack and possibly exchanges with the layers (see Fig. 4.16).

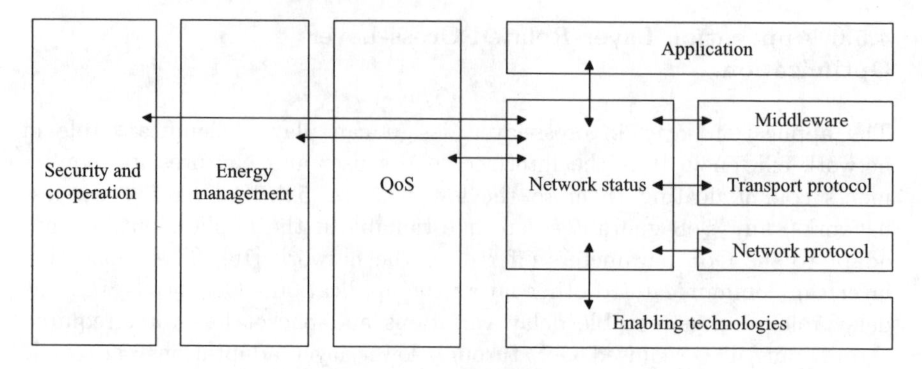

Fig. 4.16. Generic inter-layer interactions

This referent architecture allows the *cross-layer optimization for all network functions* through data sharing. Some of them (power awareness) are mandatory and some are cross-layer by nature (e.g., power management, security and cooperation). Replacing the network status oriented protocol with its legacy counterparts can only result in degraded performance. This model also improves *local and global adaptation* and provides *full context awareness at all layers*. Cross-layering can facilitate the achieving of context awareness at higher layer (middleware and application layer).

This generic approach to architecture design fits into the cross-layer coordination plans inbuilt in referent architecture proposed for future 4G systems [67, 68].

4.8 A Cautionary Perspective on Cross-Layer Design

The cross-layering approach in designing network protocols can produce unwilling consequences in the overall network performance on long-term basis [8, 69]. In the process of designing, a tension is always present between performance and architecture. Performance optimization can lead to a short-term gain, while architecture is usually based on long-term considerations. Unbridled cross-layer design must be aware of interactions with the multiplicity of already existing dynamics and some accompanying theoretical timescale framework.

Cross-layer design diminishes the advantages of modularity, which is a paradigm in building OSI and TCP/IP based systems that assures their compatibility. It can create unintentional interactions between processes and layers, so called loops, which may cause undesirable consequences on the stability of the system.

Solutions offered by cross-layered design are much more difficult to standardize and they must be adapted to particular applications resulting in increased costs. Protocols no more can be developed in isolations, and renewing of any of them must be accompanied with a reimplementation in a cross-layer fashion.

A system-wide cross-layered solution can lead to *spaghetti* implementation, when some new modification affects the overall system operation. Therefore, the cross-layer optimization should be undertaken as a holistic and not fragmented concept.

These cautions mostly go to evolutionary designs, while the revolutionary designs already have task-specific implementations. Recent projects (e.g., MobileMan [70]) and applications (e.g., WSNs) can provide some valuable answers concerning dilemma: strictly layered and modular or variations of cross-layer design.

4.9 Projects and Implementations

Questions raised towards architecture and protocol design paradigms in ad hoc environment, and more generally in wireless networks, have initiated number of projects which address different aspects of cross-layering. Few of them are listed in the following text.

MobileMAN [70]. The project aims to define and develop a metropolitan area, self-organizing, and totally wireless network called *Mobile Metropolitan Ad hoc Network* (MobileMAN). The development of the MobileMAN envisioned architecture and related protocols that will support the self-organizing paradigm can be achieved with a fruitful collaboration through information sharing (cross-layer system design) among various entities in the network.

DIANE [71]. The project develops and evaluates concepts for integrated service discovery procedures (both distributed and efficient) in ad hoc networks. Two approaches for building *overlays* as a promising means of efficient service discovery are proposed. The process of constructing these overlays incorporates the cross-layer paradigm. In some application domains, it is possible to build an overlay that is structured according to the semantics of the offered services. Examples for such semantic overlays are service rings and multi-layer clusters.

MATRICE [72]. The main objective of this project is to define and validate access and transmission concepts based on MC-CDMA technology for provision of the broadband component of future mobile cellular systems. The work encompasses layers 1–3, and specifically proposes to investigate possible *adaptive link layer techniques*, based on cross-layer optimizations, for achieving maximum throughput, and definition of criteria in assisting intelligent and fast adaptation.

4MORE [73]. The 4MORE project complements to the MATRICE project and worldwide research results and advances one step towards implementation through the design of a System on Chip (SoC) solution for a 4G terminal employing multiple antennas and based on MC-CDMA techniques. 4MORE objectives are research, development, integration, and validation of a cost effective, low power, System on Chip (SoC) solution for multi-antenna MC-CDMA mobile terminals, based on *joint optimisation of L1 and L2 functions*, i.e., cross-layer optimizations.

PHOENIX [74]. The aim of the PHOENIX project is to develop a scheme offering the possibility to let the application world (source coding, ciphering) and the transmission world (channel coding, modulation) to talk to each other over an IPv6 protocol stack (network world), so that they can *jointly* develop an end-to-end optimised wireless communication link (i.e., cross-layer system design). The design trade-offs of interactive wireless video systems are being studied and performance comparisons is provided both in the context of existing and future generations of wireless systems.

Project E — NEWCOM [75]. The goals of Project E of the IST Network of Excellence on Wireless Communications (NEWCOM) include identification of the existing gaps in European knowledge in cross-layer, and preparation of an action plan for filling them by capitalizing on project researchers' skills. The research objective of the project aims at investigating the potential benefits of cross-layer in wireless network design in relation to the methodology of separate layer design. In addition, the project intends to consider the coupling of the higher layers with the physical layer and elaborate the information to be exploited from the physical in order to optimize network performance.

ERACLIN [76]. The ERACLIN project is embedded in one of the most challenging innovation paths, affecting in a convergent mode both Mobile Communication and Information Technology. The main objective is the transfer of knowledge between the industrial and the academic partner of the consortium in fields and technologies where they are specialised. This scientific project aims to contribute to the technical innovation policies of the Community by exploiting potential of Radio Resource Management (RRM) algorithms based on *cross-layer issues* for 4G networks.

4.10 Concluding Remarks

The major idea that permits the networks proliferation was the design of layering approach, which perfectly suits the wired network architectures. Existing layering paradigm was reexamined with the idea of more intensive communication and information exchanges between the closed layers in case of wireless architectures. This was a natural consequence in attempt to improve the system performances.

In the case of ad hoc mobile networks, the cross-layering approach is proven as successful designing tool. The more traditional approach to cross-layer design allows information exchanges through the layers of OSI/TCP based protocol stacks. New network design knowledge, developed within the ad hoc networking paradigm, which primarily stems from boundaring areas (e.g., information theory, signal proccesing), shapes the development of novel approaches to cross-layering.

The benefits from cross-layering are enlightened through this chapter. A cautionary perspective on cross-layer design was also under consideration. The cross-layer paradigm will be incorporated on more/less regular basis into the systems' protocol stacks within the future standards for 4G technologies, including ad hoc mobile networks as their inevitable counterpart [67].

References

[1] ITU, Information Technology — OSI — Basic Reference Model, X.200, 1994, July.

[2] Rappaport, T. S., Annamalai, A., Buehrer, R. M., and Tranter, W. H., "Wireless Communications: Past Events and a Future Perspective," *IEEE Communications Magazine*, 40, May 2002, pp. 148–161.

[3] Raisinghani, V. T. and Iyer, S., "Cross-Layer Design Optimizations inWireless Protocol Stacks," *Computer Communications*, 27(2004), Elsevier Publishing, 2004, pp. 720–724.

[4] Shakkottai, S., Rappaport, T. S., and Karlsson, P. C., "Cross-Layer Design for Wireless Networks," *IEEE Communications Magazine*, October 2003, pp. 74–80.

[5] Li, X. and Bao-yu, Z., "Study on Cross-layer Design and Power Conservation in Ad Hoc Network," *IEEE PDCAT 2003*, 2003.

[6] Wang, Q. and Abu-Rgheff, M. A., "Cross-Layer Signalling for Next-Generation Wireless Systems," *IEEE WCNC2003*, New Orleans, USA, March 2003.

[7] Dhaou, R., Gauthier, V., Tiado, M. I., Becker, M., and Beylot, A. -L., "Cross Layer Simulation: Application to Performance Modelling of Networks composed of MANETs and Satellites," *2nd International Working Conference on Performance Modelling and Evaluation of Heterogeneous Networks (HET-NETs)*, West Yorkshire, UK, July 2004.

[8] Aune, F., *Cross-Layer Design Tutorial*, Norwegian University of Science and Technology, Department. of Electronics and Telecommunications, Trondheim, Norway, Published under Creative Commons License, November 2004.

[9] Conti, M., Maselli, G., and Turi, G., "Cross-Layering in a Mobile Ad Hoc Network Design," IEEE Computer Society, pp. 48–51, Commercial implementation, February 2004.

[10] Goldsmith, A. J. and Wicker, S. B., "Design Challenges for Energy-Contrained Ad Hoc Wireless Networks," IEEE Wireless Communications, pp. 8–27, August 2002.

[11] Obreiter, P. and Klein, M., "Vertical Integration of Incentives for Cooperation — Inter-Layer Collaboration as a Prerequisite for Effectively Stimulating Cooperation in Ad Hoc Networks," *Med-Hoc Net 2003 Workshop*, Mahdia, Tunisia, June 2003.

[12] Valenti, M. C. and Correal, N., "Exploiting Macrodiversity in Dense Multihop Networks and Relay Channels," *IEEE WCNC2003*, 4(1), March 2003.

[13] Kobayashi, M., Caire, G., and Gesbert, D., "Antenna Diversity vs. Multiuser Diversity: Quantifying the Tradeoffs," *ISITA2004*, Parma, Italy, October 2004.

[14] Ebert, J. -P. and Wolisz, A., "Combined Tuning of RF Power and Medium Access Control for WLANs," *Mobile Networks and Applications*, 6(5), Special issue on Mobile Multimedia Communications (MoMuC'99), 2001.

[15] Atanasovski, V. and Gavrilovska, L., "Throughput Performance of Contending IEEE 802.11a Systems in Rayleigh Fading Channels," *8th International Symposium on Wireless Personal Multimedia Communications — WPMC 2005*, Aalborg, Denmark, September 18–22, 2005, pp. 1925–1929.

[16] Atanasovski, V. and Gavrilovska, L., "Influence of Header Compression on Link Layer Adaptation in IEEE 802.11b," *IEEE International Conference on Wireless Networks, Communications and Mobile Computing WIRELESSCOM 2005*, Maui, Hawaii, USA, June 13–16, 2005.

[17] Gavrilovska, L. and Atanasovski, V., "Influence of Packet Length on IEEE 802.11b Throughput Performance in Noisy Channels," *Workshop on "My Personal Adaptive Global NET: Visions and Beyond" IST-FP6-IP MAGNET*, Shanghai, China, November 11–12, 2004.

[18] Gavrilovska, L. and Atanasovski, V., "Adaptive Techniques for WLAN Throughput Improvement," in *Proceedings of the 5th International Conference on 3G Mobile Communication Technologies — IEE 3G 2004*, Savoy Place, London, UK, October 18–20, 2004, pp. 73–77.

[19] Ferrus, R., Alonso, L., Umbert, A., Reves, X., Perez-Romero, J., and Casadevall, F., "Cross-Layer Scheduling Strategy for UMTS Downlink Enhancement," *IEEE Radio Communications Magazine*, June 2005, pp. S-24–S-28.

[20] Zhao, J., Guo, Z., and Zhu, W., "Power Efficiency in IEEE 802.11a WLAN with Cross-Layer Adaptation," *IEEE ICC 2003*, 2003.

[21] Maharshi, A., Tong, L., and Swami, A., "Cross layer Designs of Multichannel Reservation MAC under Rayleigh Fading," *IEEE Transactions on signal processing*, 51(8), August 2003, pp. 2054–2067.

[22] Song, G. and Li, Y., "Cross-Layer Optimization for OFDM Wireless Networks — Part I: Theoretical Framework," *IEEE Transactions on Wireless Communications*, 4(2), March 2005, pp. 614–624.

[23] Song, G. and Li, Y., "Cross-Layer Optimization for OFDM Wireless Networks — Part II: Algorithm Development," *IEEE Transactions on Wireless Communications*, 4(2), March 2005, pp. 625–634.

[24] RFC 3135 — Performance Enhancing Proxies Intended to Mitigate Link-Related Degradations. Available: www.faqs.org/rfcs/rfc3135.html.

[25] Atanasovski, V. and Gavrilovska, L., "Routing Assisted Efficient Service Discovery in Ad-Hoc Networks," *7th International Symposium on Wireless Personal Multimedia Communications — WPMC 2004*, Abano Terme, Italy, September 12–15, 2004, pp. 491–495.

[26] Xylomenos, G. and Polyzos, G. C., "Quality of Service Support Over Multiservice Wireless Internet Links," *Computer Networks*, 37(5) (2001), 2001, pp. 601–615.

[27] Methfessel, M., Dombrowski, K. F., Langendorfer, P., Frankenfeldt, H., Babanskaja, I., Matthaei, I., and Kraemer, R., "Vertical Optimization of Data

Transmission for Mobile Wireless Terminals," *IEEE Wireless Communications*, 9(6)(2002), 2002, pp. 36–43.

[28] Perkins, C., "IP Mobility Support for IPv4," *RFC3344*, August 2002.

[29] Valk, A. G., "Cellular IP: A New Approach to Internet Host Mobility," *ACM SIGCOMM Computer Communication Review*, 29(1)(1999), 1999.

[30] Wu, Y., Chou, P. A., Zhang, Q., Jain, K., Zhu, W., and Kung, S. -Y., "Network Planning in Wireless Ad Hoc Networks: A Cross-Layer Approach," *IEEE Journal on Selected Areas in Communications*, 23(1), January 2005, pp. 136–150.

[31] Toumpis, S. and Goldsmith, A. J., "Performance, Optimization, and Cross-Layer Design of Media Access Protocols for Wireless Ad Hoc Networks," *IEEE ICC2003*, Seattle, USA, May 2003.

[32] Dimic, G., Sidiropoulos, N. D., and Zhang, R., "Medium Access Control — Physical Cross-Layer Design," *IEEE Signal Processing Magazine*, September 2004, pp. 40–50.

[33] Zhang, R., Sidiropoulos, N. D., and Tsantsanis, M., "Collision Resolution in Packet Radio Networks Using Rotational Invariance Techniques," *IEEE Transactions on Communications*, 50(1), January 2002, pp. 146–155.

[34] Madueno, M., and Vidal J., "Joint Physical-MAC Layer Design of the Broadcast Protocol in Ad Hoc Networks," *IEEE Journal on Selected Areas in Communications*, 23(1), January 2005, pp. 65–75.

[35] Narayanaswamy, S., Kawadia, V., Sreenivas, R. S., and Kumar, P. R., "Power Control in Ad-Hoc Networks: Theory, Architecture, Algorithm and Implementation of the COMPOW Protocol," in *European Wireless*, 2002.

[36] Liu, Q., Zhou, S., and Giannakis, G. B., "Cross-Layer Scheduling With Prescribed QoS Guarantees in Adaptive Wireless Networks," *IEEE Journal on Selected Areas in Communications*, 23(5), May 2005, pp. 1056–1066.

[37] Fattah, H. and Leung, C., "An Overview of Scheduling Algorithms in Wireless Multimedia Networks," *IEEE Wireless Communications*, 9(5), October 2002, pp. 76–83.

[38] Martinez, I., Altuna, J., and Arnaiz, L. M., "A New Cross-Layer Design for Ad Hoc Wireless Networks with QoS Support," *4th Scandinavian Workshop on Wireless Ad-Hoc Networks*, Stockholm, Sweden, May 2004.

[39] Gronkvist, J., "Assignment Strategies for Spatial Reuse TDMA," Licentiate thesis, Royal Institute of Technology, Stockholm, Sweden, March 2002.

[40] Parekh, A. K. and Gallager, R. G., "A Generalized Processor Sharing Approach to Flow Control in Integrated Services Networks: The Single-Node Case," *IEEE/ACM Transactions on Networking*, 1, June 1993, pp. 344–357.

[41] Chen, J., Lv, T., and Zheng, H., "Joint Cross-layer Design for Wireless QoS Content Delivery," *IEEE ICC2004*, Paris, France, June 2004.

[42] Zhang, Q., Zhu, W., and Zhang, Y. -Q., "A Cross-layer QoS-Supporting Framework for Multimedia Delivery over Wireless Internet," *International Packetvideo Workshop 2002*, Pittsburgh, USA, April 2002.

[43] Zhang, J., "Bursty Traffic Meets Fading: A Cross-Layer Design Perspective," *40th Allerton Conference on Communications, Control, and Computing*, October 2002.

[44] Haleem, M. A. and Chandramouli, R., "Adaptive Downlink Scheduling and Rate Selection: A Cross-Layer Design," *IEEE Journal on Selected Areas in Communications*, 23(6), June 2005, pp. 1287–1297.

[45] Yuen, W. H., Lee, H. -N., and Andersen, T. D., "A Simple and Effective Cross Layer Networking System for Mobile Ad Hoc Networks," *IEEE PIMRC2002*, Lisboa, Portugal, September 2002.

[46] Iannone, L., Khalili, R., Salamatian, K., and Fdida, S., "Cross-Layer Routing in Wireless Mesh Networks," *IEEE ISWCS2004*, Mauritius, September 2004.

[47] Chen, K., Shah, S. H., and Nahrstedt, K., "Cross-Layer Design for Data Accessibility in Mobile Ad Hoc Networks," *Wireless Personal Communications 21*, Kluwer Academic Publishers, 2002, pp. 49–75.

[48] Borgia, E., Conti, M., Delmastro, F., and Gregori, E., "Experimental Comparison of Routing and Middleware Solutions for Mobile Ad Hoc Networks: Legacy vs Cross-Layer Approach," *2005 ACM SIGCOMM workshop on Experimental approaches to wireless network design and analysis*, Philadelphia, Pennsylvania, USA, 2005.

[49] Atanasovski, V. and Gavrilovska, L., "Efficient Service Discovery Schemes in Wireless Ad Hoc Networks Implementing Cross-Layer System Design," *27th International Conference on Information Technology Interfaces ITI 2005*, Cavtat, Croatia, June 20–23, 2005, pp. 527–532.

[50] Baker, D. J. and Ephremides, A., "A Distributed Algorithm for Organizing Mobile Radio Telecommunication Networks," *2nd International Conference on Distributed Computing Systems*, Paris, France, April 1981.

[51] Chatterjee, M., Das, S. K., and Turgut, D., "A Weighted Clustering Algorithm for Mobile Ad Hoc Networks," *Cluster Computing 5*, Kluwer Academic Publishers, 2002, pp. 193–204.

[52] Chen, L., Low, S. H., and Doyle, J. C., "Joint Congestion Control and Media Access Control Design for Ad Hoc Wireless Networks," *IEEE INFOCOM2005*, Miami, USA, March 2005.

[53] Raisinghani, V. T., Singh, A. K., and Iyer, S., "Improving TCP performance over Mobile Wireless Environments using Cross Layer Feedback," *IEEE ICPWC 2002*, 2002.

[54] Khan, S., Sgroi, M., Steinbach, E., and Kellerer, W., "Cross-Layer Optimization for Wireless Video Streaming — Performance and Cost," *IEEE ICME2005*, Amsterdam, The Netherlands, July 2005.

[55] Choi, L. -U., Kellerer, W., and Steinbach, E., "Cross Layer Optimization for Wireless Multi-User Video Streaming," *IEEE ICIP 2004*, Singapore, October 2004.

[56] van der Schaar, M., Krishnamachari, S., Choi, S., and Xu, X., "Adaptive Cross-Layer Protection Strategies for Robust Scalable Video Trasnmission over 802.11 WLANs," *IEEE Journal on Selected Areas in Communications*, 21(10), December 2003, pp. 1752–1763.

[57] Kumwilaisak, W., Hou, Y. T., Zhang, Q., Zhu, W., Jay Kuo, C. -C., and Zhang, Y. -Q., "A Cross-Layer Quality-of-Service Mapping Architecture for Video Delivery in Wireless Networks," *IEEE Journal on Selected Areas in Communications*, 21(10), December 2003, pp. 1685–1698.

[58] Nosratina, A., Hunter, T. E., and Hedayat, A., "Cooperative Communication in Wireless Networks," *IEEE Communications Magazine.*, 42, (10), October 2004, pp. 74–80.

[59] Alonso, L. and Agusti, R., "Automatic Rate Adaptation and Energy-Saving Mechanisms Based on Cross-Layer Information for Packet-Switched Data Networks," *IEEE Radio Communications Magazine*, March 2004, pp. S-15–S-20.
[60] Pham, P. P., Perreau, S., and Jayasuriya, A., "New Cross-Layer Design Approach to Ad Hoc Networks Under Rayleigh Fading," *IEEE Journal on Selected Areas in Communications*, 23 (1), January 2005, pp. 28–39.
[61] Conti, M., Gregori, E., and Turi, G., "A Cross-Layer Optimization of Gnutella for Mobile Ad hoc Networks," *MobiHoc'05*, Urbana-Champaign, Illinois, USA, May 2005.
[62] Xiao, J., Cui, S., Goldsmith, A. J., and Luo, Z. Q., "Joint Estimation in Sensor Networks Under Energy Constraints," *IEEE SECON'04*, Santa Clara, CA, October 2004.
[63] Ramstad, T., *Insights Into Mobile Multimedia Communications*, Academic Press, 1st ed., 1999. Chapter 26: Combined Source Coding and Modulation for Mobile Multimedia Communication, pp. 415–430.
[64] Ramstad, T., "Signal Processing for Multimedia: Robust Image and Video Communication for Mobile Multimedia," *Nato Science Series: Computers and Systems Sciences*, 174, ISO Press, 2000.
[65] Ramstad, T., "Shannon Mappings for Robust Communication," *Telektronikk*, 174, Information Theory and its Applications, 2002.
[66] Conti, M., Maselli, G., Turi, G., and Giordano, S., "Cross Layering in Mobile Ad Hoc Network Design," *IEEE Computer Society*, February 2004.
[67] Verikoukis, C., Alonso, L., and Giamalis, T., "Cross-Layer Optimization for Wireless Systems: A European Research Key Challenge," *Global Communications Newsletter*, July 2005, pp. 1–3.
[68] Tafazolli, R., *Technologies for the Wireless Future: Wireless World Research Forum (WWRF)*, Wiley, 2004.
[69] Kawadia, V. and Kumar, P. R., "A Cautionary Perspective on Cross-Layer Design," *IEEE Wireless Communications Magazine*, February 2005, pp. 3–11.
[70] Project MobileMAN, http://www.mobileman.projects.supsi.ch
[71] Project DIANE, http://www.hnsp.inf-bb.uni-jena.de/DIANE/en/inhalte/home.html
[72] Project MATRICE, http://www.ee.surrey.ac.uk/CCSR/IST/Matrice/
[73] Project 4MORE, http://www.ist-4more.org
[74] Project PHOENIX, http://www.ist-phoenix.org
[75] http://newcom.ismb.it/public/index.jsp
[76] ERACLIN, MKTIK-2004-517518, Marie-Curie Transfer of Knowledge.

5

Quality of Services

5.1 Introduction

Ad hoc network is a dynamic multihop wireless network established by a set of mobile nodes on a shared wireless channel, which makes the mobile ad hoc networking a challenging task. The evolution in the multimedia technology and the growing number of different applications promotes QoS in Mobile Ad hoc NETworks (MANETs [1]) to an area of great interest.

Even though some of the QoS problems are already solved in the wire-based networking, the area of mobile ad hoc networking still hides many challenges. When addressing QoS in the wireless mobile segment several unique and distinguishing issues emerge. The mature solutions for QoS in wire-based networks influenced the solutions in wireless ad hoc environments. Providing quality of service (QoS) support in ad hoc networks requires considering the:

- Limited resources (in bandwidth and in nodes' processing power),
- Time-varying topology, due to the nodes mobility,
- Fully-distributed architecture, due to the lack of infrastructure.

The dynamic nature of the ad hoc networks makes the complex QoS solutions proposed for wire-based networks inapplicable in the mobile ad hoc domain. However, some quality enforcements mechanisms can be applied.

Several overviews of the challenging task of providing QoS support in mobile ad hoc networks are found in the literature [2–10]. This chapter aims to generalize the concepts in quality of service design and features for wireless ad hoc networks and to present some novel research in this domain.

5.2 Quality of Service, What It Is?

The QoS stands for Quality of Service and is widely used in different networking scenarios taking into consideration different parameters, network topologies and variables. The CCITT's (UN Consultative Committee for International Telephony and Telegraphy) widely accepted Recommendation E.800 defines QoS as: *"The collective effect of service performance which determines the degree of satisfaction of a user of the service"*. The QoS in Internet is defined as: *"To provide a set of service requirements to the applications while routing through the network"*, RFC 2386 [4]. There are many other definitions referring to different aspects of QoS (e.g., ITU-T E.430, ISO/IEC JTC1 SC21 QoS, ETSI ETR 003) [8] and many debates about their meanings.

The QoS is the performance level of services offered by a service provider or a network to the user. QoS provisioning often requires negotiation between the host and the network about the different kinds of services, resource reservation schemes, priority scheduling and call admission control. The service can be characterized by a set of measurable predefined service requirements such as minimum bandwidth, maximum delay, maximum delay variance (jitter), maximum packet loss rate and so on.

The definition of QoS accepted in wired environment can not fit perfectly in the case of ad hoc mobile networks and it has to be redefined. A hard QoS guarantee is not plausible and it depends on the "quality" of the network, i.e., on the available resources in the wireless medium and in the mobile nodes, and their stability. The applications should adapt to these changes, so the quality of service in mobile ad hoc networks could mean *to provide a set of parameters in order to adapt the applications to the "quality" of network while routing them through the network* [2].

5.2.1 Applications and QoS Parameters

Different applications require different QoS support [8]. They can be roughly grouped according to their QoS sensitivity as QoS-sensitive *real time* and *non-real time* applications. So, the real time applications, such as voice and video transmission, require limited delay, while non-real time applications, such as file transfer and e-mail, require reliability. The real-time QoS service class can be further divided into several general classes: delay and delay-sensitive services, throughput and bandwidth-sensitive applications, error and error-sensitive applications.

A large amount of today's traffic belongs to demanding multimedia applications. Multimedia applications can be classified into three general classes:

- Real-time interactive audio and video (e.g., Internet telephony, video conferencing)
- One-to-many streaming of real-time audio and video (e.g., broadcast radio and television over the Internet)
- Streaming stored audio and video (e.g., Web broadcast, video-streaming on demand).

The most relevant QoS parameters for multimedia applications are bandwidth, delay and delay jitter. Provisioning of sufficient bandwidth for various applications affects the QoS support for all QoS parameters.

The QoS parameters in wired networks are mainly influenced by the requirements of the multimedia traffic, while in ad hoc wireless networks the QoS parameters are more influenced by the resource constraints of the nodes (e.g., battery power, processing power, buffer size).

5.2.2 Quality of Service Provisioning

The term *provisioning* refers to the allocation of resources needed at various points in the network. Wired-based networks achieve the QoS provisioning implementing one of these approaches [3]:

- Over-provisioning of the resources, or
- Network engineering.

The *over-provisioning of the resources* assumes that plentiful capacity is added in the network to make it more resistant to demanding multimedia applications. The upgraded resources can be data links, routers and network cards. It is easy to implement and allow gradualism. However, it supports only one service class and same priority for all users.

The *network traffic engineering* approach aims to classify the users (or their applications) in service classes and dedicate different priority to each class. Network traffic engineering, designed for use in combinations for different network contexts, can be:

- Reservation-based engineering
- Reservation-less engineering.

In the *reservation based engineering*, assignment of network resources is completed according to the application's QoS request and considering the network bandwidth management policy. This approach is used in ATM (Asynchronous Transfer Mode) and in RSVP — IntServ (Resource ReserVation Protocol — Integrated services) [11].

The *reservation-less engineering* assumes no reservations within the network. The QoS is achieved by the introduction of different mechanisms into the network such as Connection Admission Control (CAC), Policy Managers, Traffic Classes and Queuing Mechanisms. CAC controls the node's access to the network and whether it is served with requested QoS parameters. Policy Manager ensures that no node will violate the pre-assigned type of service. Traffic Classes (assured, controlled-load or best-effort services) differentiate the processing priority of data packets. This approach is used in DiffServ

(Differentiated Services) QoS architecture. Queuing mechanisms are responsible for dropping the packets with the lowest priority in the case of congestion or to provide explicit feedback to nodes in order to avoid congestion.

The QoS provisioning in mobile ad hoc networks depends not only on the available resources, but also on the mobility of such resources and topology and capacity changes [12, 13]. As a result, it can not rely on static traffic profiles and topology information. Many proposals have been presented in the literature to support QoS in MANETs including QoS MAC protocols, QoS routing protocols and resource reservation protocols [10].

5.3 Classification of QoS Solutions

The QoS solutions can be classified based on the *layer* in the network protocol stack, at which the appropriate mechanisms operate, or according to the *implemented approach* [9].

The layer-wise solutions are represented with:

- *MAC/DLL solutions*, which implement QoS mechanisms on MAC/link layer (e.g., 802.11e, Cluster TDMA, MACA/PR, RTMAC)
- *Network layer solutions*, represented by QoS routing protocols
  - On demand (e.g., TBR, TDR, QoSAODV, AQR, OLMQR)
  - Table-driven (e.g., PLBQR)
  - Hybrid (e.g., BR, CEDAR)
- *Cross layer solutions* for QoS provisioning, which involve the inter-layer and/or cross-layer information exchanges (e.g., INSIGNIA, INORA, SWAN, PRTMAC).

The approach-wise solutions can be based on interactions between: routing protocol and QoS provisioning mechanisms, network and MAC layers or routing information update mechanisms. More details can be found in [9].

Some other characteristics can also classify the QoS approaches. If the QoS guarantees last for the whole duration of a session, then this situation is defined as *hard QoS*. However, most of the QoS approaches are *soft QoS*, since it is very difficult to provide stable guarantees in dynamic ad hoc networks. In the case when every node maintains the information about either global or local state (link or flow specific), the approach is classified as *stateful*. In *stateless* approach, with reductions in storage and computation resources at the nodes, providing the QoS guarantees is extremely difficult.

An example of MAC layer solution for QoS support in the widely spread IEEE 802.11 WLANs will be presented in the following paragraph. It is based on an ongoing draft proposal for IEEE 802.11e standard, foreseen to introduce the QoS in WLAN environment.

5.4 MAC Layer QoS Solution — IEEE 802.11 Enhancements

IEEE 802.11 MAC specifications define two different access schemes, the mandatory Distributed Coordination Function (DCF), and the optional (rarely implemented) Point Coordination Function (PCF). DCF is distributed contention-based access scheme and can be used both in ad-hoc and infrastructure mode. It is basically listen-before-talk access scheme. According to the DCF, each station senses the medium before initiating a frame transmission. In order to determine the state of medium (free/busy), every station should perform physical carrier sensing at PHY layer and virtual carrier sensing at MAC sub-layer. The medium is considered free when both physical and virtual sensing indicate free medium. The DCF access scheme does not include any differentiation or prioritisation mechanism. All stations and traffic classes have same priority to access the wireless medium, thus different QoS requirements of applications are not supported with the use of DCF. On the other hand, PCF provides some support for time-bounded applications [14], but without any traffic prioritisation and differentiation.

The Enhanced Distributed Channel Access (EDCA) mechanism [15, 16] is simply an enhancement of the DCF access scheme with a possibility of traffic prioritisation. It allows the packets to be classified into 4 different Access Categories (ACs): AC-VO (AC-Voice), AC-VI (AC-Video), AC-BE (AC-best effort) and AC-BK (AC-background), each with different values of the contention parameters DIFS and CW. This section focuses on the main differences between DCF and EDCA mechanisms.

Instead of waiting for a DIFS interval before trying to access the medium, or continuing to decrement backoff timer after it was paused as in DCF, an interframe space called Arbitration InterFrame Space (AIFS) is used for each AC. The AIFS interval for AC i is set according to the following formula:

$$AIFS(TC_i) = DIFS + \Delta AC_i \times SlotTime \quad (5.1)$$

where ΔAC_i is an integer such that $\Delta AC_i \geq 0$. This means that a AC using large ΔAC_i (large AIFS) will have lower priority than an AC using small ΔAC_i (small AIFS), since it will wait longer before trying to access the medium or continuing to decrement backoff timer after it was paused.

In order to support further differentiation between ACs, the contention window from which the backoff timer is calculated is different for each AC. The backoff timer for AC i is calculated as follows:

$$Backoff\ Time(AC_i) = Rand(1, CW(AC_i) + 1) \times SlotTime \quad (5.2)$$

where $Rand(1, CW(AC_i)+1)$ is a pseudorandom integer drawn from a uniform distribution over the interval $[1, CW(AC_i)+1]$. $CW(AC_i)$ is current contention window size for AC i, $CW_{\min}(AC_i) \leq CW(AC_i) \leq CW_{\max}(AC_i)$, whereas $CW_{\min}(AC_i)$ and $CW_{\max}(AC_i)$ are the minimal and the maximal values of the contention window for AC i, respectively. Choosing a smaller $\mathrm{CW_{min}/CW_{max}}$ for a given AC will cause generating shorter backoff intervals for that AC, thus gaining priority over an AC with larger $\mathrm{CW_{min}/CW_{max}}$. Unlike DCF, where the backoff timer after an unsuccessful transmission attempt is calculated using doubled CW, EDCA calculates new CW with the help of the Persistence Factor (PF) which can be different for each AC. New CW for AC i is calculated as follows:

$$newCW(AC_i) \geq [(oldCW(AC_i) + 1) \times PF(AC_i)] - 1 \tag{5.3}$$

where $PF(AC_i)$ is PF for AC i, and $newCW(AC_i)$ is a value of CW that should be used in calculating backoff timer for the next attempt to transmit the frame from AC i. Of course, $newCW(AC_i)$ never exceeds the parameter $CW_{\max}(AC_i)$ which is the maximum possible value for the CW of AC i.

Since a station can transmit traffic flows which belong to different ACs, each station, that supports IEEE 802.11e [15], should have up to 4 independent transmission queues (MAC buffers). These queues behave as virtual stations which are contending for transmission opportunity within the station. If the backoff timers of two or more parallel TCs in a single station reach zero at the same time, a scheduler inside the station treats this as a virtual collision and transmits the frame which belongs to the AC with higher priority. Therefore, there exists two levels of medium access contention: internal contention among traffic of different priorities inside the same station and external contention among traffic from different stations. Collisions may happen at both levels and are resolved similarly such that higher prioritised traffic (by means of: $AIFS(AC_i)$, $CW(AC_i)$ and $PF(AC_i)$) will obtain the channel first and lower priority traffic will have to defer. A model of an EDCA WLAN station is given in Fig. 5.1.

The MAC scheme provided by the EDCA mechanism is depicted in Fig. 5.2. General explanation was previously discussed, while extensive details on EDCA performances can be found in [16].

The Dual Stage — Dynamic EDCA (DS-DEDCA) [17] scheme extends the basic EDCA by providing distributed adaptation of default $\mathrm{CW_{min}}$ and $\mathrm{CW_{max}}$ parameters for the ACs, according to the estimation of the network conditions proposed in [18]. Furthermore, it also includes two "virtual" stages of contention resolution in order to provide better channel utilization and reduce collision rate, especially in high-load network conditions, following the idea of Dual Stage Contention Resolution (DSCR) scheme [19] for improvement of DCF. These two features of DS-DEDCA provide better overall throughput and also appropriate medium access prioritization for ACs, which in turn improves the QoS support provided by EDCA.

Assuming that the priority queues which support the ACs inside the stations act as "virtual" stations with its own backoff instance, the definition of *Slot*

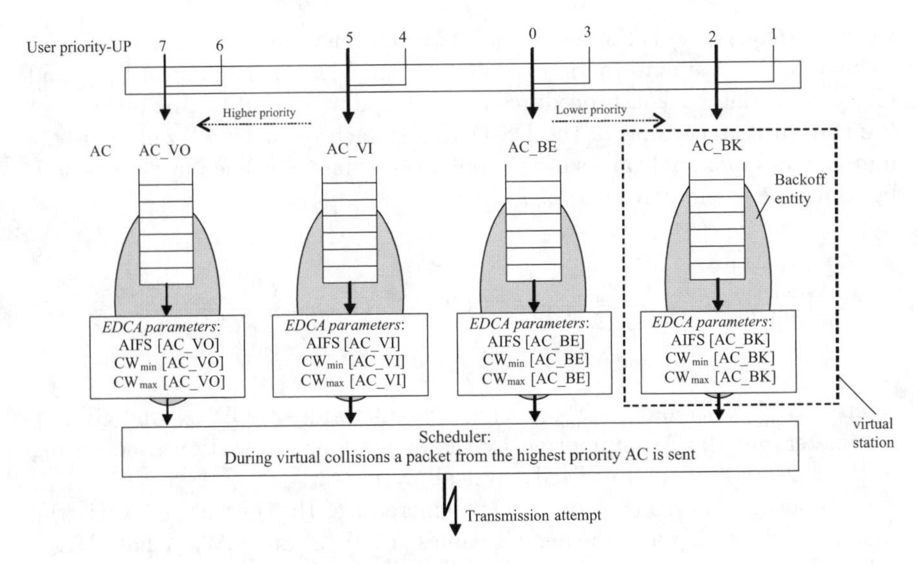

Fig. 5.1. EDCA station model

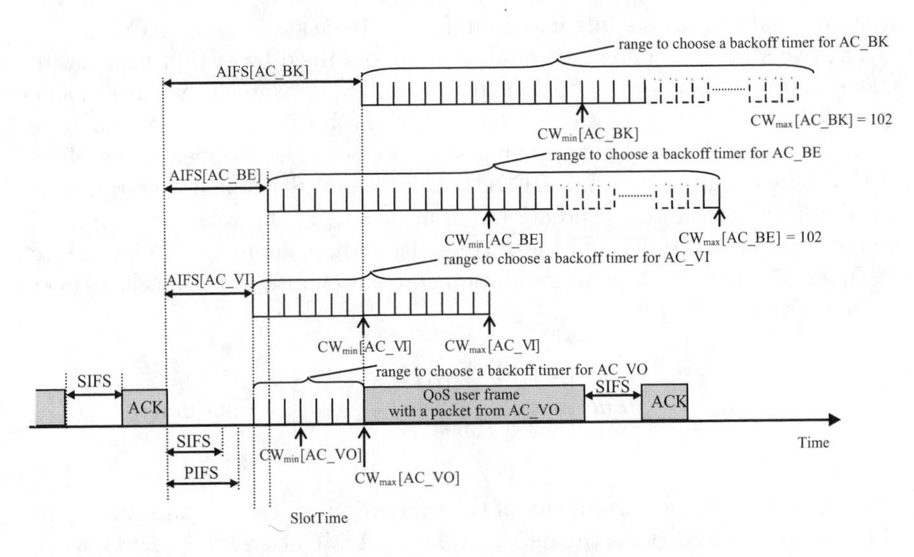

Fig. 5.2. Priroritized medium access provided by EDCA parameters

Utilization (*SU*) for stations is extended on SU for queues and a *Factor of Correction* (*FC*) for the queue k at the station i in j-th update period is defined as:

$$FC_j(i,k) = 1 - SU_{j-1}(i,k) = 1 - \frac{NBS_{j-1}(i,k)}{NAS_{j-1}(i,k)} \tag{5.4}$$

where $NBS_{j-1}(i,k)$ is the number of detected busy slots during backoff procedures in $j-1$ update period, while $NAS_{j-1}(i,k)$ is the sum of busy and idle slots during backoff procedures in $j-1$ update period for the queue k at the station i. According to the DS-DEDCA, each queue (i,k) will adapt its minimal and maximal values of the contention window in each update period, by using the calculated value of $FC_j(i,k)$:

$$CW_{\min,j}^{adapt}(i,k) = \frac{CW_{\min}^{def}(i,k)}{FC_j(i,k)}; \quad CW_{\max,j}^{adapt}(i,k) = \frac{CW_{\max}^{def}(i,k)}{FC_j(i,k)} \tag{5.5}$$

where $CW_{\min}^{def}(i,k)$ and $CW_{\max}^{def}(i,k)$ are default values of $\mathrm{CW_{min}}$ and $\mathrm{CW_{max}}$ parameters for the AC supported by the queue k at the station i, according to the 802.11e draft and utilized physical layer.

Considering that SU ($0 < \mathrm{SU} < 1$) will increase as the contention level in the channel is increasing and the default values of $\mathrm{CW_{min}}$ and $\mathrm{CW_{max}}$ have larger values for lower priority ACs then for higher priority ones, such adaptation of CW parameters provides increased contention window sizes for ACs as the network load increases while maintaining prioritization between ACs.

The two stages of contention resolution are obtained by enabling queues to choose their backoff timers from two CW intervals. Namely, when a queue is in deferring state (i.e., waiting the end of the busy period to continue its backoff timer), instead of continuing to decrement the stopped value of its backoff timer as in EDCA, DS-DEDCA proposes that the queue determines if it will "enter" the second contention resolution stage according to the stopped value of its backoff timer and the current size of its contention window. Once the queue "enters" the second resolution stage, it chooses a new backoff timer from a new CW interval as:

$$BT(i,k) = Rand\left[0, \frac{CW(i,k)}{2^{AIFSN(k)-AIFSN_{\min}}}\right] \cdot SlotTime \tag{5.6}$$

where $CW(i,k)$ is the current size of the contention window for the queue k at the station i; AIFSN(k) is the default value of Arbitration Inter Frame Space Number (AIFSN) for the AC supported by the queue k and $\mathrm{AIFSN_{min}}$ is the value of AIFSN parameter for the AC with the highest priority, according to the 802.11e draft.

Since only queues that enter the second stage can try to transmit a frame, DS-DEDCA significantly reduces the collision rate in high-load network conditions by providing only a fraction of all contending queues to enter the second stage, while the prioritization between ACs is maintained by giving a higher probability for entering the second stage to queues that support higher priority ACs.

Fig. 5.3 and 5.4 summarize the benefits of DS-DEDCA over other QoS based MAC schemes. It is easily seen that DS-DEDCA outperforms EDCA, AEDCF [21] and AFEDCF [22] in terms of achievable *total throughput* and *delay* of video packets. More details on the performance analysis on DS-DEDCA can be found in [17].

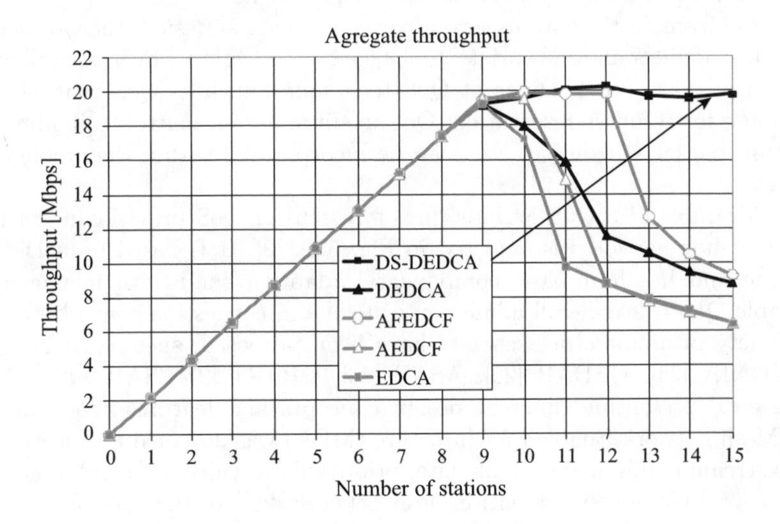

Fig. 5.3. Aggregate throughput vs. number of stations

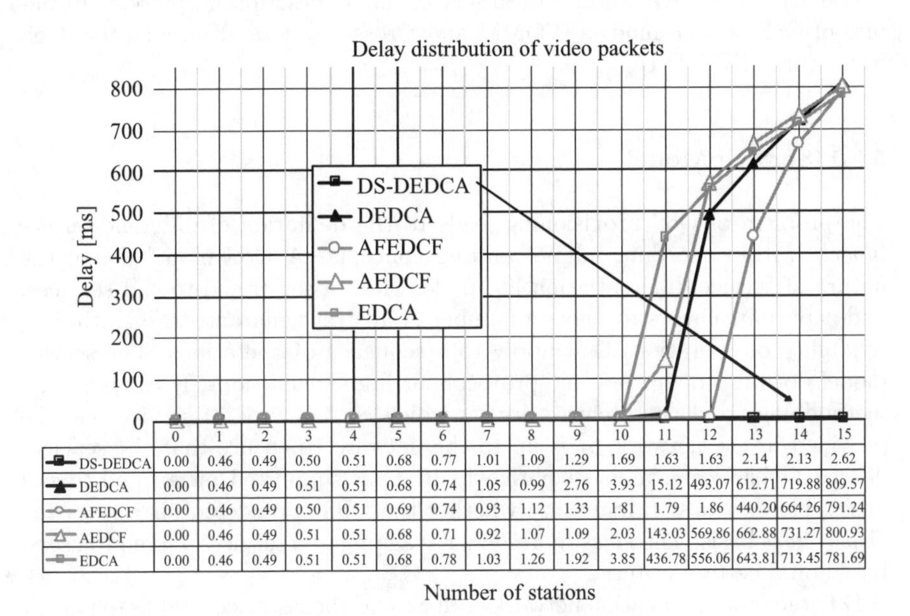

	0	1	2	3	4	5	6	7	8	9	10	11	12	13	14	15
DS-DEDCA	0.00	0.46	0.49	0.50	0.51	0.68	0.77	1.01	1.09	1.29	1.69	1.63	1.63	2.14	2.13	2.62
DEDCA	0.00	0.46	0.49	0.51	0.51	0.68	0.74	1.05	0.99	2.76	3.93	15.12	493.07	612.71	719.88	809.57
AFEDCF	0.00	0.46	0.49	0.50	0.51	0.69	0.74	0.93	1.12	1.33	1.81	1.79	1.86	440.20	664.26	791.24
AEDCF	0.00	0.46	0.49	0.51	0.51	0.68	0.71	0.92	1.07	1.09	2.03	143.03	569.86	662.88	731.27	800.93
EDCA	0.00	0.46	0.49	0.51	0.51	0.68	0.78	1.03	1.26	1.92	3.85	436.78	556.06	643.81	713.45	781.69

Fig. 5.4. Delay distribution of video packets

5.5 QoS Framework for Mobile Ad Hoc Networks

A *QoS framework* is a complete system that provides required/promised services to each user or application. All components within the system cooperate in providing the required services. A QoS provisioning within such a system is achieved by applying different mechanisms to particular *QoS service model,* which defines the way to meet the user requirements.

A QoS framework contains several key components. Besides the QoS service model, it defines an appropriate *framework architecture* which is responsible to realize that model. Different QoS techniques, such as QoS routing, QoS signaling for resource reservation, QoS medium access control, call admission control, packet scheduling, etc., can be incorporated within the architecture design.

An example of a QoS architecture, foreseen for QoS provisioning in large scale ad hoc networks is depicted in Fig. 5.5 [20]. It is based on a DiffServ scheme and has four basic components: adaptive bandwidth management, scalable QoS routing, call admission control and congestion control.

Variety of architecture schemes have been proposed, such as SWAN [23], INSIGNIA [24], CEDAR [25], ASAP [26], QPART [27], SARA [28], MEA [29], etc., presenting different design concepts and features. For example, the Mesh network Enabled Architecture (MEA) was designed for meeting the data stream requirements (bulk data, priority data, video data and interactive video streams) across the entire mesh network [29]. In the case of sensor ad hoc networks, specific design is envisioned [30].

Following text gives an insight in service model description, presents examples of QoS service models (FQMM and Cross-Layer) and an adaptive QoS framework (INSIGNIA).

5.5.1 Service Model

The problem of QoS provisioning leads to the definition of different *service models* within the different networking concepts. A QoS model define the nature of service differentiation [9]. It describes a set of end-to-end services, and it permits clients to select a number of abstract guarantees (e.g., timing, reliability of operation) that apply to a sequence of operations. The service model does not define specific protocols or implementations. It concerns the methodology and the architecture by which certain type of services can be provided in the network. It is up to the network to ensure that the services offered at each link along a path will be QoS supported on an end-to-end basis. The key design issues are whether the QoS is rendered on *per-session* or *per-class* basis. The *class* represents the aggregated services offered to users, based on a certain criteria.

Different communication networks can extend their service models to permit multiple types of services. The ATM network define service models according

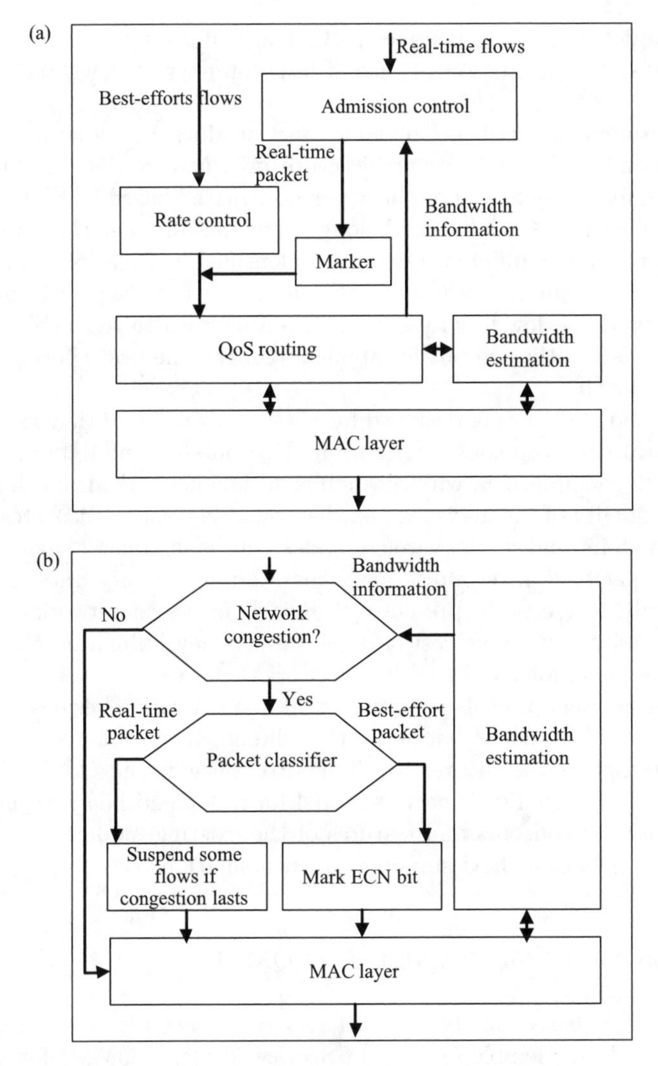

Fig. 5.5. Overview of the QoS architecture: (a) actions performed in the source nodes; (b) actions performed in the intermediate nodes

to requested bit rate, bandwidth guarantee and timing constraints (e.g., from CBR to UBR). The IETF made a tremendous effort to provide QoS in the Internet, including two QoS models that have been proposed for the Internet, which are basically aimed for wired networks: IntServ [31] and DiffServ [32].

Integrated services provide per-flow end-to-end QoS guarantees. IntServ defines two service classes: guaranteed service [27] and controlled load [33], in addition to the best effort service. The guaranteed service class guarantees to provide a maximum end-to-end delay, supporting strict delay requirements. Adaptive real-time applications are supported by controlled load service.

IntServ model is not applicable for mobile ad hoc networks, due to requested significant state information in case of large number of flows and scalability problem.

Differentiated services are scalable, but it does not guarantee services on end-to-end basis. DiffServ architecture uses core routers to differentiate between different QoS classes on a per-hop basis (based on DiffServ Code Point (DFCP) in the IP header belonging to the different classes defined in edge routers). Three different classes are identified in DiffServ: expedited forwarding (EF), Assured forwarding (AF) and best effort. Expedited forwarding provides low delay, low loss rate and assured bandwidth. Assured forwarding provides expected throughput for applications and the best effort provides no guarantees at all.

IntServ and DiffServ are designed for wired networks and they require accurate link state and topology information. The mobility and dynamic topology differs the QoS support in wireless mobile ad hoc networks from that in fixed ones. The quality of the network, i.e., the available resources that reside in the wireless medium and in the wireless nodes, varies in time. The time varying and low capacity link resources make maintaining the accurate information very difficult. It makes the present QoS models for wired networks insufficient for mobile ad hoc wireless networks. However, they influenced the design of the service model for mobile ad hoc networks.

The QoS model applicable for mobile ad hoc networks, besides considering the features of mobile ad hoc networks, should allow seamless connectivity to the Internet in the future. So, it has to consider the QoS architectures for Internet. The quality of service model for mobile ad hoc networks should benefit from the concepts and features of the existing models. The following paragraph provides a short summary of such solution.

5.5.2 QoS Model for MANETs — FQMM

The experience based on the Internet service models results in the development of a hybrid Flexible Quality of Service Model (FQMM) for mobile ad hoc networks [34], which takes advantage of the per flow granularity of IntServ and aggregation of services into classes in DiffServ.

FQMM is the first QoS model for small to medium size MANETs, with less than 50 nodes. It uses a flat, non-hierarchical, topology and defines three types of nodes (as in DiffServ): an *ingress node* which sends data, an *interior node*, which forwards data to other nodes, and an *egress node*, which is a destination. Fig. 5.6(a) depicts a possible FQMM scenario, where a data is sent from node M1 to node M6. If the nodes are static, this represents a classical DiffServ scenario, where the application is being generated in an ingress node, and terminated in the egress node. Fig. 5.6(b) depicts the scenario with two connections with notified roles of the involved nodes and illustrates the dynamic roles of the nodes.

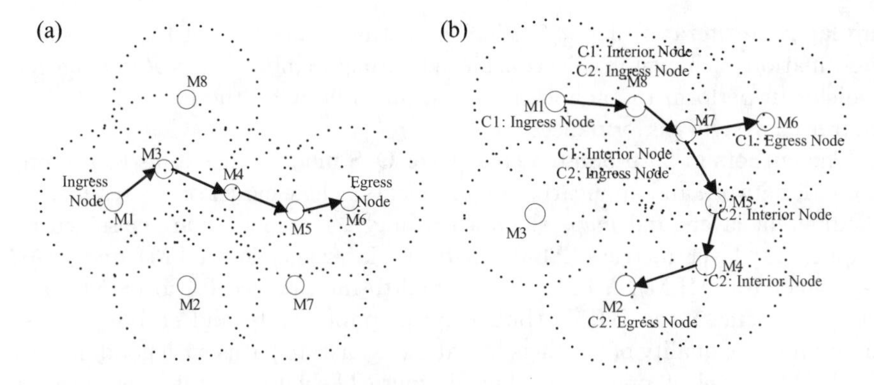

Fig. 5.6. (a) FQMM scenario 1 (static nodes); (b) FQMM scenario 2 (dynamic roles of nodes) [34]

FQMM proposes a hybrid per-flow and per-class provisioning scheme. The traffic with the highest priority is given per-flow provisioning, while other priority classes are given per-class provisioning. Since the states of per-flow granularity come from only a small fraction of the traffic, it eliminates the scalability problem. FQMM uses both IntServ and DiffServ schemes for different priority classes thus eliminating the drawbacks of both schemes.

The FQMM tried to redefine the *absolute* traffic profile to reflect the low capacity and unstable resources as *relative* percentage of the effective link capacity in order to keep the differentiation between the classes predictable and to respond to the network dynamism. The relative traffic profile contains the description of temporal properties of a flow such as bit rate and burst size. This profile should be also *adaptive* to the network's dynamics. A source node, which originates the traffic, is responsible for traffic shaping. The FQMM scheme achieves better performance in terms of throughput and service differentiation than the best-effort model.

5.5.3 A Cross-Layer QoS of Service Model

Serious performance degradations may be experienced due to the characteristics of ad-hoc networks: channel instability, spontaneous link formation, resource limited network nodes, lack of embedded infrastructure, etc. The quality of service that an application requires strictly depends on the "quality" of the networks, i.e., the available network resources that reside both in the wireless medium and in the mobile nodes. The QoS service model should include mechanisms to adapt the applications to the dynamic network changes.

Possible solution is envisioned in the model that involves extensive *cross-layer* information exchanges. As a result, the protocol stack has layers that actively interact and exchange information. Several different designs show up lately in literature [2], [35–37]. The actual realization of the cross-layer information exchange in [35] is achieved through embedded *protocol booster*, which can perform the adaptation procedures between the layers leading to improved system performances.

The authors in [2] propose a cross-layer QoS model which aims to respond to both network and application requirements. Their model separates *metrics* at different layers and *maps* them accordingly. The metrics are classified as: application layer metrics (ALM), network layer metrics (NLM) and MAC layer metrics (MLM). MLMs (link Signal-to-Interference Ratio (SINR) and coding schemes) and NLMs (buffer state, power state and stability state) determine the quality of the links in order to generate paths with good quality, while ALMs select one path which is more likely to meet the application requirements.

At the application layer requirements are classified into three *QoS classes* (class I, class II and class III) which are mapped to appropriate metrics. Class I corresponds to applications that have strong delay constraints (e.g., voice), and are mapped to *delay* metric at ALMs. Class II corresponds to applications requiring high throughput (e.g., video, transaction-processing applications), and are mapped to the *throughput* metric at ALMs. Class III is mapped with no specific constraints to best-effort at ALMs. The global view of a cross-layer QoS model is given in Fig. 5.7.

The mapping between QoS classes, ALMs, NLMs, and MLM is shown in Table 5.1. Each node checks the application requirements and appropriate mapped metrics deciding which procedure to perform on which level. The adaptation can be performed on different levels depending on the classes and metrics. Some techniques (FEC/ARQ, coding schemes, etc.) can be implemented in order to reduce the packet flow through the layer interfaces. Adaptation to network conditions can be improved with shaping mechanisms,

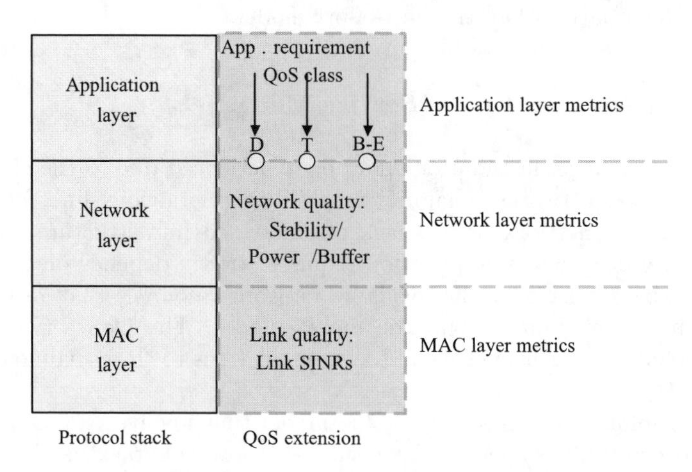

Fig. 5.7. Global view of a cross-layer QoS model [2]

Table 5.1. QoS Classes and Mapping

Priority Classes	ALMs	NLMs	MLM
Class I	Delay	buffer + hop#	SINR
Class II	Throughput	buffer + hop#	SINR
Class III	Best-Effort	stability + hop#	SINR

performing delay or dropping packets within a traffic flow. This model does not define specific protocols and implementation.

The following section presents the QoS improvements achieved with cross-layer optimizations performed through some Link Layer Adaptation techniques, such as FEC (forward Error Correction) and HC (header compression) implemented in an IEEE 802.11b system.

5.5.3.1 Link Layer Adaptation

There are numerous possibilities for inter-layer information exchanges. The wireless links exhibit location-dependent, time-varying and bursty errors, which make the channel conditions unstable. The Link Layer Adaptation (LLA) refers to the cross-layer optimizations in the lower part of the protocol stack where a channel condition's information from the physical layer (Signal-to-Noise Ratio (SNR) or Bit Error Rate (BER)) is used in order to adjust the MAC layer parameters. LLA in wireless networks will be a crucial step towards providing ubiquitous communication and consistent QoS. Future generation wireless technologies will comprise heterogeneous wireless network solutions with users expecting seamless service delivery "any time, any place". The LLA mitigates the link related degradation issues in various channel conditions through dynamic adjustments in the communication parameters between the communicating entities.

The adaptation approach can be applied to link layer techniques such as selection of packet length, use of Forward Error Correction (FEC) schemes, Header Compression (HC), scheduling, ARQ, a hybrid of ARQ and FEC, etc. It aims at ensuring more efficient network performance, measured as more reliable communication and higher throughput.

The technique of adaptive choice of the transmitting packet length can efficiently cope with the throughput degradation issues in wireless environments. Namely, introduction of an adaptive MAC Service Data Unit (MSDU) length selection guarantees maximum throughput under different channel conditions. FEC is an efficient way of enabling detection and correction of bit errors at the receiver based on sending a portion of redundant bits which is used for error correction. There is a variety of codes used for generating error correction bits (BCH codes, RS codes, convolutional codes, etc.). FEC schemes significantly improve the throughput and even allow the communicating entities to change their transmission rate for a higher value, though the channel conditions (SNR) are unchanged. Fig. 5.8 illustrates the benefits of using RS-based FEC scheme in IEEE 802.11b based systems under the assumption of fixed length MSDUs of 1500 bytes. It shows that the SNR axis can be segmented into definition zones which provide maximum throughput for a particular operational mode (neglecting FEC). The MSDU length does not significantly influence the width of the definition zones. However, it has a strong impact on the maximally achievable throughput in the zones. Furthermore, it is clear that the FEC scheme allows expansion of the working definition zones.

Using header compression techniques can significantly improve the application seen throughput under various channel conditions. The concept of header compression discussed here refers to the compression of UDP/IP headers. Header compression can be used in order to increase the upper bound of WLAN capacity for supporting voice streams (VoIP). Namely, header reduction makes more room for payload information, so the actual number of supported voice streams can be increased. Header reduction can also increase the application QoS since it makes more available bandwidth for the application packet units. Fig. 5.9 summarizes the throughput dependence on the SNR value in IEEE 802.11b systems with and without header compression.

Figs. 5.8 and 5.9 are based on an analytical performance analysis of IEEE 802.11b based systems. Extensive details on the adaptive choice of the packet length and the transmission rate, the benefits of using RS-based FEC scheme in WLANs, as well as the advantages of HC over IEEE 802.11b wireless links, can be found in [35, 38–40].

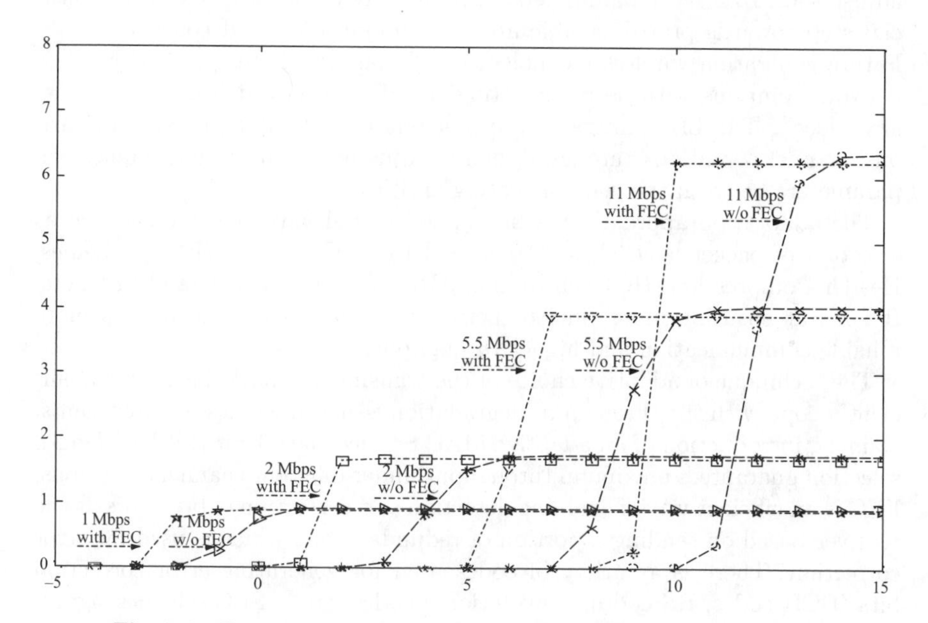

Fig. 5.8. Throughput vs. SNR with and without RS-based FEC scheme

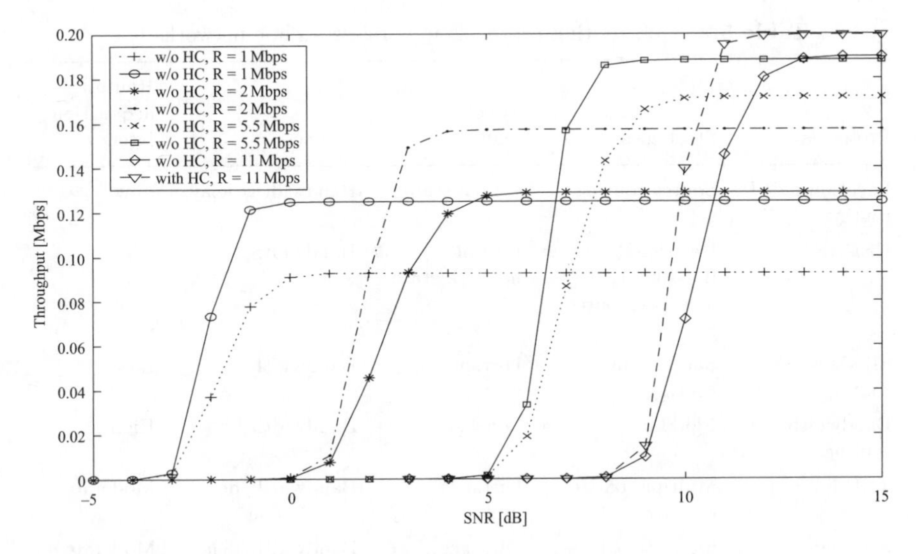

Fig. 5.9. Throughput vs. SNR with and without HC for different IEEE 802.11b operational modes; Application unit size = 20 bytes

The previous paragraphs have presented partially the performance analysis results that involve cross-layer interactions. The idea of cross-layer system design in mobile ad hoc networks is fully elaborated in Chapter 4.

5.5.4 Adaptive QoS Frameworks

The necessity to adapt to the unstable and mobile nature of wireless ad hoc networks has created a number of adaptive QoS frameworks. They mainly concentrate on QoS signalling [24, 28], QoS routing [2, 6, 9, 11, 31, 41–43] or QoS resource management [20, 26] or their combinations [9]. A comprehensive survey and comparisons of QoS routing protocols implemented in wireless ad hoc networks are given in [44] (see also Table 5.2) and more details on the particular routing protocols can be found in Chapter 3.

The next section briefly describes INSIGNIA (an example of a QoS framework) which was developed to provide adaptive services in wireless ad hoc networks.

5.5.4.1 INSIGNIA

INSIGNIA is a first signalling protocol explicitly designed for MANETs. The QoS signalling has to fulfill two prerequisites: reliable transfer of signals between routers and correct interpretation of the appropriate mechanisms to handle the signals [3]. The transfer of signals between routers can be divided into *in-band signalling* and *out-of-band signalling.* In-band signalling means that any network control information is encapsulated into data packets (easy

Table 5.2. QoS routing protocols in wireless ad hoc networks [44]

Protocols	Strategies	Topology management	Metrics	Route acquisition latency
Location-aided routing	Source routing	Global state	Bandwidth/delay	Low
CEDAR	Restricted flooding + localized source routing	Partial network state	Bandwidth	High
Min-hop routing	Shortest path routing	Distance vector	Bandwidth	Low
Bandwidth routing	Flooding	Local state	Bandwidth	High
TBP	Multiple paths routing	Distance vector	Bandwidth/delay	Moderate
Alternate routing	Shortest path + alternate routing	Distance vector	Bandwidth/delay	Moderate

and lightweight signalling), while out-of-band signalling uses explicit control packets (heavyweight signalling that involves extra packets).

INSIGNIA can be characterized as an in-band RSVP signalling protocol since it encapsulates control signals in the IP option of every IP packet (see Fig. 5.10), in this case named INSIGNIA IP option. The INSIGNIA IP Option efficiently supports fast reservation, restoration and adaptation schemes to deliver the adaptive services through the information content of different fields. The proposed signalling system can rapidly respond to changes in the network topology and to end-to-end QoS conditions.

INSIGNIA IP Option field (Fig. 5.10(a)) in field called *reservation mode* indicates if the packet is seeking for a reservation (REQ) or has already received resources (RES). The INSIGNIA differentiates two service types: real time (RT) and best effort (BE). Bandwidth request indicates minimum and maximum amount of bandwidth requested by a packet. Indicator fields are used to indicate if the route and receiver can fulfil the requested service type and bandwidth (see Fig. 5.10(b)). So, the payload indicator indicates the type of packets being transported (*base-BL* and *enhanced-EL*). The bandwidth indicator plays an important role during the flow setup and adaptation processes. It indicates the resource available at intermediate nodes during the setup. Resetting this indicator to MIN means that along the route, a node experiences unsufficient resources to meet the service requirements.

The adaptive service model allows packet audio, video and real-time data applications to specify their maximum and minimum bandwidth needs. The key design issues in providing adaptive services concentrates on how to operate on a particular service level (how to sustain the variations in the QoS level)

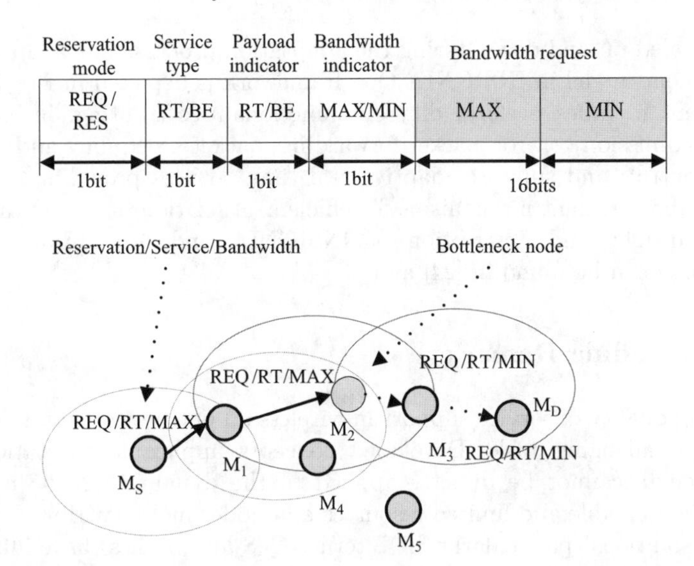

Fig. 5.10. (a) INSIGNIA IP Option field; (b) A flow setup and use of Bandwidth Indicator

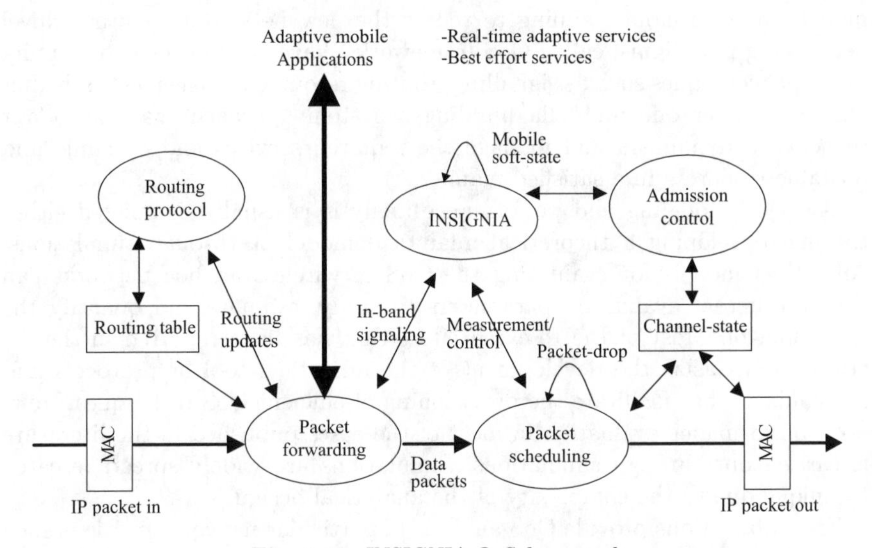

Fig. 5.11. INSIGNIA QoS framework

and how to switch from one service level to the other following the network topology and channel conditions. In order to provide adaptive quality of service support for real-time services in mobile ad hoc networks, *flow-state* (i.e., reservation state at nodes associated with flows) needs to be managed. Flows are represented as having a discrete base layer, with minimum bandwidth and an enhanced layer, which requires maximum bandwidth. Flows can be scaled up and down during the adaptation process. The adaptive flow management model in INSIGNIA QoS framework is depicted in Fig. 5.11.

Different modules perform different functionalities (routing, in-band signalling, admission control, packet forwarding, packet scheduling and medium access control) and support adaptive real time services providing soft state resource management mechanisms for efficient utilization of resources.

More insight and description of INSIGNIA and other different QoS frameworks can be found in [24] and [9].

5.6 Concluding Remarks

Achieving quality of service support in wireless ad hoc networks is a demanding and challenging task. Developed concepts applicable to static wired environments cannot be directly applied to the dynamic ad hoc networks due to the unstable and limited resources and nodes mobility. However, these existing solutions (particularly the Internet QoS approaches) have influenced the newly designed QoS solutions for wireless ad hoc networks.

Ad hoc wireless networks have modified the existing QoS definitions to adapt to the specifics of the quality of service provisioning. Specific QoS models were developed aiming to reflect the new networking status. Novel networking paradigms, called QoS frameworks, have been developed to unify different techniques such as signalling, routing resource management, scheduling, etc. under one umbrella building a system approach that can adapt to network dynamism and manage the required services and dynamic and unstable resources in a satisfied manner.

The QoS modeling and performance analysis is usually completed either through developing a theoretical analytical model or through simulations. Analytical models for evaluating the QoS in wireless ad hoc networks can choose different evaluation parameters to study, evaluate and quantify the QoS [45, 46]. Most of the research effort that has been reported in the literature emphasize the development of the analytical tool or protocols and mechanisms that facilitate the provisioning of efficient QoS techniques. However, performance evaluation in most of the cases published in the literature is carried out through simulations. Simulations are widely spread research technique due to the complexity of the analytical design.

Few publications provide QoS analysis of particular network functions such as handover algorithm [47], effects of Rayleigh fading [48], beamforming antennas [49], etc. QoS provisioning in wireless ad hoc networks stays an open issue, still attracting significant research potentials.

References

[1] IETF Mobile Ad-Hoc NETworks (MANETs) Charter, http://www.ietf.org/html.charters/manet-charter.html

[2] Nikaein, N. and Bonnet, C., "A Glance at Quality of Service Models for Mobile Ad Hoc Networks," *DNAC*, Paris, France, December 2002.

[3] Demetrios, Z. -Y., "A Glance at Quality of Services in Mobile Ad-Hoc Networks," *Final Research Report for CS260 — Seminar in Mobile Ad Hoc Networks*, University of California — Riverside, USA, Fall 2001.

[4] Crawley, E., Nair, R., Rajagopalan, B., and Sandick, H., "A framework for QoS Based Routing in the Internet," RFC 2386, August 1998.

[5] Perkins, D. D. and Hughes, H. D., "A Survey on Quality-of-Service Support for Mobile Ad Hoc Networks," *Wireless Communications and Mobile Computing*, 2002, pp. 503–513.

[6] Wu, K. and Harms, J., "QoS Support in Mobile Ad Hoc Networks," http://www.cs.ualberta.ca/~wkui/research/QoSReview.ps

[7] Chakrabarti, S. and Mishra, A., "QoS Issues in Ad Hoc Wireless Networks," http://www.ececs.uc.edu/~guptanis/research/papers/QoS/QoSAdHoc.Mishra.pdf

[8] Aggelou, G., *Mobile Ad Hoc Networks: From Wireless LANs to 4G Networks*, McGraw Hill, 2005.

[9] Siva Ram Murthy, C. and Manoj, B. S., *Ad Hoc Wireless Networks: Architectures and Protocols*, Prentice Hall Communications Engineering and Emerging Technologies Series, 2004.

[10] Ilyas, M., *The Handbook of Ad Hoc Wireless Networks*, CRC Press, 2003.

[11] Braden, R., Zhang, L., Berson, S., Herzog, S., and Jamin, S.,"Resource Reservation Protocol (RSVP) — Version 1 Functional Specification," IETF RFC 2205 September 1997.

[12] Comaniciu, C. and Vincent Poor, H., "QoS Provisioning for Wireless Ad Hoc Data Networks," *42nd IEEE Conference on Decision and Control*, Hawaii, December 2003.

[13] Barry, M. and McGrath, S., "QoS Techniques in Ad Hoc Networks," *1st International ANWIRE Workshop*, Glasgow, UK, April 2003.

[14] Gu, D. and Zhang, J., "QoS Enhancement in IEEE 802.11 Wireless Local Area Networks," *IEEE Communications Magazine*, 41(6), pp. 120–124, June 2003.

[15] IEEE 802.11 WG, IEEE 802.11e/D8.0, "Draft Amendment to Standard for Telecommunications and Information Exchange between Systems — LAN/MAN Specific Requirements — Part 11: Wireless Medium Access Control (MAC) and Physical Layer (PHY) Specifications: Medium Access Control (MAC) Enhancements for Quality of Service (QoS)," February 2004.

[16] Dimitrovski, Z. and Gavrilovska, L., "Evaluation of QoS Enhancements Provided by EDCF Medium Access Scheme in IEEE 802.11 WLAN," *39th International Conference on Information, Communication and Energy Systems and Technologies (ICEST 2004)*, Bitola, Macedonia, June 2004, pp. 567–570.

[17] Dimitrovski, Z. and Gavrilovska, L., "DS-DEDCA: Providing Better QoS in Ad-Hoc IEEE 802.11 WLAN," *Poster Presentation at the 12th IEEE International Conference on Network Protocols — ICNP 2004*, Berlin, Germany, October 2004.

[18] Bononi, L., Conti, M., and Donatiello, L., "Design and Performance Evaluation of a Distributed Contention Control (DCC) Mechanism for IEEE 802.11 Wireless Local Area Networks," *JPDC*, 60(4), April 2000, pp. 407–430.

[19] Yang, X. and Vaidya, N., "DSCR: A More Stable MAC Protocol for Wireless Networks", *Technical Report*, (Electrical and Computer Engineering

Department, and Coordinated Science Laboratory University of Illinois at Urbana-Champaign), August 2002.

[20] Xu, K., Tang, K., Bagrodia, R., Gerla, M., and Bereschinsky, M.,"Adaptive Bandwidth Management and QoS Provisioning in Large Scale Ad Hoc Networks," *MILCOM 2003*, Boston, MA, USA, October 2003.

[21] Romdhani, L., Ni, Q., and Turletti, T., "Adaptive EDCF: Enhanced Service Differentiation for IEEE 802.11 Wireless Ad Hoc Networks," *IEEE WCNC'03*, New Orleans, Louisiana, USA, March 2003.

[22] Malli, M., Ni, Q., Turletti, T., and Barakat, C., "Adaptive Fair Channel Allocation for QoS Enhancement in IEEE 802.11 Wireless LANs," in *Proceedings of IEEE International Conference on Communications (ICC) 2004*, Paris, June 2004.

[23] Ahn, G., Campbell, A., Veres, A., and Sun, L., "SWAN: Service Differentiation in Stateless Wireless Ad Hoc Networks," *Proceedings of IEEE INFOCOM 2002*, June 2002.

[24] Lee, S., Ahn, G., Zhang, X., and Campbell, A., "INSIGNIA: An IP-Based Quality of Service Framework for Mobile Ad Hoc Networks," *Journal of Parallel and Distributed Computing (JPDC)*, 60(4), April 2000.

[25] Sinha, P., Sivakumar, R., and Bharghavan, V., "CEDAR: A Core-Extraction Distributed Ad Hoc Routing Algorithm," *Proceedings of IEEE INFOCOM 1999*, August 1999.

[26] Xue, J., Stuedi, P., and Alonso, G., "ASAP: An Adaptive QoS Protocol for Mobile Ad Hoc Networks," *14th IEEE International Symposium on Personal, Indoor and Mobile Radio Communications (PIMRC2003)*, Beijing, China, September 2003.

[27] Yang, Y. and Kravets, R., "Distributed QoS Guarantees for Realtime Traffic in Ad Hoc Networks," *1st IEEE International Conference on Sensor and Ad hoc Communications and Networks (SECON)*, Santa Clara, CA, USA, October 2004.

[28] Yeh, C. -H., Mouftah, H. T., and Hassanein, H., "Signaling and QoS Guarantees in Mobile Ad Hoc Networks," *IEEE International Conference on Communications*, New York, USA, April 28–May 2, 2002.

[29] Mesh Networks, *White paper: Quality of Service for Ad Hoc Networks*, http://www.meshnetworks.com

[30] He, T., Stankovic, J. A., Lu, C., and Abdelzaher, T., "SPEED: A Stateless Protocol for Real-Time Communication in Sensor Networks," *International Conference on Distributed Computing Systems (ICDCS 2003)*, Providence, RI, May 2003.

[31] IETF Integrated Services (IntServ) Charter, http://www.ietf.org/html.charters/intserv-charter.html

[32] IETF Differentiated Services (DiffServ) Charter, http://www.ietf.org/html.charters/diffserv-charter.html

[33] Sweeney, J. D., Grupen, R., and Shenoy, P., "Active QoS Flow Maintenance in Robotic, Mobile, Ad Hoc Networks," *University of Massachusetts Amherst Computer Science technical report #04-20*, April 2004.

[34] Xiao, H., Seah, K. G., Lo, A., and Chua, K. C., "A Flexible Quality of Service Model for Mobile Ad-hoc Network," *Proceedings of IEEE VTC2000 — Spring*, Tokyo, May 2000.

[35] Power Aware Communications for Wireless OptiMised personal Area Network, PACWOMAN (IST-2001-34157), http://www.imec.be/pacwoman
[36] Kawadia, V. and Kumar, P. R., "A Cautionary Perspective on Cross-Layer Design," *IEEE Wireless Communications Magazine*, 12(1), February 2005, pp. 3–11.
[37] Shakkottai, S., Rappaport, T. S., and Karlsson, P. C., "Cross-Layer Design for Wireless Networks," *IEEE Communications Magazine*, 41(10), October 2003, pp. 74–80.
[38] Atanasovski, V. and Gavrilovska, L., "Throughput Performance of Contending IEEE 802.11a Systems in Rayleigh Fading Channels," *8th International Symposium on Wireless Personal Multimedia Communications (WPMC) 2005*, Aalborg, Denmark, September 2005, pp. 1925–1929.
[39] Gavrilovska, L. and Atanasovski, V., "Adaptive Techniques for WLAN Throughput Improvement," *IEE-3G 2004*, London, UK, October 2004, pp. 73–77.
[40] Atanasovski, V. and Gavrilovska, L., "Influence of Header Compression on Link Layer Adaptation in IEEE 802.11b," *IEEE International Conference on Wireless Networks, Communications and Mobile Computing WIRELESSCOM 2005*, Maui, Hawaii, USA, June 2005.
[41] Aarthi, V. and Siromoney, A., "Quality of Service Routing in DSR," *Trusted Internet Workshop*, Bangalore, India, December 2002.
[42] Huang, C., Dai, F., and Wu, J., "On-Demand Location-Aided QoS Routingin Ad Hoc Networks," in *Proceedings of the 33rd International Conference on Parallel Processing*, Montreal, Canada, August 2004, pp. 502–509.
[43] Gupta, R., Jia, Z., Tung, T., and Walrand, J., "Interference-aware QoS Routing (IQRouting) for Ad-Hoc Networks," *Submitted to IEEE International Conference on Communications 2005*, Seoul, Korea, May 2005.
[44] Zhang, B. and Mouftah, H. T., "QoS Routing for Wireless Ad Hoc Networks: Problems, Algorithms, and Protocols," *IEEE Communications Magazine*, October 2005, pp. 110–117.
[45] Jia, X., Li, D., and Du, D., "QoS Topology Control in Ad Hoc Wireless Networks," *IEEE INFOCOM 2004*, Hong Kong, March 2004.
[46] Zhu, C. and Corson, M. S., "QoS Routing for Mobile Ad Hoc Networks," *IEEE INFOCOM 2002*, New York City, USA, June 2002.
[47] Lo, C. -C. and Lin, M. -H., "QoS Provisioning in Handoff Algorithms for Wireless LAN," in *Proceedings of 1998 International Zurich Seminar on Broadband Communications*, February 1998, pp. 9–16.
[48] Lee, K. -W., Cheng, M., and Chang, L. F., "Wireless QoS Analysis for A Rayleigh Fading Channel," in *Proceedings of ICC 1998*, vol. 2, 1998, pp. 1089–1093.
[49] Ramanathan, R., "On Performance of Ad Hoc Networks with Beamforming Antennas," in *Proceedings of MobiHoc 2001*, October 2001, pp. 95–105.

6

Service Discovery

6.1 Introduction

Wireless market today is overcrowded with large number of different devices and technologies. New type of devices, ranging from tiny sensors to extremely powerful devices, provide a variety of information and services. The problem of managing these services, configuring different applications and dynamically discovering the available services becomes a difficult task in the design of ad hoc networks and pervasive computing. The *service discovery* targets these issues enabling networks usability.

Design of mechanisms proposed for service discovery in ad hoc mobile networks significantly differs from traditional wired solutions. Service discovery protocols in traditional wired networks can rely on powerful infrastructure nodes and reliable links, while in infrastructureless ad hoc mobile networks, service discovery must cope with node mobility, unstable channel conditions and nodes with limited energy and availability. Service discovery in such networks has to be completed in an automatic and self-organizing manner. The appropriate protocols for service discovery have to be efficient and lightweighted, with decentralized and distributed approach, which makes the design a demanding task. Many different solutions were proposed in research communities and different industrial consortiums. They all differ in implemented features and targeted network type.

This chapter gives an overview and definitions of basic entities and functionalities of service discovery in wireless ad hoc networks. It presents possible architectures and explains implemented mechanisms in different service discovery protocols. In addition, new aspects of service discovery in wireless ad hoc networks and ongoing research projects which deal with novel solutions are highlighted.

6.2 Definition and Goals of SD

Service represents a hardware or a software resource that can perform specific function/functions on behalf of users and applications over a network (e.g., printing services, location information, wireless network connections). A service has a name, a list of attributes and user privileges [1].

Service discovery is a process of discovering location of software entities/ agents that can provide access to network resources such as devices, data and services [2]. Major goal of the service discovery mechanism is to make devices and networks smart/intelligent and capable of being aware of the available services.

In order to facilitate the *discovery* process, the service discovery protocols have to perform number of functions such as: *use of a description language* to semantically describe the services and service requests, *storage of service information* in appropriate manner and *search for services* according to service requests expressed in description language (by sending service requests to directory nodes or disseminating them into the network) [3].

The service discovery protocol has to function with no administration, i.e., without human intervention. It means that it should dynamically adapt to changes in service descriptions and topology and should provide the user with reliable information about available services. The solutions for service discovery depend on network type, its topology and architecture, available resources, incorporated routing and communication mechanisms. Four different communication techniques are used in service discovery protocols: unicast, anycast, multicast and broadcast [1]. Relevant network features are also: size, data rates, dynamics and type of devices participating in the network.

Recent pervasive computing and ad hoc networking have identified service discovery as one of the major design components [1, 4]. The service discovery in ad hoc networking should facilitate:

- Services to announce their presence to the network;
- Automatic discovery of local and remote services regardless of the type of network and technology used;
- Automatic adaptation to mobile and sporadic availability;
- Services to describe their capabilities as well as query and understand the capabilities of other services;
- Self-configuration without administrative intervention.

Challenges imposed by ad hoc mobile networking require solutions to network dynamism (mobility of nodes), instability (node and links failure) and limited resource constraints (cash memories and energy), resulting in number of initiatives in this area. In ad hoc dynamic networking environment it is also important to maintain a consistent view of components in a network. Authors in [1–3, 5] offer a good overview of the existing service discovery protocols.

6.3 Service Discovery Entities

Service discovery functionalities are regulated through appropriate service discovery protocols. The structure of service discovery protocols identifies the building blocks and the links between the participating components (entities). Each service discovery protocol consists of at least two basic components: *client* and *server*. *Client* (*user* or *user agent* — *UA*) represents the entity that is interested in finding and using a service and hosts certain applications which access specific services. *Server (server provider* or *service agent* — *SA)* represents the entity that hosts and offers the service. The process of service discovery is actually a mapping between *service description* (that helps identify a service) and *service location* (that helps identify the location where the service can be found).

In order to facilitate the mappings, the service discovery protocols can involve another entity, named *directory*. *Directory (server coordinator, service broker, directory agent* — *DA* etc.) presents a node in the network that hosts partially or entirely the service description information [3, 6]. That node can act as a registry or broker for the discovery and provision processes, improving the performance of the service discovery. It is also often called *3rd party*. *Resources*, such as storage, bandwidth, database etc., represent the entities or tools that support the servers or clients in their activities. Different service discovery protocols involve some of previously mentioned entities under different names. Table 6.1 presents the active structure's components for some service discovery protocols.

Almost all service discovery protocols either include the *client-server* operation mode, or involve the *service brokers* (or *directories*), which reside between clients and servers as a logical entity. In the client-server paradigm, service requesting nodes (clients) send out service *request messages* and servers listen to such messages at a predetermined network interface and port. If the requested service is supported, then a *reply message* is sent back to the client. In the alternate scheme, clients direct their requests to well-known service brokers whereas servers *register* their services with these brokers. In return, the service brokers send their reply messages to the clients and *registration acknowledgements* to the servers. Accordingly, the service discovery can be completed in two-party or three-party scenarios, depicted in Fig. 6.1.

In two-party scenario multicasts are necessary for every discovery process, whereas in three-party scenario multicasts from service users (clients) and service managers (servers) are only necessary for initial discovery of the service broker (directory) node. The amount of overhead is limiting the scalability of the client-server solutions. For ad hoc mobile networks, where the nodes have similar capabilities and the availability of nodes and links is often questionable, neither approach can be implemented in a straight manner without modifications towards ad hoc networks' peculiarities. Both approaches have some benefits and a number of disadvantages for application in ad hoc wireless environment. The classification of service discovery architectures will be addressed later in this chapter.

Table 6.1. Active structure's components of some service discovery protocols

Protocol	Bluetooth SDP	SLP	Jini	UPnP	Salutation
Client	Client	User Agent (UA)	Client	Control Point	Client Functional Unit (FU)
Server	Server	Service Agent (SA)	Service	Device	Server Functional Unit (FU)
3rd party	N/A	Directory Agent (DA)	Lookup	N/A	SLM
Responsible for protocol development	Bluetooth SIG	IETF	Sun Microsystems	Microsoft	Salutation Consortium
Implementation examples	Any Bluetooth device	OpenSLP, Novel Netware	Macromedia's application server — Jrun, NIST's Aroma projector	UPnP Digital Media Receiver by Arcadyan Technology Corporation	IBMs NuOffice (Salutation enhancements of lotus notes)

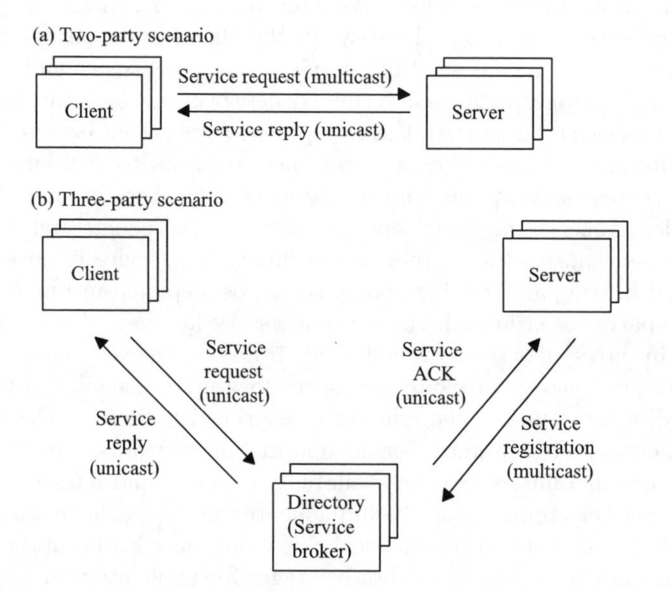

Fig. 6.1. Two-party and three-party scenarios for service discovery

6.4 Service Discovery Phases and Mechanisms

The service discovery process is accomplished through several phases:

- *Advertisement* of services and their properties makes a service discoverable. The new services that come into a network require some form of advertisement to make them available to consumers. The advertisement may be required to cover some state changes, expiration of life time, etc.
- *Locating* a service is a process of discovering service location. It involves querying the network through a broadcast or directory query.
- *Utilizing* a service (optional) is the actual use of services and may be included or not in the service discovery process.

Announcements/advertisements (push method) and *queries* (pull method) are two basic mechanisms for clients, services and directories to exchange information. Each technology uses one or both concepts.

From client's perspectives, service discovery involves bootstrapping, querying and obtaining service handle(s) as a result of a query [5]. The service first bootstraps and then registers itself with a directory agent if present. Fig. 6.2 depicts different steps/mechanisms in service discovery process.

The *bootstrapping* specifies how users and services establish contact with the discovery system. It requires some pre-configured knowledge by most discovery approaches and can use different communication techniques such as:

- *Unicast* communication to directory (IP address obtained by DHCP server — adopted by INS, Twine, SLP, P2P);

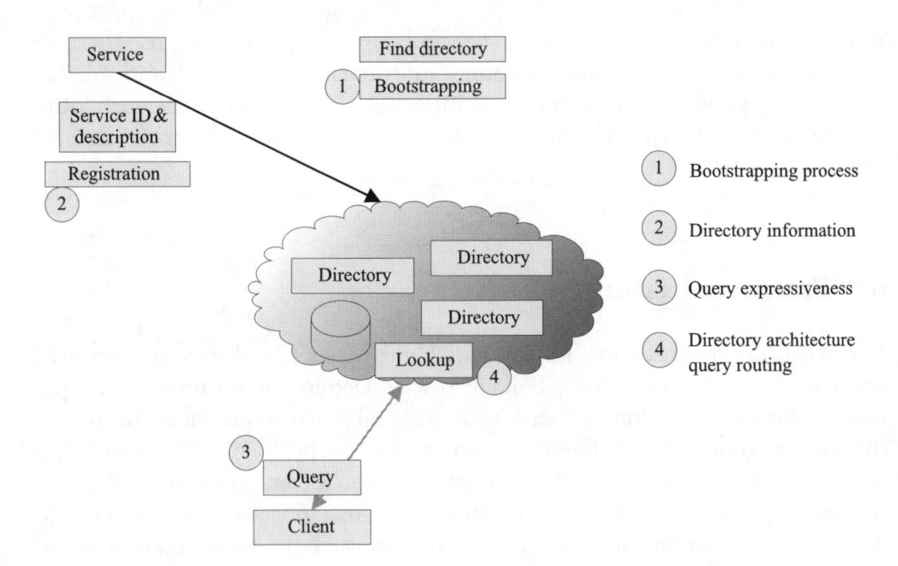

Fig. 6.2. Steps mechanisms in service discovery

- *Multicast by client/server* (client/server announces its presence using a well-known multicast address — adopted by SLP, Jini, Splendor);
- *Periodic multicast advertisement by directory* (client/server listens to periodic multicast of directory advertisements on a well known multicast address — approach accepted by SLP, Splendor).

The usage of a well-known unicast address performs better in sense of bandwidth. However, it requires that the directory resides in a known location. The multicasting approaches are more bandwidth demanding, but provide higher flexibility.

The *directory information* is usually generated by service advertisements and is stored in the directory with some other maintenance information. Service registers with some directory, which helps to offer that service on the network. Directory information presents the server needs, from server providers' side, into an abstract form (e.g., service records in INS, UDDI in Web services, index entry in P2P systems, etc.). The directory information has to support *formalism* and *expressiveness*. The formalism examines the structure of directory information, while the expressiveness concerns the competence and contents of directory information. The client transforms its service request into a *query*. A query is matched against the directory information. The directory information can directly affect the effectiveness, correctness and scalability of the service discovery.

The *query expressiveness* defines the way the query is shaped in various service discovery approaches. Client formulates its request for a particular service through a query and performs a lookup process.

Directory approach in service discovery can be influenced by appropriate *directory architecture* and chosen method for *query architecture*. Directory architecture defines the way directory information are stored and accessed in the system and how directory components coordinate with each other. The query architecture specifies the way a query is routed across directory agents.

6.5 Functional Issues

Various functionalities are implemented into a service discovery process in order to design it in the most beneficial way. Depending on protocol purpose and implementation domain, various functional features are incorporated in a different manner into different service discovery protocols. Table 6.3 (presented in the Appendix of this chapter) gives an extensive list of service discovery protocols and their specifics, relevant to ad hoc networking [3]. The following paragraphs give a short overview of the most relevant service discovery functionalities.

6.5.1 Service Identity

The service must be identified among all advertised services. It can be done by globally unique ID (as in Salutation where the identifier is generated locally by the service), locally unique ID combined with the network address (as in INS and Twine, where the identity is presented by IP:port pair) or with URI/URL (as in SLP, Web/Grid services).

6.5.2 Service Description

Service description is crucial for the existence of service discovery protocols. Many discovery protocols, based on service description allow *yellow pages service*, i.e., service lookup based on given description. Different protocols use different description formats (data structures) to enable search for a particular service according to appropriate format's fields and values. A service description must contain *service identity* and *service type*.

The service discovery protocols usually specify the service types by classification of services into *classes* (e.g., *print* is a class of service). The service, as an instant of the appropriate class, is called *object*. Each object contains a list of *attributes*. Attributes, specified with assigned *values*, identify the object. The service description is usually stored as a set of *attributes/values pairs* [3].

Generally, different service discovery protocols use the same approach, however, the naming and conventions are different. So, some protocols like SLP and Jini store the attribute/value pairs as lists and do not relate attributes to each other. Others, like INS and TWINE relate attributes based on dependency relationship and store them in a tree-like hierarchy. Predefined set of attributes can be used for each service type in Salutation and SLP. The XML based representation is used by many approaches to provide extensibility and openness (e.g., UPnP). Table 6.2 illustrates different data structures for describing services in some of the most popular service discovery protocols today.

Service description is closely related to features such as naming, invocation, and status query. Many protocols solve the *naming* problem by defining a service naming standard, in order to avoid the naming conflicts. Different service discovery domains, which are designed for different SD protocols, may use different vocabularies for the same service (e.g., print or printing). The new services must be identified by a new name.

Service attributes have also standard *naming conventions* in order to avoid conflicts. They are used to identify the client's request by matching its query requirements. If the attribute is not specified, it is generally considered that it can take any value and appropriate procedure must be used to discover the service (e.g., wild-card matching in INS [7]).

Service *invocation* is performed by a client who invokes the service through a service interface. Different protocols define this interface in a different way

Table 6.2. Data structures for describing the services

Protocol	Bluetooth SDP	SLP	Jini	UpnP	Salutation
Service Class	Class	Service Template	Service Template	Template	Functional Unit
Service Object	Service Record	Addressed by Service Location	Java Object (proxy)	Description of the service	FU Description record
Attribute Name	Identifier	Attribute Name	Attribute Name	Attributes (XML Tags)	Attribute ID
Attribute Value	Value	Attribute Value	Attribute Value	Values	Value

(e.g., Bluetooth SDP — applications define a service interface, Salutation — use of Remote Procedure Call (PRC), Jini — use of downloadable Java code). Some service discovery protocols (Jini and UPnP) need TCP/IP protocol, HTTP servers and JVMs (Java Virtual Machines) to perform service invocation.

Service *status inquiry* is used to follow the services' events status changes. It can be performed by polling the service (when service status changes frequently), service event notification (service notify the clients who are registered and have shown interest) or agents' event filtering and aggregation (when agent perform services' events management).

6.5.3 Maintenance

Permanent adjustment of stored service information is crucial to efficient service discovery and overall network efficiency. In case when service information is stored on directory nodes, the process of maintenance should adjust to changes in service description and topology (e.g., FRODO [8], Cheng and Marsic [9], UPnP [10]). Topology state information can be updated periodically through re-registration or on request. Overlay networks have to implement algorithms for preserving consistency in case of network partition and integration and nodes leaving or entering the network (Service Rings [11] and LANES [12]).

6.5.4 Search Methods

Storage mechanisms strongly influence the discovery processes. The information about the servers and their offer can be obtained through either *passive discovery (push model)*, when services or brokers announce their presence, or *active discovery (pull model)*, when application needs information about services or brokers and sends discovery messages. The object of discovery can be directory nodes or services. Most *discovering mechanisms* can be either directory based or peer-to-peer.

6.5.5 Service Selection

Process of best service selection is required if service is offered by multiple servers in the system. The user can *manually* make a choice or it can be performed by means of *optimization algorithm* implemented on client's side or selected by one of the directory agents present in the system [1]. The appropriate *best offer metric* plays an important role in case of *automatic protocol selection.* These metrics could be: lowest hop count [13], best Randezvous Point (RP) either with smallest response time or smallest database [14], etc. The Broker Agent (BA), which is responsible for the service registry in CSP [15], evaluates the context attributes (distance to server, server load and channel conditions) in order to make the best choice. In service selection, context awareness, scope awareness (e.g., location awareness, administrative domains), QoS awareness (e.g., application metric-based load balancing), etc. may have an important role. The following text in this chapter addresses some of these issues.

6.5.6 Service Usage

After the service discovery phase, the client usually connects directly to the device offering the service. Special protocols offer methods for using the services.

The access to a service is facilitated through its *service handle*, which presents the form of reference to the service. They can be service identifiers used to identify and invoke the service, network address of the service, a URL for the service, XML service description and proxy stub to the service that can be used for service invocation. For example, INS, Twine, Splendor and most peer-to-peer systems use network addresses; SLP returns the URLs of the matching service; JXTA and Jini return proxy objects for the matched services, etc.

6.5.7 Security and Privacy

Security and privacy are important issues concerning ad hoc networks [16–18]. Despite the unstable availability of services caused by nodes dynamics, ad hoc networks can be attacked by many possible misuses, both from fraudulent servers and from misbehavior clients. Security is addressed in different domains, including design of service discovery protocols. Deploying security in service discovery protocols adds more administrative overhead. More details on general security aspects in ad hoc networking are presented in Chapter 8.

Major security constraints include: *authentication, authorization, trust, confidentiality, integrity,* and *non-repudiation.* Different service discovery protocols support some or none of these features [19]. For instance, INS [7], TWINE [20] and UPnP [10] protocols do not include any security mechanisms. Authentication is the most often implemented security feature (in Salutation [21], SLP [22], SSDS [23]), which includes a username and a password or even digital signature (in case of Salutation [21]). Jini [24] implements authorization (including access control lists), as well as authentication, confidentiality and integrity in its own standardized security package. SLP provides message integrity. There are very few protocols which implement most of the security features (see Table 6.3 in the Appendix). SSDS and Splendor [25] are among the protocols with built-in security mechanisms. Bluetooth has its built-in request-response authentication, authorization and also an encrypted mode of communication.

Incorporating security features in service discovery protocols leads to new solutions [3]. SPDP is a fully distributed protocol, based on anarchy trust model of public key infrastructure (PKI) and on existing protocols [19]. Authors in [26] offer solution for combining service discovery with a proxy-based approach that uses other existing networks channels to set up a secure and trust relationship and facilitate ad hoc wireless communications. SPINS [27] is a security service discovery protocol designed for sensor networks.

User privacy is also important. It allows the user to keep his information private and to determine the degree to which it will interact with its environment. Security is an issue that has not yet been solved in a completely satisfactory way in any of the existing protocols. Future service discovery protocols should inevitably include the security features in their design.

6.6 Service Discovery Architectures

Service discovery architecture is the framework correlating with different domains such as storage of service information, directory design, topology, information flow, routing, etc. There are many approaches to definition of service discovery architectures and classification of SD protocols [1, 3, 28, 29]. The service discovery architectures applicable to ad hoc design can be classified into two general groups: *query-based* (or *directory-less*) and *directory-based* [30]. Fig. 6.3 presents the classification of service discovery architectures.

Query-based architectures are represented by: traditional *client-server* (two-party) architectures, based on master-slave mode of operation; *unstructured* (distributed peer-to-peer) architectures, where all nodes have equal functionalities or by *multi-tier* architectures, where nodes are layered according to their capabilities into heterogeneity levels.

The directory-based architectures can operate with *one directory* (acting as a service broker or a coordinator, i.e., a simple three-party architecture) providing *centralized* approach. The architectures with more directories can be

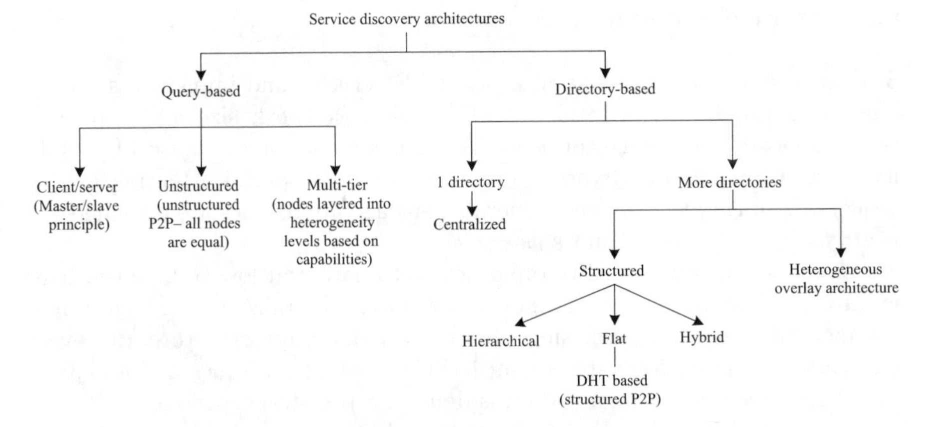

Fig. 6.3. Classification of service discovery architectures

organized as *structured (hierarchical, flat or hybrid)* or *heterogeneous overlay* architectures, designed for heterogeneous environment with more discovery domains. Hierarchical architectures adopt parent-child relations between the nodes, often leading to tree-like structures or implementing some node clustering. Clustering architectures differentiate nodes functionality within the nodes in a cluster and between the nodes in different clusters. They can be combined with the multi-tier approach resulting in hierarchical multi-layer architectures, such as in [31]. Flat architectures are usually distributed DHT based structures (structured peer-to-peer), while hybrid architectures combine the elements of flat and hierarchical approaches as well as of query-based and directory-based elements.

Peer-to-peer networks recently gain on popularity. Especially applicable to networks with a large number of nodes, they seem to offer appropriate mechanisms to ad hoc mobile network design, capable of following the nodes dynamism and mobility. They will be addressed in one of the following subsections.

Network type and imposed requirements influence the service discovery architecture in ad hoc domain. Absence of any infrastructure, node dynamism and mobility declare the directory-less designs as more appropriate. However, the large amount of overhead in the client-server approach, distinguishes the distributed principle in both service discovery architectures as the most efficient for ad hoc design (light service discovery protocols. Proposed solutions should cope with mobility and node failures, low caches and resources constraints, scalability. Table 6.3 in the Appendix specifies the service discovery architecture of listed protocols appropriate for ad hoc networking.

The architecture design of service discovery process can be recognized, possibly with some modifications, within organization of storage of service information and particularly directory design. The following subsection gives an overview of these topics providing examples within existing service discovery protocols.

6.6.1 Storage of Service Information

Storage system enables retrieval of available services and implements mechanisms to rapidly access and provide service. Network size and type are very important designing concerns. Ad hoc networking, because of mobility of nodes and limited storage resources, needs a special attention in the design of storage mechanisms. Generally, storage approaches are classified as: *centralized, unstructured* and *structured.*

Centralized approach allows rapid access to data and low traffic even it is failure sensitive creating single point of failure. Centralized approach is not common for ad hoc networks, since they exist without infrastructure. However, elements of this approach are present in SLP (with directory agents) and Jini, even though actual communication is done in a peer-to-peer mode.

Unstructured distributed storage systems base communications on broadcasting and multicasting mechanisms. They are often used in local area networks and ad hoc networks. In the majority of protocols (UPnP, Konark, DEAPspace, Wu and Zitterbart, Varshavsky et al., GSD [10, 13, 32–35]), interested nodes maintain their own information of services and devices in the network. For instance, UPnP maintains a consistent view of all devices and services present in the network, while GSD proposes that a node caches the service descriptions available in a maximum number of hops (the diameter of its knowledge). The complexness that has to be maintained in each node is an important and critical feature.

Structured distributed storage methods are applicable for large networks. This approach allows directory nodes to organize themselves in *overlay* structures, so that the discovery messages are routed through a limited number of hopes. Structured distributed storage systems recognize several categories: *hierarchical, flat* and *hybrid* [3].

In *hierarchical* type of storage, the information, advertisements and queries are propagated through hierarchical structure in parents–children manner. The root nodes can become a bottleneck. Examples of protocols which implement hierarchical storage are SSDS [23] and CSP [36].

Flat type of protocols mostly rely on peer-to-peer overlay networks. Unlike the attribute/value approach, they are built upon the *stored key/value* approach. They are constructed by means of *distributed hash tables* (DHTs). DHT protocols provide efficient lookup mechanism. Examples are CAN [37], Chord [38], Pastry [39] and Tapestry [40]. Messages enter the hash function before routing through a bounded number of hops (O(log(N), where N is a number of nodes) to the nodes responsible for resulting key. The routing tables, with identifiers and network addresses of other nodes, are maintained in each node. An example of service discovery protocols based on DHT is INS/Twine (relies on Chord) [20]. More details can be found in [3].

Hybrid solutions combine the approaches of hierarchical and flat storage mechanisms with some additional optimization techniques. They are often based on hierarchical models for node organization and try to optimize the benefit of DHT approaches (i.e., efficient lookup mechanisms). Several examples can be found in the literature. Service Rings [11] maintains hierarchical ring architecture with service access points (SAPs) and optimizes the structure monitoring network traffic. LANES builds up two-dimensional CAN structure organized in lanes [12]. Service announcements are propagated through a lane in a top-bottom manner and each node has full information about the services offered by that lane. Anycast routing is used to propagate service requests to another lane. JXTA [41] establishes a virtual network overlay on top of the physical network and combines DHT methods with a random walker approach (to check a consistent distributed hash index). INS [7] is based on a spanning tree architecture, with a central component in the network (domain space resolver — DSR), which helps in establishment of neighbor relationship (according to the round trip latency). Interesting proposal for overlay networks comes from Kozat and Tassulas in [28]. They propose a distributed directory solution based on a network *backbone* node structure, formed as a *dominating set* (see Chapter 2).

6.6.2 Directory Design (Architecture)

Directories cache service information and answer client's lookup requests. They can be organized in a different way depending on environment according architectures, service information cache strategies and hierarchies. Several classifications, concerning different features, are possible:

- Centralized vs. decentralized directories;
- Number of service information copies (*single copy*: failure sensitive, *multiple copies* provided by multiple directories: more reliable, but greater overhead; *fully replicated* directories: a service search only goes to the directory to which a client is attached);
- Flat (peer-to-peer) vs. hierarchical directory structure (directories have parent/child relationships);
- Service state (soft vs. hard);
- Directory address;
- Number of directory hierarchies.

The directory architecture adopted by different service discovery protocols can be broadly classified as centralized or decentralized (see Fig. 6.4) [42]. In the *centralized* architectures, the directory information is stored on a central location in the network. *Decentralized* architectures can be categorized as *replicated* (the entire directory information is stored at multiple network locations — as in INS), *distributed* (directory information is portioned and parts are stored in different network locations) or *hybrid* (both replication and distribution methods are used — as in Twine).

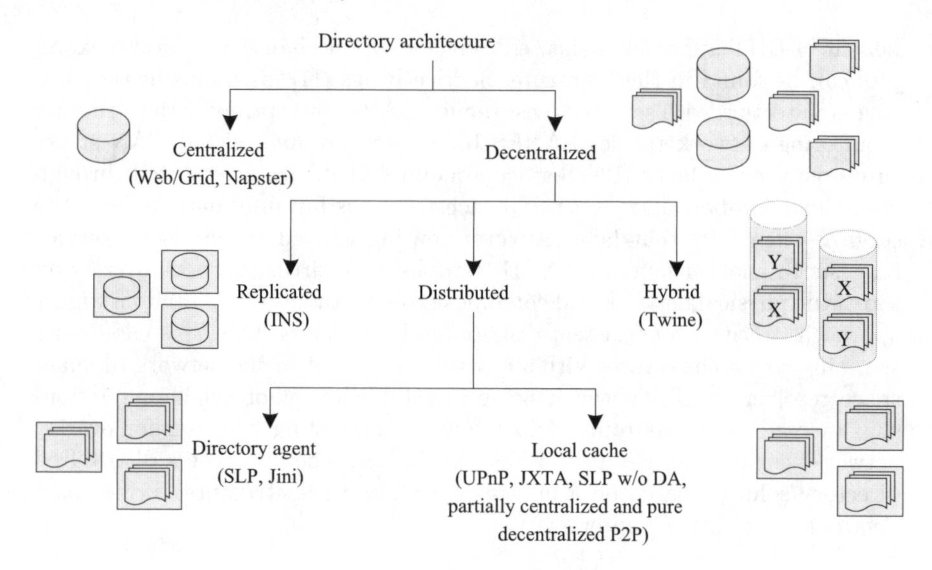

Fig. 6.4. Directory architectures

The directory information in distributed directory architecture can be stored in dedicated servers, i.e., *directory agents* (DAs) (as in SLP, Jini and Salutation) or can be *cached locally* by the service providers in the system (as in UPnP, JXTA, SLP without DA, pure decentralized and partially-centralized P2P systems).

Centralized directory architecture is not suitable for large systems and is failure sensitive. The major problem in replicated systems is consistency of the replicas. Distributed and hybrid architectures are scalable and provide better level of fault-tolerance. Centralized approach does not fit the dynamic ad hoc nature while decentralized distributed approaches can perform satisfactorily. Since each node acts as client and router at the same time, the network nodes have limited resources to perform the service discovery. So, the small amount of service information cached locally is more convenient to low power ad hoc nodes than directory based approaches, which occupy more resources of directory nodes.

6.6.3 Peer-to-Peer Architectures

The classical client/server paradigm is hardly applicable in modern networks and is increasingly replaced by peer-to-peer (P2P) approaches, allowing constant changes in network topology and making fixed infrastructure obsolete. P2P networks distinguish from traditional client/server or master/slave networks in sense that there is neither a central point of control nor centralization of data. In recent trends, more and more research efforts are dedicated towards deployment of peer-to-peer networks in the context of mobile ad hoc networks (MANETs). The pervasive computing and emerging sensor networks also contribute to focus research activities on these architectures and their features. P2P technology is recently successfully deployed in several file sharing systems, such as Gnutella [43], Napster [44], Freenet [45]. The music file sharing revolution started by them led to considerable interest in ad hoc networking issues. However, they are not scalable. Second generation of P2P systems (Chord [38], Pastry [46], Tapestry [47], CAN [37]) are based on DHT and provide reliable content location (with persistency and availability). They guarantee a definite answer to a query into a bounded number of hops and are more scalable. Their major characteristic is that all nodes tend to participate and contribute equally to the system. Improvements considering nodes heterogeneity and matching to performance metrics (load balance, system utilization, reliability and trust) are offered in [48]. Srivatsa et al. in [49] propose to structure P2P overlay topology using a capacity-aware multi-tier topology to better balance between peers' heterogeneous capabilities. Good overview of P2P systems can be found in [50].

What is the difference between (mobile) ad hoc networking and peer-to-peer? The basis of both is self-organisation, scalability, decentralized information dissemination and discovery, independence from centralised servers and support for constantly changing network topology [51]. Whereas ad-hoc-networking refers more to the lower network layers, i.e., radio transmission instead of wired connections, dynamic host address assignment, special routing mechanisms and so on, the peer-to-peer paradigm refers to the application design and is an antipode to the client server paradigm. Thus, applications in ad hoc networks are very likely to use the peer-to-peer paradigm, but conversely, peer-to-peer applications are not dependent on the network architecture underneath and are currently gaining popularity even in traditional TCP/IP networks.

P2P networks are overlay networks that abstract away the underlying physical network [52]. They impose a *virtual* network topology that usually does not consider the underlying physical topology in its construction. It means that two physical neighbor nodes are not necessarily the overlay neighbor nodes.

Virtual P2P topology, as independent of physical node layout, often can be identified with known topology network structures, such as *random graph* or *Erdos-Renyi graphs*, *small world graphs* and *scale-free graphs* (see Chapter 2). Using these *graph-theoretic* techniques makes analytical approach to network performance analysis [53, 54] easier.

P2P systems are broadly classified into *unstructured* (search based) and *structured* (DHT based) systems.

Most of the research effort in unstructured P2P systems focus on improving the search algorithms (by introducing the flooding based and guided search procedures based on query latency, messaging bandwidth, node degree, etc.) and the overlay topologies (e.g., multi-tier topology, capacity aware topologies, etc.) [55]. *Multi-tier topologies*, unlike the hierarchical topologies, can be realized using simple bootstrapping and do not require high construction or maintenance costs. They are built according to the classification of nodes into *heterogeneity levels (HPs)*, based on nodes capabilities. Connections between two nodes are allowed only if differences in their heterogeneity levels satisfy certain conditions preventing week nodes to become a hot spots or bottlenecks in the system. Powerful nodes would overtake more connections and queries, achieving significant performance gains in the system.

Much research effort on P2P systems has been devoted to *distributed hash tables (DHTs)*. P2P approaches overcome the *scalability* problem and provide an *upper bound* on the number of routing hopes that have to be taken to locate the object (i.e., a given *key*), bounding it to $O(\log N)$, where N is the number of nodes. Although DHT helps sufficiently to locate the target node on the overlay network, the discrepancy between the number of hopes and actual distances can result in increasing the actual network traffic and latency. Due to limited bandwidth and vulnerable transmission media, the probability of dropping a packet clearly increases with the path length. Also, mobility of nodes can result in increasing probability of broken route. That means that the existing P2P systems can not be merely deployed to ad hoc mobile networks without some modifications and improvements.

A possible solution is offered in [52], where the authors propose an *efficient clustering* scheme, based on Random Landmarking (RLM) group to actively take advantage of physical proximity in DHT P2P networks. Their method is based on grouping of physically close nodes into common sections (clusters) of the overlay ID spaces, achieving that two nodes physically close are also likely to be "close" to each other in the virtual overlay network. It can significantly decrease the physical path length of overlay hops and consequently the network traffic in the context of ad hoc mobile networks. Clustering can be semantic-based, providing clustering of peers with semantically closed data and mapping the high-dimensional clusters into a one-dimensional *small world network* (see Chapter 2) [56]. Ramaswamy et al. in [57] propose connectivity based distributed node clustering scheme with completely decentralized approach and based only on knowledge of neighboring nodes. The clustering is performed automatically with reduced messaging cost and improved QoS of the running applications.

Since future networks will be much more dynamic than traditional (wired) ones, peer-to-peer approaches, together with service discovery, will gain more and more importance. The goal of service discovery mechanisms is to enable software components to find each other on such highly dynamic networks (e.g., mobile ad hoc networks). Apart from arranging advertising and discovery, *consistency maintenance* and *failure handling* are also important challenges to service discovery in dynamic peer-to-peer approaches.

Distributed approach in ad hoc networks imposes importance of having consistent and up-to-date service information in the whole network [58]. Consistency maintenance can be provided by polling (query based) and notification (registry based). Ad hoc networks have to cope with changes due to failures. The service discovery architecture must provide mechanisms to detect failures and to recover. Two basic mechanisms exist: *soft state persistence* by monitoring periodic announcements and *application–level persistence* by bounded retries and remote exceptions [58].

Integration of existing peer-to-peer, service discovery and ad hoc technologies provides a robust infrastructure for deploying *agents* within an ad hoc network [59, 60] leading to a standard development (FIPA — Foundation for Intelligent Physical Agents) for mobile ad hoc environments and its own protocol for service discovery.

6.7 New Service Discovery Paradigms

Importance of service discovery for the overall functioning of ad hoc mobile networks, where nodes are highly dynamic and new services appear and disappear from the network, initiated intensive research activities towards improvements in service discovery efficiency [61]. The enhancements of service discovery protocols are offered through cooperation with content awareness, cross-layer approaches and combination with different routing protocols, interoperability and variety of different proposals.

6.7.1 Context Awareness

Context awareness is one of the key elements for successful service provisioning [62]. Context-aware service discovery means provisioning of the most appropriate service/services to mobile users by exploiting any meaningful contextual information (e.g., user's preference, location and present environment). Context awareness, combined with service discovery, can offer more intelligence in the network and result in significant improvements in discovery efficiency [63]. MAGNET offers a solution of such cooperation, where both service and content discovery entities are inbuilt in MAGNET architecture and interact between each other [31]. This cooperation is supported by appropriate naming system [64]. Extensive ongoing research in this area is illustrated in references [65–69].

6.7.2 Cross-Layer Approach in Service Discovery

Mobility is a major problem in achieving efficient service discovery in mobile ad hoc networks. It causes unpredictable changes in availability of services and other resources. The effective way of providing updated information into all system layers and boosting the network performance may be through *cross layer information exchanges and optimizations* [31, 70–72]. The cross-layer approach outperforms the traditional approaches [70–72]. A separate chapter in this book (Chapter 4) is dedicated to this important issue.

A special case of cross-layer cooperation is *routing-based service discovery* (see Chapter 4), where service discovery information is combined with routing information in order to achieve more efficient system in sense of number of hops and time necessary for discovery process, as well as overhead information [4, 13, 30, 70–74].

6.7.3 Interoperability

Service discovery protocol heterogeneity is a key challenge in the ad hoc mobile computing domain. The mobile user can be isolated in heterogeneous environment if it is unable to discover services advertised with service discovery protocols different than the ones he recognizes. In order to facilitate the service discovery in wireless and mobile computing environments, across various domains, heterogeneous devices and communication protocols must be made interoperable (either a single service discovery or most commonly used technologies should dominate). The middleware is usually responsible to dynamically *detect* actual service discovery protocols and *adapt* their protocol interactions to interoperate with services available in the environment. Offered solutions have to minimize resource consumption and introduce lightweight mechanisms that may adapt easily to any platform (see [6] for more details).

Currently there are five auspicious protocols, each one eligible in different environments: SLP2, Jini, Salutation,UPnP and Bluetooth SDP. Probably all of them will continue to co-exist, since each one has unique characteristics indispensable for certain applications and limited support for interoperability. It is very probable though, that new architectures will be developed for more sophisticated and critical scenarios (like military operations).

Another solution to interoperability is achieved with *service broker approach* [75]. This approach discovers services not based on network addresses, but based on *service usage intention.* Different service discovery domains form a single logical service discovery domain *federation.* The service broker architecture consists of a broker engine integrating service discovery domains with the aid of *domain adapters.* This approach may be implemented according to a *service query translation* (extends the reach of service queries to multiple domains) or *service registration translation* (enables distribution of service registrations to multiple domains). Service broker node performs *bridging* in order to provide interoperability within the federation.

6.8 Projects Addressing Service Discovery

Many recent and ongoing projects address the service discovery paradigm tackling different features such as architecture, location and content awareness, integration, security, etc. They deal with different networking domains such as personal networks (MAGNET), sensor networks (SPINS), ad hoc environment in general (NOMAD, DIANE), heterogeneous networks (UBISEC), etc. The following lines shortly present some of these projects.

- *NOMAD.* Project NOMAD deals with integration of *location aware service discovery* mechanisms, handover procedures and service/user profiling [76]. It develops technology that allows users to freely roam across existing and future heterogeneous network infrastructures.
- *DIANE.* This project develops and evaluates concepts that allow *integrated, efficient and effective usage of the services offered in an ad hoc network.* It proposes mechanisms to describe, find and select services and methods to efficiently process these queries. DIANE proposes a new approach to service discovery in ad hoc networks based on a multi-layered clustering that results from a natural combunation of semantic issues (based on ontology) and structural issues (based on radio connectivity) [77]. The emphasis is on the information services, i.e., the services allowing an integrated access to digitally available information.
- *UBISEC.* This project [78] addresses integration aspects in new business areas and technologies. It primarily addresses security aspects. However, the problem of personalized content delivery, flexible service announcements (directory services), discovery, provisioning, and delivery support for mobile user, while moving across the heterogeneous networks, are inevitably addressed.
- *MAGNET.* MAGNET [31] is a project dedicated to personal network domain. The problem of service discovery is addressed in a truthfully and systematic manner. A proposed MSDP (MAGNET service discovery protocol), which deals with service discovery across different personal area entities is a peer-to-peer protocol and is based on existing peer-to-peer protocols (such as JXTA [41] or INS/Twine [20]). MAGNET defines layered architecture which recognizes five tier levels. Different nodes and entities participate in each tier level [31]. Tier level 4 recognizes SMN (service management nodes) nodes, which communicate with MSDP protocol, and are responsible for service discovery. The SMN nodes form a peer-to-peer overlay network, where they act as super-peer nodes. Details of such peer-to-peer based service discovery approach in hybrid architectures can be found in [79].

6.9 Concluding Remarks

This chapter deals with problems and solutions related to service discovery in ad hoc wireless networks. Extensive research and standardization activities are lately performed trying to address different issues and offer appropriate solutions. Ad hoc dynamic environment imposes limitations and requirements resulting in variety of solutions [16]. However, there are very few implemented service discovery protocols in the ad hoc domain.

The design of the service discovery process is still an open issue. New emerging technologies and variety of wireless devices impose new challenges. *Scalability* requires attentions due to growing number of heterogeneous devices that offer variety of services. Those services need to be discovered and used. *Resource awareness* imposes a need for a lightweight protocol that is able to run on a small and resource limited devices (such as the ones in sensor networks). *Security* has to be included in any service discovery protocol. *Inter-operability* deals with collaboration among service discovery protocols and permits discovery across heterogeneous networks which use different protocols. It will support building service discovery integration platform [4, 31, 80, 81] towards seamless communication.

New emerging network concepts, such as emergency and disaster communications, additionally concentrate interest on service discovery paradigms [81, 82]. The evolving service discovery protocols have to manage a large scale environment and resource constraints of present and future devices [83]. Efficient and flexible service discovery is crucial feature of ad hoc networks towards seamless communications and their integration into future wireless systems.

6.10 Appendix

Table 6.3. Comparative table of service discovery protocols

	Architecture	Storage of service information	Search methods	Event notification	Service description	Service selection and usage	Fault tolerance and mobility support	Network scalability	Security
Jini	Centralized	On Lookup Service	Both active and passive discovery for finding the Lookup Service	Distributed events	Java proxy objects	*Usage* -Java RMI via proxy objects	Lease mechanism for granting access to services	-Service grouping -Use of federation	-Authentication -Authorization (access control lists) -Confidentiality -Integrity
UPnP	P2P	On every control point	Both active and passive discovery for finding services	Eventing mechanism	XML description based on UPnP template language	Usage -Messages encapsulated in SOAP	-Expiry time for advertise-ments -“Device unavailable” notification	-	-
SLP	-Centralized (with DA) -P2P (without DA)	-On DA -On UA and SA	-Both active and passive discovery -Active discovery of services	-	Service templates registered with IANA	-	- Lifetime for service registrations	-More DA -Scope mechanism for service grouping	Optional authentication of DA and SA (using digital signatures)
Bluetooth SDP	Client - Server	On every server	Request/ Response messages between client and server	-	Service attributes (ID and value)	-	Implicit (no caches are maintained)	-	Determined in the Bluetooth connection negotiation phase
Salutation	Flexible (can be P2P or centralized)	Service Registry on every Salutation Manager	Salutation Manager Protocol between two SLMs	Long-term requests	Service description records	Usage through RPC	Periodic availability check of services	-	User authentication (username and password)
INS/Twine	Overlay network of resolvers which form a DHT	Each resolver holds a range of keys and their values	Service discovery messages are routed in O(logN) hops	-	Hierarchies of attribute-value pairs	-	-Each strand is stored on multiple nodes -Hybrid state management scheme	Hash-based partitioning of resource descriptions among resolvers	-

Continued

	Architecture	Storage of service information	Search methods	Event notification	Service description	Service selection and usage	Fault tolerance and mobility support	Network scalability	Security
INS	Spanning-tree overlay network	Overlay network formed by INRs	-Domain Space Resolver for INR discovery -Passive discovery for services -Early and late binding	-	Name-specifiers consisting of attributes and values	-	-Replication of service information among resolvers -Soft-state name exchanges and updates	Spawning INRs to candidates resolvers	-
CDS	Hash-based overlay network	Small set of rendezvous points for each content name	Applying the hash function to AV pairs in the query	-	Attribute-value pairs	*Selection* -Query optimization algorithm: node with the smallest response time or the node with smallest database is selected	Names are stored in a soft state fashion	Load Balancing Matrix to share registration and query load	-
Splendor	Four components: clients, services, directories and proxies	Directories	-Active and passive discovery of directories -Clients query directories for services	-	-	-	-Soft state storage of service information -Hard-state storage of services represented by proxies	Aggregation and filtering of service registrations	-Authentication -Authorization -Confidentiality -Integrity -Non-repudiation
Service Rings	Overlay structure of hierarchical rings that groups services which are geographically and semantically close	Service Access Points present on every ring	Queries are routed through the ring hierarchy	-	-Languages that support "dist" and "sum" functions -Descrip-tions are summarised at SAP points	-	-RingCheck message periodically checking for consistency -Algorithms for broken rings and network partition and reintegration	-Scaling through the hierarchical structure -Algorithms for ring restructuring	-

LANES	Loosely-coupled lanes overlay structure	Nodes within a lane share the same service information	Discovery messages are transmitted from lane to lane using anycast routing	-	-	-	-Proactive maintenance -Algorithms for node login/logoff, broken connections, network partition and reintegration	Algorithms for optimizing the lanes structure	-
Kozat and Tassiulas	Virtual backbone (a subset of network nodes forms a dominating set)	Distributed directory located on the virtual backbone	Source based multicast tree algorithm	-	-	-	-Periodic service registration -Algorithms for backbone maintenance	Through the use of multiple service broker nodes	-
FRODO	Centralized	Central acts as a repository for service information	Clients ask the central for services	Subscribers are notified about any change in the state of a service	-	*Selection* -Devices can request for the best matched service *Usage* - Through XML parsing	-Central election and re-election -Recovering from Central/Backup failure -Soft state for 300D devices, periodic poll of 3D devices	-	-
DEAP-space	P2P	Entire world view on every device	Modified passive search	-	Includes a TTL and the address	-	Nodes broadcast their entire view – rapid convergence	-	-
Konark	P2P	Service registry on every device	Both active and passive discovery	Clients may subscribe to events	XML	Usage -SOAP	Servers periodically announce their services	-	-
Allia	P2P (nodes are organized in alliances)	Every node caches advertisements from nodes in its alliance	-Passive discovery -Active discovery by multicast or broadcast service requests	-	-	-	Adjustment of advertisement rates and alliance diameter based on mobility of nodes	-	Policies for access rights and credential verification

Continued

	Architecture	Storage of service information	Search methods	Event notification	Service description	Service selection and usage	Fault tolerance and mobility support	Network scalability	Security
GSD	P2P	Every node caches advertise-ments to maximum N hops away (advertise-ment diameter)	-Passive discovery -Active discovery by selectively forwarding the request to a set of nodes based on semantic information	-	-DAML + OIL -Facilitates grouping of services based on service functionality	-	Adjustment of the advertisement time interval and advertisement diameter based on mobility of nodes	-	-
Wu and Zitterbart	P2P (based on DSR routing protocol)	Every node caches advertise-ments	Both active and passive discovery	-	-	-	Reactive: -Implicit service confirmation -Reception of a service poll reply -Reception of a "no service" indication	-	-
Varshavsky et al.	P2P (based on DSR and DSDV routing protocols)	Each node maintains a service table	Both active and passive discovery	-	-	*Selection* -Clients choose the service with the lowest hop count	*Reselection*: -None, route breaks, any change *Rediscovery*: -Proactive, reactive (route breaks, no route to server, no route to any server)	-	-
Cheng and Marsic	P2P (based on ODMRP routing protocol)	On every node which is interested in the service	-Active discovery -Passive discovery in the bootstrap	Nodes subscribe for receiving service updates	-	*Usage* -Service invocation through mobile objects	Implicit through active dissocvery	-	-

References

[1] Zhu, F., Mutka, M., and Ni, L., "Classification of Service Discovery in Pervasive Computing Environments," *MSU-CSE-02-24,* Michigan State University, East Lansing, 2002.

[2] Alex, H., Kumar, M., and Shirazi, B., "Service Discovery in Wireless and Mobile Networks," *Wireless Information Highways*, Idea Group Publication, December 2004.

[3] Marin-Perianu, R., Hartel, P., and Scholten, H., "A Classification of Service Discovery Protocols," *Technical report TR-CTIT-05*-25, Centre for Telematics and Information Technology, University of Twente, The Netherlands, June 2005.

[4] Kozat, U. C. and Tassiulas, L., "Network Layer Support for Service Discovery in Mobile Ad Hoc Networks," *IEEE INFOCOM 2003*, San Francisco, USA, 2003.

[5] Ahmed, R., Boutaba, R., Cuervo, F., Iraqi, Y., Li, D. T., Limam, N., Xiao, J., and Ziembicki, J., "Service Discovery Protocols: A Comparative Study," *IFIP/IEEE International Symposium on Integrated Network Management (IM'2005) Application Sessions*, Nice, France, 2005.

[6] Bromberg, Y. -D. and Issarny, V., "Service Discovery Protocol Interoperability in the Mobile Environment," *International Workshop Software Engineering and Middleware (SEM)*, September 2004.

[7] Adjie-Winoto, W., Schwartz, E., Balakrishnan, H., and Lilley, J., "The Design and Implementation of an Intentional Naming System," *Symposium on Operating Systems Principles*, Charleston, SC, December 1999.

[8] Sundramoorthy, V., Scholten, J., Jansen, P. G., and Hartel, P. H., "Service Discovery at Home," *4th International Conference on Information, Communications & Signal Processing and 4th IEEE Pacific-Rim Conference on Multimedia (ICICS/PCM)*, IEEE Computer Society Press, December 2003.

[9] Cheng, L. and Marsic, I., "Service Discovery and Invocation for Mobile Ad Hoc Networked Appliances," *2nd International Workshop on Networked Appliances (IWNA 2000)*, New Brunswick, New Jersey, USA, November 30–December 1, 2000.

[10] Microsoft Corporation, White Paper: Understanding Universal Plug and Play, June 2000. http://www.upnp.org/resources/UpnPbkgnd.htm

[11] Klein, M., Konig-Ries, B., and Obreiter, P., "Service rings — a semantic overlay for service discovery in ad hoc networks," *DEXA Workshops*, 2003.

[12] Klein, M., Konig-Ries, B., and Obreiter, P., "Lanes — A Lightweight Overlay for Service Discovery in Mobile Ad Hoc Networks," *Technical Report 2003-6*, University of Karlsruhe, 2003.

[13] Varshavsky, A., Reid, B., and de Lara, E., "The Need for Cross-Layer Service Discovery in MANETs," *Submitted for conference publication*, January 2004. Available at: http://www.cs.toronto.edu/~delara/papers/crosslayer.pdf

[14] Gao, J. and Steenkiste, P., "Rendezvous Points-Based Scalable Content Discovery with Load Balancing," *Fourth International Workshop on Networked Group Communication (NGC'02)*, Boston, MA, October 2002.

[15] Lee, C. and Helal, S., "A Multi-Tier Ubiquitous Service Discovery Protocol for Mobile Clients," *2003 International Symposium on Performance Evaluation of Computer and Telecommunication Systems (SPECTS 2003)*, Canada, 2003.

[16] Zhu, F., Mutka, M., and Ni, L., "Facilitating Secure Ad hoc Service Discovery in Public Environments," *2003 IEEE Computer Software and Applications Conference (Compsac 2003)*, Dallas, Texas, USA, November 2003.
[17] Yuan, Y. and Agrawala, A., "A Secure Service Discovery Protocol for MANET," *Computer Science Technical Report CS-TR-4498 and UMIACS Technical Report UMIACS-TR-4498*, Computer Science Department, University of Maryland, 2003.
[18] Handorean, R. and Roman, G. -C., "Secure Service Provision in Ad Hoc Networks," *The First International Conference on Service Oriented Computing*, Trento, Italy, December 2003.
[19] Almenarez, F. and Campo, C., "SPDP: A Secure Service Discovery Protocol for Ad-hoc Networks," *EUNICE 2003*, Budapest, Hungary, September 2003.
[20] Balazinska, M., Balakrishnan, H., and Karger, D., "INS/Twine: A Scalable Peer-to-Peer Architecture for Intentional Resource Discovery," *International Conference on Pervasive Computing 2002*, August 2002.
[21] Salutation Consortium, Salutation Architecture Specification Version 2.1, Part 1, 1999.
[22] Guttman, E., Perkins, C., Veizades, J., and Day, M., "Service Location Protocol, Version 2," RFC 2608, IETF, June 1999.
[23] Hodes, T. D., Czerwinski, S. E., Zhao, B. Y., Joseph, A. D., and Randy H. Katz, "An Architecture for Secure Wide-Area Service Discovery," *Wireless Networks*, 8(2/3):213–230, 2002.
[24] Sun Microsystems. Jini architecture specification version 2.0, June 2003.
[25] Zhu, F., Mutka, M., and Ni, L., "Splendor: A Secure, Private, and Location-Aware Service Discovery Protocol Supporting Mobile Services", in *First IEEE International Conference on Pervasive Computing and Communications (Percom'03)*, Dallas-Fort Worth, Texas, USA, March 2003.
[26] Zhu, F., Mutka, M., and Ni, L., "Facilitating Secure Ad hoc Service Discovery in Public Environments," *2003 IEEE Computer Software and Applications Conference (Compsac 2003)*, Dallas, Texas, USA, November 2003.
[27] Perrig, A., Szewczyk, R., Wen, V., Culler, D., and Tygar, J. D., "SPINS: Security Protocols for Sensor Networks," *ACM Mobile Computing and Networking*, Rome, Italy, 2001.
[28] Kozat, U. C. and Tassiulas, L., "Service Discovery in Mobile Ad Hoc Networks: An Overall Perspective on Architectural Choices and Network Layer Support Issues," *Ad Hoc Networks Journal,* 2(1): 23–44, Elsevier, June 2003.
[29] Jacobsson, M., Hoebeke, J., Heemstra de Groot, S., Lo, A., Moerman, I., Niemegeers, I., Munoz, L., Alutoin, M., Louati, W., and Zeghlache, D., "A Network Architecture for Personal Networks," *IST Summit 2005*, Dresden, Germany, 2005.
[30] Engelstad, P. E., Zheng, Y., Koodli, R., and Perkins, C. E., "Service Discovery Architecture for On-Demand Ad Hoc Networks," *International Journal of Ad Hoc and Sensor Networks*, Old City Publishing (OCP Science), 1 (3), March 2005, pp. 27–58.
[31] My personal Adaptive Global NET, MAGNET (IST-507102), http://www.ist-magnet.org
[32] Helal, S., Desai, N., Verma, V., and Lee, C., "Konark — A Service Discovery and Delivery Protocol for Ad-Hoc Networks," *3rd IEEE Conference on Wireless Communication Networks (WCNC)*, New Orleans, Louisiana, March, 2003.

[33] Nidd, M., "Service Discovery in DEAPspace," *IEEE Personal Communications Magazine*, 8(4): 39–45, August 2001.
[34] Wu, J. and Zitterbart, M., "Service Awareness in Mobile Ad Hoc Networks," *11th IEEE Workshop on Local and Metropolitan Area Networks (LANMAN)*, Boulder, Colorado, USA, March 2001.
[35] Chakraborty, D., Joshi, A., Finin, T., and Yesha, Y., "GSD: A Novel Group-Based Service Discovery Protocol for MANETs," *4th IEEE Conference on Mobile and Wireless Communications Networks (MWCN)*, September 2002.
[36] Lee, C. and Helal, S., "A Multi-Tier Ubiquitous Service Discovery Protocol for Mobile Clients," *2003 International Symposium on Performance Evaluation of Computer and Telecommunication Systems (SPECTS 2003)*, Canada, 2003.
[37] Ratnasamy, S., Francis, P., Handley, M., Karp, R. M., and Schenker, S., "A Scalable Content-Addressable Network," *Technical Report TR-00-010*, University of California, Berkeley, Berkeley, CA, August 2001.
[38] Stoica, I., Morris, R., Karger, D. R., Frans Kaashoek, M., and Balakrishnan, H., "Chord: A Scalable Peer-to-Peer Lookup Service for internet applications," *2001 ACM SIGCOMM Conference*, August 2001.
[39] Rowstron, A. and Druschel, P., "Pastry: Scalable, Decentralized Object Location, and Routing for Large-Scale Peer-to-Peer Systems," *Lecture Notes in Computer Science*, 2218: 329–350, November 2001.
[40] Zhao, B. Y., Kubiatowicz, J. D., and Joseph, A. D., "Tapestry: An Infrastructure for Fault-Tolerant Wide-Area Location and Routing," *Technical Report UCB/CSD-01-1141*, UC Berkeley, April 2001.
[41] Traversat, B., Abdelaziz, M., and Pouyoul, E., "A Loosely-Consistent DHT Rendezvous Walker," *White Paper*, May 2003. Available at: http://www.jxta.org/docs/jxta-dht.pdf
[42] Ahmed, R., Boutaba, R., Cuervo, F., Iraqi, Y., Li, D. T., Limam, N., Xiao, J., and Ziembicki, J., "Service Discovery Protocols: A Comparative Study," *IFIP/IEEE International Symposium on Integrated Network Management (IM'2005) Application Sessions*, Nice, France, 2005.
[43] Gnutella. The gnutella home page: http://gnutella.wego.com/, 2002.
[44] Napster. Napster home page. http://www.napster.com/, 2001.
[45] Clarke, I., Sandberg, O., Wiley, B., and Hong, T. W., "Freenet: A Distributed Anonymous Information Storage and Retrieval System," *ICSI Workshop on Design Issues in Anonymity and Unobservability*, San Diego, CA, USA, July 2000.
[46] Rowstron, A. and Druschel, P., "Pastry: Scalable, Decentralized Object Location and Routing for Largescale Peer-to-Peer Systems," *IFIP/ACM International Conference on Distributed Systems Platforms*, November 2001.
[47] Zhao, B. Y., Kubiatowicz, J. D., and Joseph, A. D., "Tapestry: An Infrastructure for Fault-Tolerant Wide-Area Location and Routing," *Technical Report UCB/CSD-01-1141*, U. C. Berkeley, April 2001.
[48] Gedik, B. and Liu, L., "PeerCQ: A Decentralized and Self-Configuring Peer-to-Peer Information Monitoring System," *23rd International Conference on Distributed Computing Systems*, Providence, Rhode Island, USA, 2003.
[49] Xiong, L. and Liu, L., "A Reputation-Based Trust Model for Peer-to-Peer eCommerce Communities," *4th ACM conference on Electronic commerce*, San Diego, California, USA, 2003.

[50] Nakauchi, K., Morikawa, H., and Aoyama, T., "Design and Implementation of a Semantic Peer-to-Peer Network," *7th IEEE International Conference on High Speed Networks and Multimedia Communications*, Toulouse, France, 2004.

[51] Dyrna, M., "Peer2peer Network Service Discovery for Ad hoc Networks," *Seminar on Ad Hoc Networking — Technical University of Munich*, 2003/2004.

[52] Zahn, T., Winter, R., and Schiller, J., Vladimir A., "Simple, Efficient Peer-to-Peer Overlay Clustering in Mobile, Ad-Hoc Networks," *IEEE International Conference on Networks (ICON 2004)*, Singapore, November 2004.

[53] Kapur, A., Gautam, N., Brooks, R., and Rai, S., "Performance and Design of P2P Networks for Efficient File Sharing," *Industrial Engineering Research Conference*, Portland, USA, 2003.

[54] Kapur, A., Gautam, N., Brooks, R., and Rai, S., "Design, Performance and Dependability of a Peer-to-Peer Network supporting QoS for Mobile Code Application," *Tenth International Conference on Telecommunications Systems, Modeling and Analysis*, Monterey, Califonia, USA, October 2002.

[55] Xiong, L. and Liu, L., "A Reputation-Based Trust Model for Peer-to-Peer eCommerce Communities," *4th ACM conference on Electronic commerce*, San Diego, California, USA, 2003.

[56] Zhou, D., "Semantic Clustering for Peer to Peer Network," *Colloqium Report CSE 590*, The Pennsylvania State University, September 2004.

[57] Ramaswamy, L., Gedik, B., and Liu, L., "A Distributed Approach to Node Clustering in Decentralized Peer-to-Peer Networks," *Accepted for publication in IEEE Transactions on Parallel and Distributed Systems (TPDS)*, 16(9), September 2005.

[58] Castro, M., Costa, M., and Rowstron, A., "Performance and Dependability of Structured Peer-to-Peer Overlays," *International Conference on Dependable Systems and Networks (DSN'04)*, Florence, Italy, 2004.

[59] Lawrence, J., "LEAP for Ad-hoc Networks," *Workshop on Agents in Ubiquitous and Wearable Computing AAMAS 2002*, Bologna, Italy, 2002.

[60] Pirker, M., Berger, M., and Watzke, M., "An Approach for FIPA Agent Service Discovery in Mobile Ad Hoc Environments," *Workshop on Agents for Ubiquitous Computing*, Columbia University, USA, July 2004.

[61] Farkas, K., Ruf, L., May, M., and Plattner, B., "Framework for Service Provisioning in Mobile Ad Hoc Networks," *First International Conference on Telecommunications and Computer Networks*, San Sebastiàn, Spain, December 2004.

[62] Olesen, H. Jiang, B., Thongthammachart, S., and Butkus, A., "User-Centric Factors of Context Aware Services," *Workshop on "My personal Adaptive Global NET: Visions and beyond" IST-FP6-IP MAGNET*, Shanghai, China, November 2004.

[63] Khedr, M. and Karmouch, A., "Context Based Service Discovery Protocol," *21th Biennial Symposium on Communications*, Kingston, Ontario, Canada, June 2002.

[64] Murakami, H., Olsen, R. L., Schwefel, H. -P., and Prasad, R., "User-Centric Name Services for Personal Networks," *7th International Symposium on Wireless Personal Multimedia Communications — WPMC 2004*, Abano Terme, Italy, September 2004.

[65] Lee, C. and Helal, S., "Context Attributes: An Approach to Enable Context-Awareness for Service Discovery," *2003 ACM Symposium on Applications and the Internet*, 2003.
[66] Riva, O., Nadeem, T., Borcea, C., and Iftode, L., "Mobile Services: Context-Aware Service Migration in Ad Hoc Networks," Submitted for publication, Available at: http://www.cs.rutgers.edu/~iftode/mobservice.pdf
[67] Mostefaoui, S. K. and Mostefaoui, G. K., "Towards a Contextualisation of Service Discovery and Composition for Pervasive Environments," *Workshop on Web Services and Agent-based Engineering AAMAS'2003*, Melbourne, Australia, July 2003.
[68] Chen, G. and Kotz, D., "Context-Sensitive Resource Discovery," *First IEEE International Conference on Pervasive Computing and Communications*, March 2003.
[69] Tchakarov, J. B. and Vaidya, N. H., "Efficient Content Location in Mobile Ad Hoc Networks," *IEEE International Conference on Mobile Data Management (MDM)*, January 2004.
[70] Atanasovski, V. and Gavrilovska, L., "Efficient Service Discovery Schemes in Wireless Ad Hoc Networks Implementing Cross-Layer System Design," *27th International Conference on Information Technology Interfaces ITI 2005*, Cavtat, Croatia, June 2005, pp. 527–532.
[71] Atanasovski, V. and Gavrilovska, L., "Cross-Layer Optimizations in Wireless Ad-Hoc Networks," *Poster Presentation at the 12th IEEE International Conference on Network Protocols — ICNP 2004*, Berlin, Germany, October 2004.
[72] Atanasovski, V. and Gavrilovska, L., "Routing Assisted Efficient Service Discovery in Ad-Hoc Networks," in *Proceedings of the 7th International Symposium on Wireless Personal Multimedia Communications — WPMC 2004*, Abano Terme, Italy, September 2004, pp. 491–495.
[73] Ververidis, C. N. and Polyzos, G. C., "Routing Layer Support for Service Discovery in Mobile Ad Hoc Networks," *Third IEEE International Conference on Pervasive Computing and Communications Workshops (PERCOMW'05)*, 2005.
[74] Oh, C. -S., Ko, Y. -B., and Roh, Y. -S., "An Integrated Approach for Efficient Routing and Service Discovery in Mobile Ad Hoc Networks," *CCNC 2005*, Las Vegas, USA, January 2005.
[75] Koponen, T. and Virtanen, T., "A Service Discovery: A Service Broker Approach," *37th Hawaii International Conference on System Sciences*, 2004.
[76] Integrated NetwOrks for seaMless And transparent service Discovery, NOMAD (IST-2001–33292), www.ist-nomad.net/index.php
[77] Klein, M. and Konig-Ries, B., "Multi-Layer Clusters in Ad-Hoc Networks — An Approach to Service Discovery," *International Workshop on Peer-to-Peer Computing*, Pisa, Italy, May 2002.
[78] Ubiquitous Networks with a Secure Provision of Services, Access, and Content Delivery, UBISEC (IST- 506926), http://jerry.c-lab.de/ubisec/
[79] Atanasovski, V. and Gavrilovska, L., "Providing Efficient Service Discovery in Wireless Personal and Ad Hoc Networks," *TELFOR2005*, Belgrade, Serbia and Montenegro, November 2005.
[80] Detken, K. -O., Fikouras, I., and Phillipopoulos, P., "Service Discovery Integrated Network Platform," *Interworking 2002*, October 2002.

[81] Blange., M. J., Karkowski, I. P., and Vermeulen, B. C. B., "Service Discovery in Heterogeneous Wireless Networks," *International Workshop on Wireless Ad-Hoc Networks*, Oulu, Finland, 2004.
[82] Bodanese, E., Gavrilovska, L., Rakocevic, V., and Stewart, R., "Eliminating the Communication Black Spots in Future Disaster Recovery Networks," *8th International Symposium on Wireless Personal Multimedia Communications — WPMC 2005*, Aalborg, Denmark, September 2005, pp. 1930–1934.
[83] Munoz, M. and Rubio, C. G., "A New Model for Service and Application Convergence in B3G/4G Networks," *IEEE Wireless Communications Magazine*, October 2004, pp. 6–12.

7

Mobility

7.1 Introduction

An ad hoc wireless network is a multi-hop wireless configuration without a fixed infrastructure or central administration. It consists of mobile nodes, which create a highly dynamic environment that poses some of the major challenges in network design and performance analysis. *Mobility modelling*, that specifies the dynamic characteristics of nodes' movement, is a crucial mechanism in evaluation and study of such networks.

Intensive research activities were dedicated to investigation of mobility in the area of wireless ad hoc networks. Some ideas were adjusted from the cellular mobile networking, others were originated within the ad hoc networking paradigm. In cellular networks, mobility models are mainly focused on individual movements respecting the point-to-point connectivity and cell organization. The motion patterns in ad hoc networks, in order to reflect the realistic movements in different ad hoc applications, must efficiently model the behavior of a mobile host (node), as well as a group of nodes in cooperative ad hoc environment (e.g., rescue operations, disaster recovery, moving group in a museum, battlefield, etc.), with no reference to particular "cell" as in cellular systems [1].

Different mobility patterns will affect the performance of different network protocols in different ways. Therefore, it is of great importance to study the impact of mobility patterns on different network protocols in order to achieve the best performance in each scenario. Appropriate metrics are defined to capture the mobility influence. Some scenarios exploit the node movement to achieve better performance, such as in the case of capacity, security, QoS, etc.

This chapter covers mobility issues relevant to ad hoc networking. It gives a short survey of the most important existing mobility models and presents several relevant metrics. In addition, the impact of mobility on different aspects of ad hoc networking concludes this chapter.

7.2 Mobility Models in Ad Hoc Networks — General Remarks

The movement pattern of nodes (users) plays an important role in performance analysis of mobile and wireless networks. Moreover, various mobility patterns affect the performance of different network protocols in different ways [1]. So, it is necessary to develop a flexible mobility framework, which allows modelling of different applications and network scenarios and identification of the impact of mobility on different scenarios. Generally, the modelling of movement can be completed by using analytical or simulation approaches [2].

Analytical mobility models are usually based on simple assumption regarding the movement behavior of the nodes (users). However, they allow the calculation of the mathematical expressions with respect to system performance. The analytical modelling, even with some simplifications, can be quite complicated and does not always follow the realistic behavior. Combined with corresponding traffic models it can allow estimation of important system performance parameters such as channel holding time, handovers and location update events, etc. Examples can be found in: Brownian mobility model [3, 4], often referenced Guerin model [5] and many others [2]. Varieties of generalizations of analytical models are used in different simulation studies.

Simulation-based studies introduce variety of different mobility models which can describe the movement of nodes (users) in a more detailed and realistic manner [6]. It makes the simulation approach in investigation of mobility a preferred research method today. This method is also accepted by ETSI in different test scenarios for system simulation as a recognized methodology for an indoor office, outdoor pedestrian and a vehicular environment [7, 8]. Currently, there are two types of mobility models used in the simulation of networks: *traces* and *synthetic models* [9]. Traces are the mobility patterns that are observed in real life systems. They usually involve a large number of participants and long observation period resulting in accurate information. However, ad hoc networks are not easily modelled with traces. The synthetic models, on the other hand, attempt to realistically represent the behavior of mobile nodes in ad hoc environment. General disadvantage of simulation methodology is that it does not allow the derivation of analytical expressions.

In wireless ad hoc networks, the mobility models focus on the individual motion behavior between *mobility epochs* (the smallest time periods in the simulation) in which a mobile host (node) usually moves in a constant direction at a constant speed [1]. The most simple and often used mobility model, random mobility model [10], assumes that the speed and the direction of motion in a new time interval has *no relations* to the values in the previous intervals and can be chosen between a uniformly distributed values. *Repeating pause and motion intervals* are introduced in random waypoint mobility model [11, 12]. The *correlation* between the speed/direction values in previous and current interval are introduced in incremental models [e.g., 13]. The *relationship and cooperation* between the nodes, which move with a common objective (such as disaster recovery, military deployment) give new flavour to mobility modelling (e.g., Pursue model [14], Column model [15]) initiating various group mobility models [1, 16]. A new movement patterns, related to new application scenarios [17], and different enhancements [2, 18, 19, 20–22] are continuously appearing. Some of them consider the obstacles that restrict movement and signal propagation [21, 22], offer a generic mobility model (e.g., random trip model in [23]) or propose the hybrid [22] and two-tiered [24] approaches to shape the realistic movement scenario. These approaches are influencing many different mobility modelling techniques.

Design of mobility model considers some general features such as level of details, dimension, border behavior, degree of randomness. The mobility modelling in wireless domain considers different *levels of detail* (level of description): microscopic, mesoscopic (kinetic) and macroscopic. A *microscopic* model describes the movement of a single node by its space and speed coordinates at a given time (e.g., random waypoint model [11, 12], smooth random mobility model [2], etc.). Most of the researchers that investigate the mobility modelling in ad hoc networking work in this area. A *mesoscopic* level reflects the homogenized movement behavior of several nodes (e.g., a distribution function that describes the number of nodes/vehicles at a certain location (x, y) or speed v at time t). A *macroscopic* modelling is interested in features such as density, mean speed and speed variance, traffic flow, etc. The mobility model in this group can range from city scale to international scale and can treat an individual or aggregated description of movement. Examples can be: fluid flow model, gravity models, random walk models, etc. [2, 25–27].

The mobility modelling can also deal with three *dimensional movements*: 1D, 2D and 3D (e.g., vertical movements in stairs and elevators) [28, 29]. The impact of the *border behavior* on the spatial node distributions considers three strategies (see Fig. 7.1):

- The node leaving the area is bounced back to the system area according to a certain rule;
- The leaving node is deleted and a new node is initiated according to the node initialization distribution;
- The leaving node is wrapped around the other side of the simulation plane.

These methods guarantee that the number of nodes in the system area remains constant, changing the angle, direction or speed after reaching the edge. In the third case, a leaving node enters the system area on the opposite

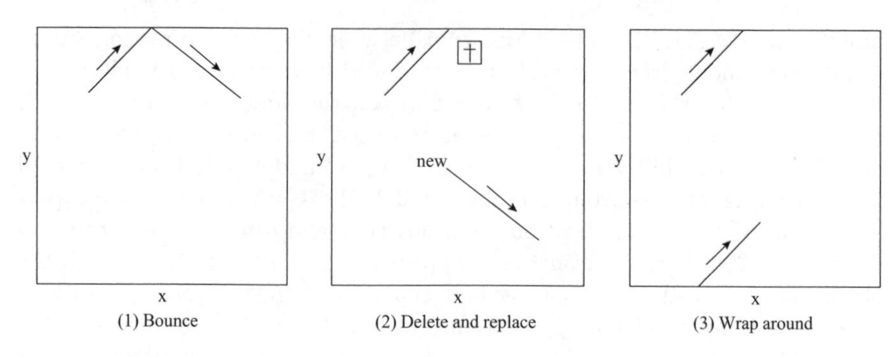

Fig. 7.1. Border behavior strategies

side, keeping its current speed and direction, modelling the system area as a torus. Border rules must be carefully used to avoid the *effect of border behavior,* which can result in higher node density in the middle of the area [2].

Different mobility models consider different *degree of randomness.* These include:

- Models that allow nodes (users) to move anywhere in the system plane following a pseudo-random process for speed and direction (e.g., random waypoint model, random direction model, etc.);
- Models that bound the movement of nodes by street, building, obstacles and so on, not using a pseudo-random process for speed and direction choice at crossing points (e.g., Manhattan-like street model);
- Models that bound the movement of users to a predefined path (either with determined speed and direction, or with fixed direction trace and randomly chosen speed).

The relevant parameters in randomized approaches are *speed* and *direction* (or *destination*). At any moment of changes, they can be treated as independent of their previous values. This results in sharp variations (sharp turning and sudden stopping), which is not a realistic movement behavior. The inclusion of some degree of correlations with the previous values (speed and/or direction) and incremented changing can result in more realistic mobility models [1, 13].

Different criteria are used in variety of mobility models based either on simulation or on analytical approaches [2]. Fig. 7.2 illustrates some criteria used for categorization of mobility models. New features introduced in upcoming mobility models can be added to this chart.

Many other enhancements are suggested recently in mobility modelling of wireless ad hoc networks. The following text gives a categorization view and short description of the most relevant mobility models.

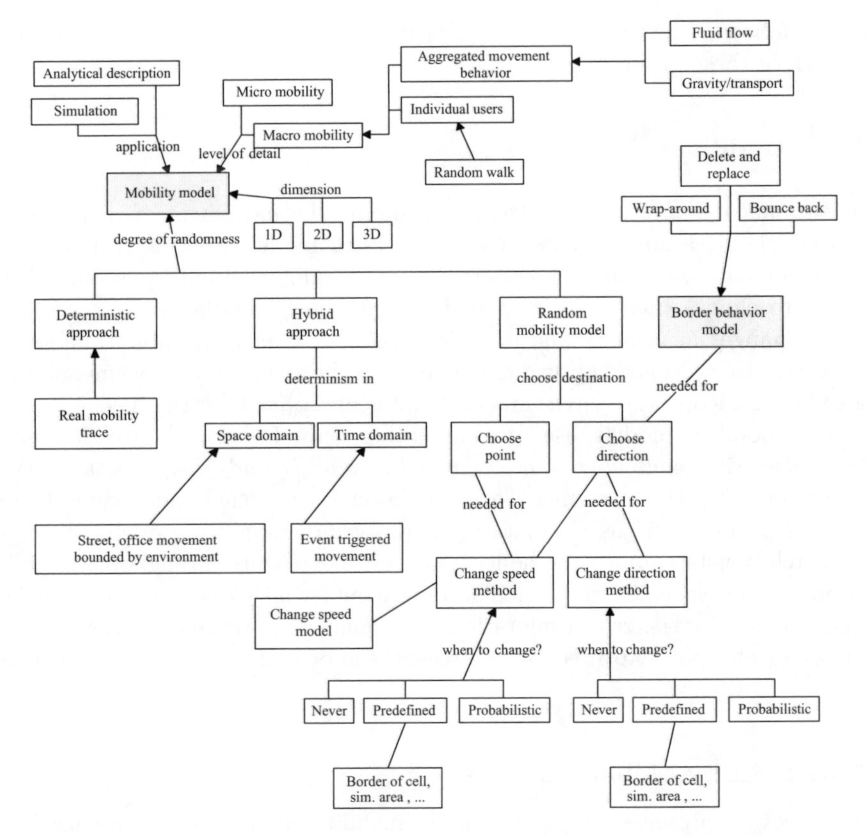

Fig. 7.2. Concepts used in simulation and analytical based mobility models

7.3 Classifications' Varieties

Nodes' movements in ad hoc networks significantly affect their performance and also invoke protocol mechanisms to react to such dynamism. The mobility model specifies the dynamic characteristics of node movement creating a base for investigations of links between the mobility and performance.

The growing variety of mobility models often changes their classification, depending on the different modelling considerations. A good survey of mobility models in ad hoc networks is offered in [2, 9, 17, 20, 30–34].

It is difficult to provide a unique classification of mobility models for ad hoc networks. However, they can be grouped as:

- Traditional mobility models;
- Enriched mobility models;
- Interdisciplinary mobility models

The following text presents examples and explains the major characteristics of each of these groups.

7.3.1 Traditional Models

Traditional mobility models attempt to mimic the movement of real mobile objects [1, 9] as an entity or as a group. Such models are relatively simple to implement and analyze. Most of them use randomized approach and allow nodes to choose their *velocity* and *direction independently*, with no restriction. Changes in velocity and direction must occur in reasonable time slots. However, they do not capture the correlation between node movements and also fail to incorporate environmental and geographical restrictions.

The mobility models used in simulations can be roughly divided into two categories: *individual (entity) mobility models and group-based.* In the entity models, the movement of each node is modelled independently of any other nodes in the simulation. In the group mobility models, there is some relationship among the nodes and their movements throughout the cells or field. The specifics of particular movements which may exist on battlefield, disaster areas, crowd migration, movement of a group of visitors into a museum, etc., are captured by the recently appeared *group mobility models* [9, 16].

7.3.1.1 Entity Mobility Models

This section presents several synthetic mobility models which are used in the performance evaluation of ad hoc network protocols. The most common mobility models used in research purposes are: *random walk mobility model, random waypoint mobility model* [11, 12] and *random direction mobility model* [2]. Each of them is designed to produce particular motion behavior.

Random Walk The Random Walk Mobility Model (RWMM), originally described by Einstein in 1926, was developed to mimic the extremely unpredictable movements of many entities in nature [10]. In ad hoc networks, it models the node movement from current location to a new one by randomly choosing the direction and speed (velocity), from the pre-defined ranges $[0, 2\pi]$ and $[v_{min}, v_{max}]$ respectively. The changes may occur either at the end of a *constant time interval* t, or, once a *constant distance* d is travelled (values for t and d should be adjusted to the network dynamism). Fig. 7.3 depicts the variations of the model in these two cases [9]. If the node reaches the boundary of the simulation area, it bounces off the simulation border with an angle determined by the incoming direction.

The Random Walk Mobility Model is sometimes referred to as *Brownian motion.* Many derivatives of Random Walk Mobility Model include the *1D*, *2D*, *3D* and *dD* walks [9, 35]. It is widely used [32, 36] and

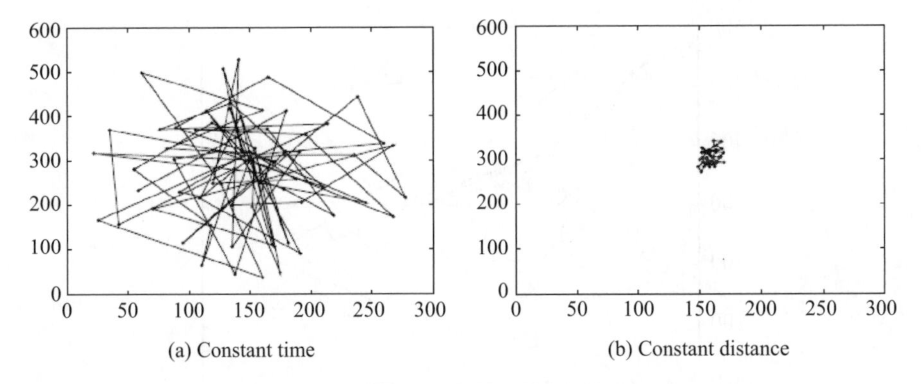

Fig. 7.3. Motion pattern of a node using the Random Walk Mobility Model with: (a) constant time; (b) constant distance

presents the *memory-less mobility pattern*, since it does not keep knowledge of previous speeds and directions. The current speed and direction are also independent of the previous ones. This can generate discrepancy, resulting in sudden stops and sharp turns. In order to avoid these problems, many authors modify the Random Walk Model by changing the calculation of speed, direction or both. This model was extended to various other models such as: Random Waypoint Model, Random Gauss-Markov model, Markovian Model.

Random Waypoint The Random WayPoint Mobility Model (RWP) breaks the entire movement of a mobile node into a sequence of pause and motion period [11, 12]. During the pause periods, the node stays in a certain location for a specified time period. Then, it travels towards the newly chosen destination with a selected speed. The speed is uniformly chosen between $[0, v_{max}]$. If the pause interval is zero, this model turns into a Random Walk Mobility Model [9].

Fig. 7.4 depicts the motion pattern of a node which uses the Random Waypoint Mobility Model. RWP model is a well designed model, but it is unable to capture the spatial dependence of movement among nodes, temporal dependence of node movement over time and existence of obstacles or barriers.

The RWP model has also some peculiarities. In most cases, when using RWP model, the initial node distribution is considered random around the simulation area. However, the average node neighbor percentage (cumulative percentage of total nodes that are a particular node's neighbors) can change over time, resulting in high variability in performances.

There is a complex relationship between node speed and pause time in this model, which can cause the average *nodal speed consistently decrease* over time [9]. Authors in [37] offer a modified solution which enables to

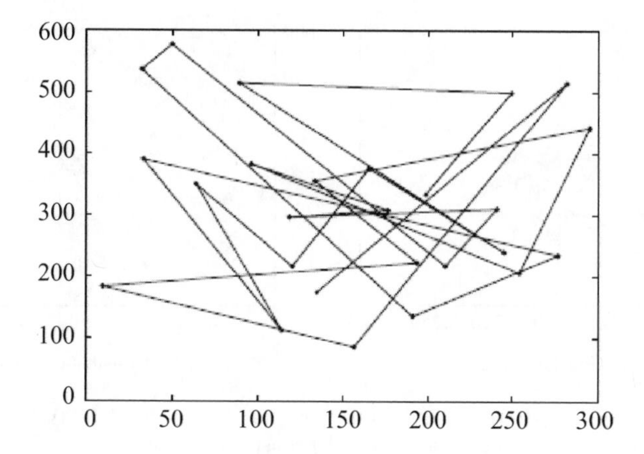

Fig. 7.4. Motion pattern of a node using the Random Waypoint Mobility Model

reach a *steady state* condition. Another general technique, based on renewal theory, which enables an accurate derivation of the steady state distribution function for node movement parameters (such as distance and speed), is proposed in [38]. The stationary behavior, which experiences the *density waves* in average number of neighbors, was only recently overcome in [11, 12, 38].

Random Direction A Random Direction Mobility Model forces the node to travel to the edge of the simulation area before changing a direction and speed (see Fig. 7.5), in order to alleviate the stationary mis-behavior of nodes in RWP, and promote a semi-constant number of neighbors.

The Random Direction Mobility Model is also a memory less mobility model. However, it differs from the RWP model in one critical issue, e.g., how the users choose the next segment to traverse. Under the RWP model, the user chooses a point (destination) within the space with equal probability and a speed from some given distribution. Both models operate in a finite two dimensional plane, usually a square. However, under the Random Direction Mobility Model, a user chooses a *direction* to travel in, and a *speed* at which to travel [2, 9]. The *boundary effect* (when nodes hits the boundary) is resolved either with *wrap around* or *reflection* (see previous subsection and [34]).

A RWP model can be considered as a modified Random Direction Mobility Model, if the nodes are not forced to travel to the simulation boundary before stopping to change the new direction and speed.

In a slightly modified version of Random Direction Mobility Model, the mobile node can choose random directions, without being forced to travel to simulation boundary before stopping and changing to new direction [32].

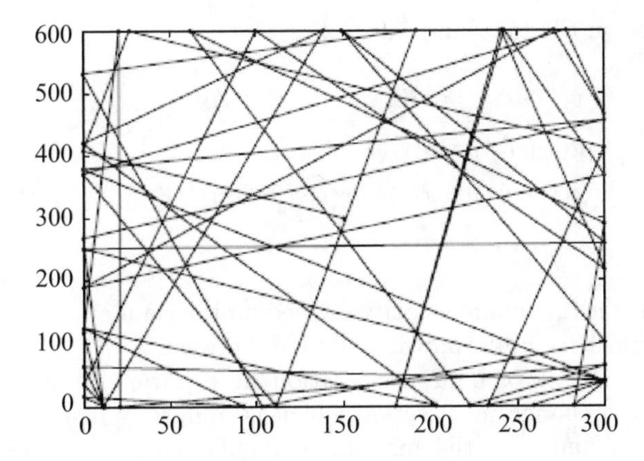

Fig. 7.5. Motion pattern of a node using the Random Direction Mobility Model

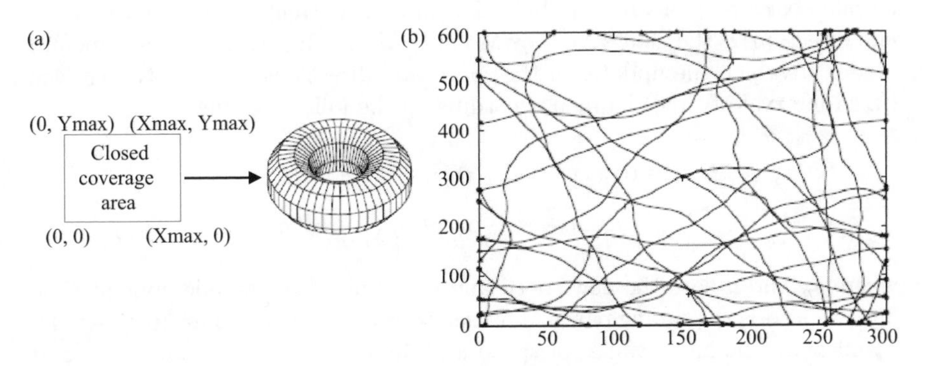

Fig. 7.6. Boundless Simulation Area Mobility Model: (a) mapping of simulation area to a torus; (b) motion pattern

A Boundless Simulation Area Mobility models that restrict nodes to move only within the boundary of a network (e.g., simulation area) have characteristic that, over time, nodes tend to converge towards the centre of the network area and thus present a skewed node distribution. To remove this limitations, Haas proposed the Boundless Simulation Area Mobility Model [9, 13], in which, once reaching its boundary, the node wraps around to the other side of the area. The rectangular simulation area is then mapped to a torus (see Fig. 7.6(a) and (b)), allowing the mobile node to travel unobstructed.

This model includes the dependences between the *previous direction* of travel and *speed* with the *current* ones [34]. The velocity vector $\overline{v} = (v, \theta)$ describes the node's velocity v and its direction θ, while the node position is represented as (x, y). The velocity vector and the position are updated every

Δt time steps, in the following way [32]:

$$v(t + \Delta t) = min[max(v(t) + \Delta v, 0), v_{max}];$$
$$\theta(t + \Delta t) = \theta(t) + \Delta\theta;$$
$$x(t + \Delta t) = x(t) + v(t)^* cos\theta(t);$$
$$y(t + \Delta t) = y(t) + v(t)^* sin\theta(t).$$

The v_{max} is the maximum velocity defined for the simulation, v is the change in velocity which is uniformly distributed between $[-A_{max}{}^*\Delta t, A_{max}{}^*\Delta t]$. The A_{max} is the maximum acceleration/deceleration of a given mobile node, $\Delta\theta$ is the change in direction which is uniformly distributed between $[-\alpha^*\Delta t, \alpha^*\Delta t]$, and α is the maximum angular change in the direction a mobile node is travelling.

Gauss-Markov Other model which considers the relationship between a mobile host's previous motion behavior and the current movement (in speed and direction) is the Gauss-Markov Mobility Model. It presents an incremental model, in which the updates of the speed and directions at the n-th instance (fixed interval of time) happen according to the following rule:

$$s_n = \alpha s_{n-1} + (1 - \alpha)\overline{s} + \sqrt{(1 - \alpha^2)} s_{x_{n-1}}$$
$$d_n = \alpha d_{n-1} + (1 - \alpha)\overline{d} + \sqrt{(1 - \alpha^2)} d_{x_{n-1}}$$

where s_n and d_n are the new speed and direction of the mobile node at time interval n, α is the tuning parameter used to vary the randomness ($0 \leq \alpha \leq 1$), $\overline{s}$ and $\overline{d}$ are the mean values of speed and direction as n$\rightarrow \infty$ and $s_{x_{n-1}}$ and $d_{x_{n-1}}$ are random variables which have Gaussian distribution. A node's next location is predicted (or generated) by its past location and velocity. To assure that the node is pushed away from the edge of a simulation area, the value for the mean direction $\overline{d}$ is changed according to a certain rule [9, 32] when the node approaches the edge within some distance. The travelling pattern of a node moving according to this model is depicted on Fig. 7.7. Depending upon parameters set, this allows modelling along a spectrum from Random Walk to Fluid-Flow. Total randomness is obtained for $\alpha = 0$ (e.g., Brownian motion), while linear motion is obtained for $\alpha = 1$ [9].

The Gauss-Markov model eliminates the sudden stops and sharp turns introducing correlation between the past and future velocities and directions. There are several variations of this model, which target different features (e.g., position equation) [9].

A Probabilistic Version of Random Walk A Probabilistic Version of Random Walk Mobility Model uses a transition probability matrix P, where each entry $P(a, b)$ represents the probability of node transition from state

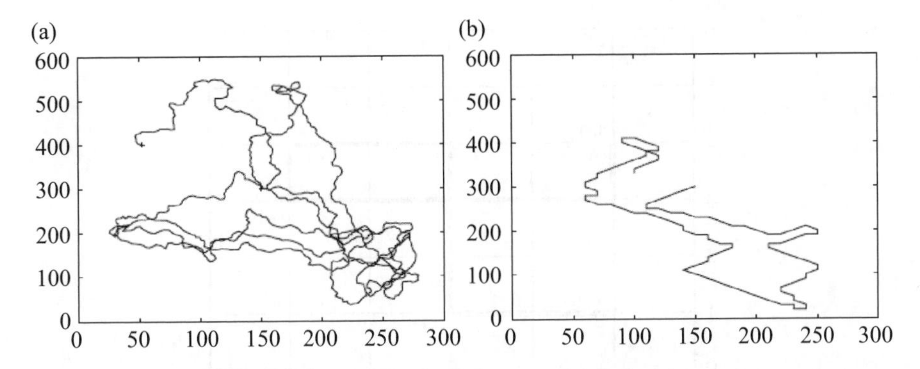

Fig. 7.7. Motion patterns of a node using: (a) Gauss-Markov Mobility Model; (b) a probabilistic Version of Random Walk

a to state *b* [9, 32]. Chiang in [11] proposed three different states: previous, present and future. In this model, the meaning of a "slot" is based on distance, not on time. This model produces probabilistic rather than purely random movements, resulting in more realistic movement patterns of nodes in ad hoc mobile networks (see Fig. 7.7(b)). Examples for similar movement behavior can be found in movement in *semi-constant direction* of people, during their daily activities. Unless the traces are available for a given movement scenario, it is difficult to choose appropriate values for $P(a,b)$ in individual simulations.

City Section Mobility Model All previously mentioned mobility models do not represent any type of real node motion. There have been some attempts to design mobility models that reflect more realistic scenario [31]. For instance, analyzing joint nodes' movements (as in group models), introducing obstacles and restriction of wireless transmission, or tracing the particular node (e.g., bus) trajectory are examples for such efforts.

The City Section Mobility Model provides realistic movements for a section of a city where the ad hoc networks exist [9]. It severely restricts the travelling behavior of the mobile nodes, forcing them to follow predefined paths and behavior guidelines (e.g., traffic laws). They cannot roam freely without regards to obstacles and traffic regulations. The travelling pattern of such movement is presented in Fig. 7.8.

In this model, the mobile node, after starting at a defined point at some street, chooses a new destination (also a point on some street), and travels towards it corresponding to the shortest time between the two points. Additional characteristics (e.g., speed limit, other traffic specifics) are usually taken in consideration. Enforcing all mobile nodes to follow predefined paths, the average number of hops increase compared to other models. Further improvements of this model are possible by including pause

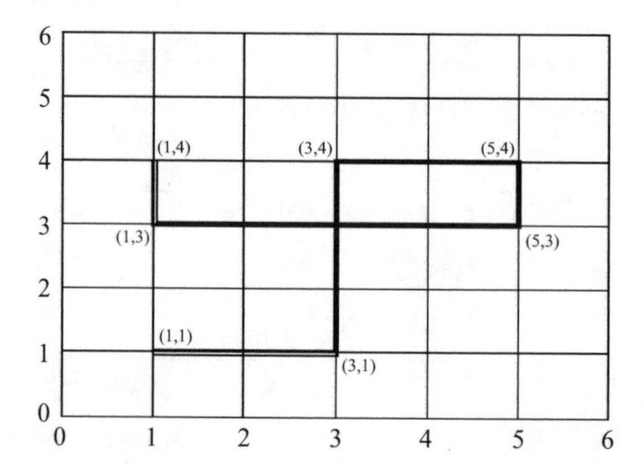

Fig. 7.8. Motion pattern of a node using the City Section Mobility Model

times at certain intersection and destinations, introducing acceleration and deceleration, daily-based time variations of node concentration in certain sectors, etc. Three different section models are defined (*city area, area zone* and *street unit* mobility models) by combining information for population, geographical area organized into regions and time (as input parameters) together with transportation theory [32]. Even more realistic approach can be achieved if the predefined paths and moving patterns are based on the actual street (town, country) maps [31]. A mobility models specified for *vehicular* ad hoc networks are using these approach [20, 21, 31, 39, 40].

They can be integrated to achieve a realistic simulation environment. However, these theoretical approaches can increase significantly computational efforts and complexity. The hierarchy of mobility models can additionally be increased in scale, so there exists metropolitan (METMOD), national (NATMOD) and international (INTMOD) mobility models [32].

Obstacle Mobility Model The Obstacle Mobility (OM) model aims to provide mechanisms for modelling movement in variety of real world environment where ad hoc networks deployment is expected (such as cities, campuses, highways, conferences and battlefields) [21, 22]. The OM model has been designed to model the movement of mobile nodes in terrains that resemble the real-world topographies (*modelling of terrain*). Unlike the *free-space* models, the OM model is considering the movement paths and obstructed signals due to natural and man-made obstacles. The *movement graphs,* defined as a set of pathways along which the mobile nodes move, are modelled by Voronoi diagram [41]. As a result the pathways tend to lie "halfway in between" adjusted buildings [21]. The OM models performs the *route selection,* using the shortest path routing policy to move the nodes between the locations in the movement graph. In addition, the OM model considers the *channel propagation* characteristics (diffraction, reflection, scattering, multi-path propagation and attenuation) in an obstructed environment. Variations of this model can be achieved with combination with group mobility models, in order to model realistic movements of group of nodes or to build the two-tier modelling approaches [24].

Mobility Vector Model The Mobility Vector Model [30], simulates natural and realistic mobility for various applications (especially in heterogeneous networks). In order to overcome the unrealistic movements in most random models (sudden stops, sharp turns, etc.), it allows only *partial changes* in the current mobility state. The mobility of a node is expressed by a vector (x_y, y_y), which represents the two dimensional velocity component. The scalar value of the mobility vector is the speed, presented as the distance between the current and the next position of a node after a unit time. The mobility vector $\vec{M} = (x_m, y_m)$ or (r_m, θ_m) is the sum of two sub-vectors: the *base vector* and the *deviation vector*. They are keeping the current mobility information and providing partial changing in motion. The base vector, $\vec{B} = (b_{x_v}, b_{y_v})$ or (r_b, θ_b), defines the major direction and speed of a node. A deviation vector, $\vec{V} = (v_{x_v}, v_{y_v})$ or (r_v, θ_v), stores the mobility deviations from the base vector. The model shows that $\vec{M} = \vec{B} + \alpha \times \vec{V}$, where α is the acceleration factor. The acceleration factor can be properly adjusted together with the variation of speed in the range of $[v_{min}, v_{max}]$. This can provide more realistic movement patterns, which is an important feature of this model. The Mobility Vector Model was used as a framework to other models (e.g., gravity model, location dependent model, targeting model, group mobility model) [33]. Besides these models, where movement of each object is independent from the others, there are mobility models that treat mobility of a group of objects. They provide smoothed and more uniform mobility pattern over a group of nodes. The following text presents some of these *group mobility models.*

7.3.1.2 Group Mobility Models

In ad hoc networks, there are many situations when the group of nodes, under some cooperation rules, is moving together. Examples of cooperative behavior of mobile nodes (users) are soldiers in some searching scenario, group of people moving together at a tourists' sightseeing, etc. A group mobility model is necessary to model these cooperative characteristics. The most general model is the Reference Point Group Mobility (RPGM) model. Column, Nomadic and Pursue model can be considered as special cases of the RPGM.

Exponential Correlated Random Mobility Model The Exponential Correlated Random Mobility Model is considered as one of the first group models which theoretically describes all the other mobility models. A *motion function* is used to create movements of a mobile node or group of nodes. Their position

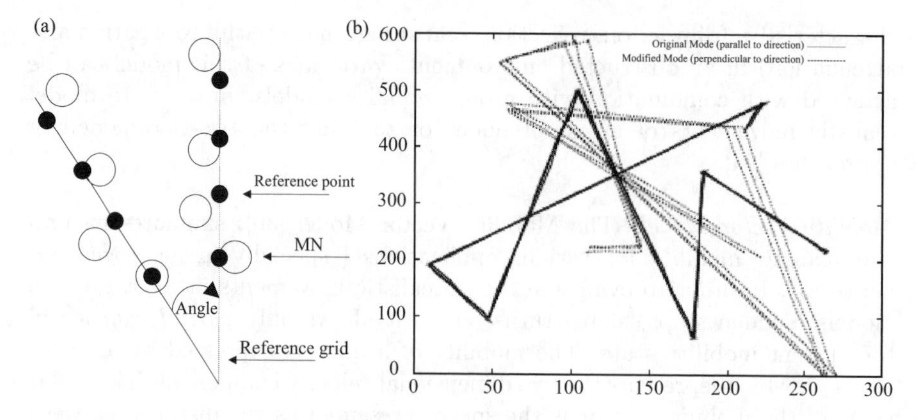

Fig. 7.9. Column Mobility Model (a) movements of nodes and (b) motion patterns

at time $t+1$, $\vec{b}\,(t+1)$ is defined with the position at time t, $\vec{b}\,(t)$:

$$b(t+1) = b(t)e^{-\frac{1}{\tau}} + (\sigma\sqrt{1-(e^{-\frac{1}{\tau}})^2})r$$

where τ reflects the range of change from the node's previous to its new location and r is a random Gaussian variable with variance σ. It is difficult to create a motion pattern for particular values of τ and σ.

Column Mobility Model Column Mobility model is used to present a movement of a group of nodes around a given line (column). An initial reference grid, which is forming a column of mobile nodes, defines the positions of the reference points. Each mobile node is placed in relation to its reference point in the reference grid. The mobile nodes are allowed to move randomly around the reference points according some entity mobility model (e.g., Random Walk Mobility Model). The new reference point for a given node is defined as:

$$new\_reference\_point = old\_reference\_point + advance\_vector$$

where *old reference point* is the previous reference point of a mobile node, and the *advance vector* is the predefined offset that moves the reference grid (calculated with a random distance and a random angle between $[0, \pi]$). The reference grid is a 1D line. The movement of a group of children into one line, or marching group of soldiers are examples of this movement patterns. The movements of mobile nodes using the Column Mobility Model and the appropriate motion patterns are depicted in Fig. 7.9(a) and (b).

Nomadic Community Model The Nomadic Community Model represents groups of mobile nodes that *collectively* move from one point to another [9].

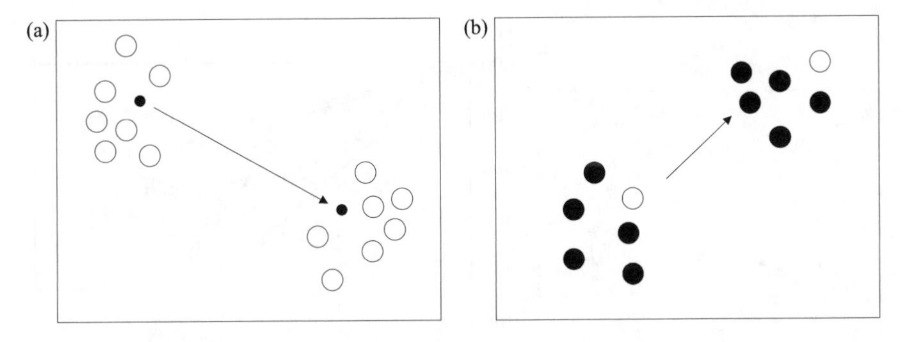

Fig. 7.10. Movements of nodes using (a) Nomadic and (b) Pursue Mobility Model

Within each group of mobile nodes, the individual nodes maintain their own personal spaces where they move in random ways according to an entity mobility model (e.g., the Random Walk Mobility Model) and roaming around the *reference point.* When the reference point changes, all mobile nodes in the group jointly move to the new area defined by the new common reference point (see Fig. 7.10(a)). Example of the Nomadic Community model comprises an organized group of visitors at the exhibition. They move together from one exhibition space to another. However, within the exhibition space, each of the visitors can move individually according to their interests.

Pursue Mobility Model The Pursue Mobility Model represents group of mobile nodes tracking a particular target. It consists of a single update equation for the new position of each mobile node:

$$\begin{aligned} new\_position =& old\_position + acceleration(target - old\_position) \\ &+ random\_vector \end{aligned}$$

where *acceleration(target − old_position)* is information of the movement of the mobile node being pursued and *random vector* is a random offset for each mobile node. The *random vector* is obtained via an entity mobility model (e.g., Random Walk Mobility Model). Fig. 7.10(b) illustrates six mobile nodes moving with the Pursue Mobility Model. The black node represents the pursued node while the white nodes represent the pursuing nodes.

Reference Point Group Mobility Model The Reference Point Group Mobility (RPGM) model represents the random motion of a group of mobile nodes allowing each node to move individually within the group [1, 9]. Each group defines a *logical centre* for the group that is used to calculate group motion via a *group motion vector*, GM (vector). The motion of the group centre completely characterizes the movement of its corresponding group of mobile nodes including their direction and speed. Individual nodes randomly move around

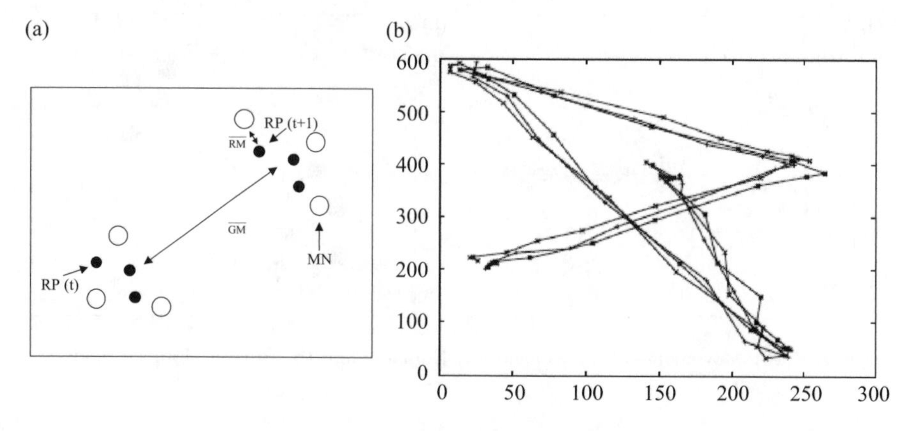

Fig. 7.11. RPGM Model: (a) movements of 3 nodes and (b) motion patterns of 1 group with 3 nodes

their own predefined reference points. The movement of these reference points depends on the group movement. Example of movements and motion patterns for group of 3 nodes are depicted in Fig. 7.11. The RPGM model can be used to model different mobility applications such as *in-place mobility model*, *overlap mobility model* and *convention mobility model*. As the most general mobility concept, with some restrictions and consideration towards the initial group locations, it can be used to illustrate different models (e.g., Column Mobility Model — [14, 15]). By proper selection of check point path and initial group location and parameters in the RPGM model, various mobility applications are modelled (InPlace Mobility Model, Overlap Model, Conventional Model, Flies on a Cake model, etc.).

Other Group-Based Mobility Models Combination of group mobility models with entity mobility models was proposed in [24] in order to enhance the realism in movement patterns. For instance, group of students moving towards school, before entering the building follows the group mobility behavior. However, once inside the building, each student goes individually towards the appropriate class room, acting according to some entity mobility pattern. The *two-tier mobility model* is based on correlation of mobility states and is more general comparing to other group mobility models.

In some applications (moving group of n nodes) the *worst case scenario* is of interest. The authors in [39] consider pedestrian mobility assuming a maximum speed v_{max} (*velocity bounded model*) and vehicular mobility assuming a maximum acceleration a_{max} (*acceleration bounded model*). They investigate the way to maintain persistent routes with nice communication network properties such as hop-distance, energy consumption, congestion and number of interferences.

The field of group mobility is an open issue. New applications and moving paradigms provoke intensive research and combination of various movement patterns adjusted to realistic scenarios.

7.3.2 Enriched Mobility Models

Lu et al. in [17] proposed different classification and tried to introduce a systematic mechanism for classification of mobility models proposing *enriched mobility models.* They identify the mobility models as being *recurring* or *terminal.* They introduce new features, such as *contraction, expansion* and *circling* and also differentiate between *single* and *hybrid* mobility models, trying to capture some characteristics of mobility neglected in the previous classifications. They also introduced new metrics, such as average node degree and link duration (see subsection 7.4.2) to compare the proposed mobility models.

Recurring Mobility Models In recurring models, the same synthetic model is repeated through the whole simulation without terminating or changing (in terms of parameters) [17]. It results in a not drastically changed performance on long term basis. Even though nodes may pause temporarily, mobility will not be terminated at a specific time point. Examples of these models are RWP, RPGM, Freeway and Manhattan [40].

Terminal Mobility Models In terminal mobility models, the rules of movement change over time meaning that some nodes may stop. The time-related characteristics of the performance may change significantly under these circumstances. Examples include contraction and expansion models.

Hybrid Mobility Models In hybrid mobility models, the movements of a node may switch from one mobility model to another, based on node's location in the network or on the simulation time. It can be illustrated by movement of a node, which could follow the RWP model in the simulation area and switch to Manhattan model after crossing the demarcation line. Other examples are *Hybrid Contraction & RWP model* and *Hybrid Manhattan & RWP model* [17].

In order to capture specific node behavior, authors in [17] proposed mobility models which emulated contraction, expansion or cyclic movements with respect to a logical point.

Contraction Mobility Model Contraction mobility model emulates the movement of mobile nodes towards a logical center from all directions (see Fig. 7.12(a)). The nodes are uniformly distributed in the area and move towards the center in a straight line. At the end of each time interval (e.g., 10s), a node calculates a new speed randomly within the interval $[1, v_{max} - 1]$. It may pause with respect to pause probability and maximum pause time.

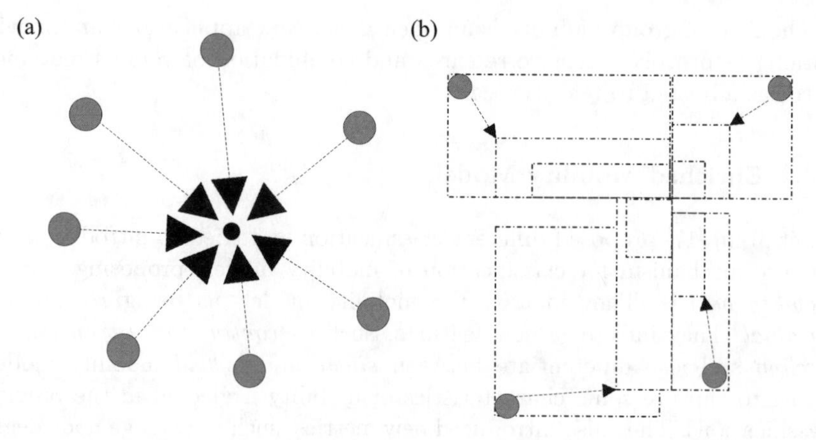

Fig. 7.12. Movements of nodes using (a) Contraction Mobility model and (b) Modified Contraction Mobility model

When nodes reach the center, they stop. An example of this movement pattern can be found in pursuit model. Also, similar behavior exhibit students gathering for class.

To avoid moving along straight lines, the *Modified Contraction Mobility model* (Fig. 7.12(b)) randomly selects the next destination of a node in the square defined by the current position and the center point. It results in node movement closer to the center in each interval, eventually reaching the center.

Expansion Mobility Model Opposite to the previously discussed enriched mobility models, the expansion mobility model emulates the movement towards the edges away from the center in a line (Fig. 7.13(a)). Initially, the nodes are uniformly distributed in the area. For each interval, the nodes move to next destination and randomly choose a speed within $[1, v_{max} - 1]$. Nodes may pause with a certain probability for some time. When nodes reach the edges, they stay there. Examples for such behavior can be found in emergency evacuations to a safe zone or students moving away from the class.

Circling Mobility Model Circling mobility model emulates movements of mobile nodes circling around the center (Fig. 7.13(b)). Each node has its unique radius value and circles around the center in a specific direction. For each time interval the nodes pick the next direction position and speed within $[1, v_{max} - 1]$. Each node can also pause with some probability when it reaches destination. Nodes circle around the same center through the simulation. Such movement behavior can be observed in cars circling around to park the car.

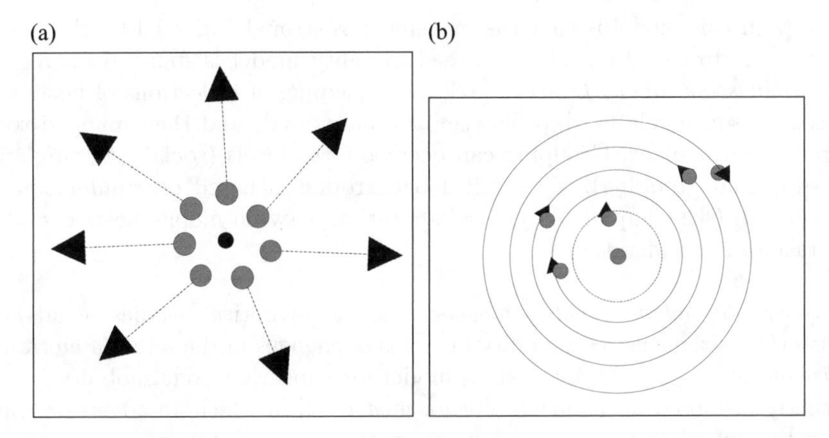

Fig. 7.13. Movements of nodes using (a) Expansion Mobility Model and (b) Circling Mobility Model

The movement patterns in reality are more complex and can not be emulated in a proper manner by a single mobility model. *Hybrid mobility models* combine different patterns providing more realistic mobility modelling. For instance, combination of RWP mobility outside the circle area and contraction model inside that area can model more realistically the movement of a student on campus. The combination of Manhattan (the node behavior on the street) and RWP (the node behavior inside the building) is a good example of modelling movement in a city.

Comparison study of proposed models showed that *contraction-based* models (contraction, modified contraction and RWP & contraction) lead to increase in average node degree and link duration resulting in increased throughput with increased velocity. The expansion model reduces the average node degree and link duration with increased velocity. Circling model has the maximum relative speed. However, it does not incur as much disconnections as does the expansion model.

7.3.3 Interdisciplinary Mobility Models

Some mobility models applicable to ad hoc networking stem from different areas (e.g., social theories). The movement of nodes and their individual or group behavior can show similarities with other areas entities' and/or groups' behavior which allows to implement different research mechanisms in investigation of mobility in ad hoc networks.

Social Network Theory Based Mobility Models In the absence of traced data, the proposal in [20] offers solution of a synthetic group mobility model. However, the synthetic models are unrealistic. The basic idea behind the social aspects in this model is that the movement is strongly affected by the needs of humans to socialize. The proposed mobility model is founded on recent results in *social network theory* [42]. The grouping of collections of hosts are based on social relationship between the individuals and then mapped to a topographical space. Grouping can occur on two levels (social grouping and geographical grouping). Theoretical background is based on *random graph theory* [43] (see Chapter 2). The collaboration between mobile hosts can also be taken in consideration.

Topology dependent mobility models Recent theoretical studies of ad hoc networks have measured mobility in terms of changes in the *underlying transmission graph* [38]. An interesting model for capturing node mobility is the recently proposed *adversarial network model* [44] in which an adversary may alter the underlying graph in an unpredictable manner. Arbitrary node movement can be represented by adversarial changes in topology. Topology-based mobility model can also use the idea of *kinetic data structures (KDSs)* [39, 45]. The mobility in these models is not characterized in terms of velocity and direction. In the proposed model, the mobility of objects is described by *pseudo-algebraic functions of time* and is fully or partially predictable. Another approach that captured unpredictable mobility is the concept of *soft kinetic data structures (SKDS)* [46]. KDS can be applied to investigate the cluster-based networks [39].

7.4 Relevant Metrics

Recent studies on mobility models focus on capturing mobility characteristics and metrics in order to build a framework for performance comparison. A set of meaningful metrics is introduced in order to measure the effect of mobility and better differentiate between the different mobility models. The mobility characteristics are application specific. They have a significant impact on the performance of ad hoc networking protocols (e.g., routing). It is important to define and choose appropriate metrics, which will reflect the mobility pattern in a most realistic way. The effect of mobility is also influenced by the traffic pattern, network topology and environment.

Various metrics were defined for comparison of particular mobility models. For instance, the efficiency of routing protocols is often measured by end-to-end throughput and control overhead, which depend on mobility of nodes [40].

Comparison can consider performance parameters such as data packet delivery ratio, end-to-end delay and average hop count, which achieve different values under different mobility models [9]. The results of numerous studies underline the importance of choosing the appropriate mobility model for the performance evaluation of a given ad hoc network.

Many new mobility models appear in order to achieve accurate modelling of realistic movements of entities in real world, and variety of individual and

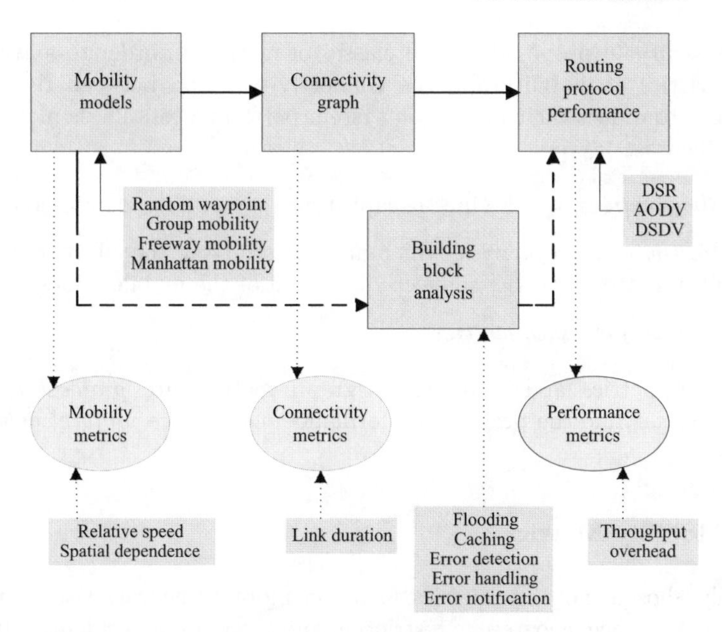

Fig. 7.14. IMPORTANT framework

group behaviors. They request different metrics which are capable to catch up specifics of movement patterns. Many other features (such as *relative velocity, spatial dependence, temporal dependence, geographical restrictions*, etc.) are used to distinguish among different mobility models. It is impossible to define a universal metric sufficient enough to capture the mobility for all possible scenarios. However, some of them (*average node degree* and *link duration*) are very useful regarding capturing mobility characteristics and dynamics [17].

In order to analyze the impact of mobility in a framework-based approach, Bai et al. proposed the IMPORTANT framework in [40]. The functional blocks concentrate different features (e.g., mobility models, mobility metrics, connectivity metrics, performance metrics, etc.) and provide a systematic approach to investigation of their interdependences. Fig. 7.14 depicts the general idea. Within this framework, the communicating traffic pattern consists of fixed, randomly chosen, source-destination pairs (with long enough session time). The major goal in this framework is to quantitatively and qualitatively analyze the impact of mobility on performance of unicast routing ad hoc protocols.

Another generic framework mobility model called Weighted WayPoint Mobility Model (WWP) is proposed in [18]. It is based on the RWP concept and highlights the impact of choosing "popular locations" as destination in a universal campus environment. It dedicates higher weights to popular places such as cafes, libraries, etc.

Authors in [40] proposed several classes of metrics in order to extract the characteristics of mobility and the connectivity graph between the mobile nodes and to reflect their impact on system performances. These metrics can be classified as:

- *Mobility metrics*, which aims at capturing the mobility characteristics;
- *Connectivity graph metrics,* which aims at studying the effect of different mobility patterns on the connectivity graph of the mobile nodes;
- *Protocol performance metrics.*

Mobility metrics and connectivity graph metrics are *protocol independent metrics*, while the protocol performance metrics are *protocol dependent metrics* [47].

7.4.1 Mobility Metrics

Carefully chosen mobility metrics focus on mobility characteristics such as spatial dependence, geographic restriction and temporal dependence [40]. The *spatial dependency* expresses the influence and correlation among nodes in neighborhood. The *temporal dependency* expresses the velocity dependences with time constraints, while the *geographic restrictions* define the possible restricted movements (such as motion along the street or highway) and established boundaries. Additional mobility metric (*relative speed* [48]) should differentiate mobility patterns based on relative motion. The following text gives the formal metrics definition of most relevant mobility metrics.

Degree of spatial dependence:

$$D_{spatial}(i,j,t) = RD(\vec{v_i}(t),\, \vec{v_j}(t)) * SR(\vec{v_i}(t),\, \vec{v_j}(t))$$

where $\vec{v_i}(t)$ and $\vec{v_j}(t)$ are the velocity vectors of nodes i and j at time t. *Relative direction (RD)* and the *speed ratio (SR)* between the two velocity vectors are defined as:

$$RD(\vec{v_i}(t), \vec{v_j}(t)) = \frac{\vec{v_i}(t) \cdot \vec{v_j}(t)}{|\vec{v_i}(t)| * |\vec{v_j}(t)|}$$

$$SR(\vec{v_i}(t), \vec{v_j}(t)) = \frac{min\, |\vec{v_i}(t)|\, ,\, |\vec{v_j}(t)|}{max\, |\vec{v_i}(t)|\, ,\, |\vec{v_j}(t)|}$$

$D_{spatial}(i,j,t)$ expresses the similarity of the velocities of two nodes that are not too far apart. Its value is high when the nodes i and j travel in more or less the same direction, and in almost similar speed. The $D_{spatial}(i,j,t)$ decreases if the relative direction or the speed ratio decreases. Distant nodes (out of transmission range) are excluded. The averaged value of $D_{spatial}(i,j,t)$ over node pairs and time instant satisfying certain conditions gives the *average degree of spatial dependence.*

Degree of temporal dependence:

$$D_{temporal}(i,t,t') = RD(\vec{v_i}(t),\ \vec{v_i}(t'))^{*}SR(\vec{v_i}(t),\ \vec{v_i}(t'))$$

This metric expresses the similarity of the velocities of a node at two time slots that are not too far apart, similarly to $D_{spatial}$ *(i,j,t)*. It has high value if the node travels in more or less same direction and almost the same speed over a certain time interval (which is limited). This metric can also be averaged over nodes and time instants (satisfying certain conditions) resulting in similar *average* metric [40].

Relative speed:

$$RS(i,j,t) = |\vec{v_i}(t) - \vec{v_j}(t)|$$

This metric expresses the speed of relative motion and is limited with the following condition:

$$D_{i,j}(t) > c * R => RS(i,j,t) = 0$$

Here, c is constant, R is a transmission range of a mobile node, and *RS(i,j,t)* is the average relative speed, averaged over node pairs and time instants satisfying certain condition:

$$\overline{RS} = \frac{\Sigma_{i=1}^{N}\,\Sigma_{j=1}^{N}\,\Sigma_{t=1}^{T} RS(i,j,t)}{P}$$

where P is the number of triples *(i,j,t)* such that $RS(i,j,t) \neq 0$.

Numerous analyses involve some of these metrics in performance differentiation of ad hoc networking protocols under appropriate mobility model. As illustration, Fig. 7.15(a) and (b) show the appropriate values for spatial and temporal dependences for different mobility models as a function of maximum node speed. It is clear that these metrics are capable of differentiating the analyzed mobility. So, expansion and contraction mobility methods have high spatial dependence, while hybrid and circular have medium to low. The contraction and expansion model exhibit medium dependencies, while Manhattan&RWP model exhibits the strongest temporal dependencies. Understanding the mobility behavior and its consequences is a powerful tool in protocol design for appropriate ad hoc networks and application scenarios.

Geographic restrictions metrics:
This metric is not quantitatively well defined. Geographic restriction in [40] is expressed through *degree of freedom of a point,* which is the number of

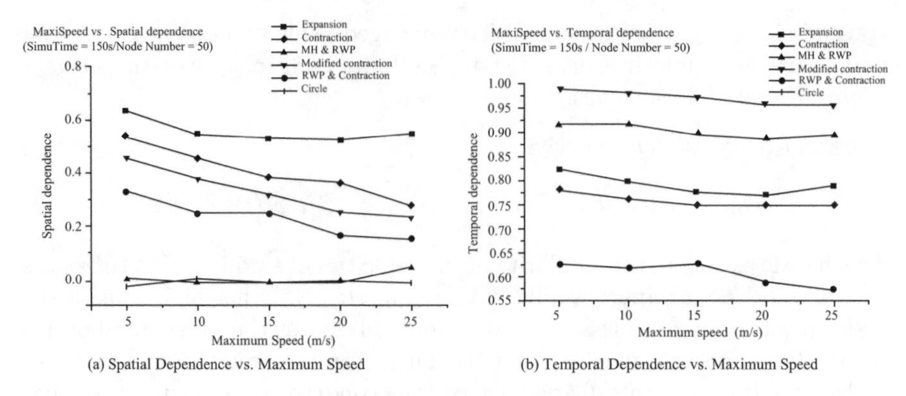

(a) Spatial Dependence vs. Maximum Speed

(b) Temporal Dependence vs. Maximum Speed

Fig. 7.15. (a) Spatial and (b) Temporal dependence vs. Mobility

directions a node can move after reaching that point and is usually connected to a map (e.g., in Manhattan and Freeway mobility model).

Other metrics:
Kwak et al. in [49–51] propose a mobility measure for mobile ad hoc networks through *remoteness function*. As a simplest case, this function represents the distance between two nodes at time t. For ad hoc mobile nodes, this definition includes the transmission range of a node and different weights to the movements of the nodes near the communication range. It is flexible and consistent for a wide range of scenarios, and provides a unified means of measuring the degree of mobility. It has the same linear relationship to the link change rate for a wide range of mobility scenarios (mobility model, physical dimension of network and the number of nodes). The flexibility of the scheme makes it useful for more different applications (e.g., MANET).

Intensive ongoing research of mobility pattern in new networking scenarios, invents metrics trying to catch up the movement specifics and provide mechanisms to investigate the impact of mobility on different networking features.

7.4.2 Connectivity Graph Metrics

In ad hoc wireless networks with dynamic network topology, it is useful to have metrics which can analyze the effect of mobility on the connectivity graph between the mobile nodes (see Chapter 2). The connectivity graph metrics reflects this effect and helps in relating the mobility metrics with protocol performance.

The *connectivity graph* is the graph $G = (V, E)$ such that $|V| = N$ and at time t, a link $(i, j) \in E$ if $D_{i,j}(t) \leq R$. Here N is the number of nodes and $D_{i,j}(t)$ is the Euclidean distance between nodes i and j at time t. The *indicator random variable X(i,j,t)* has value 1, if there is a link between nodes i and j at time t, otherwise it is zero. The value $X(i,j,t) = \max_{t=1}^{T} X(i,j,t) = 1$ indicates that a link existed between nodes i and j at any time during simulations. Based on these definitions, several connectivity graph metrics are defined: number of link changes and average number of link changes, link duration and average link duration and path availability and average path availability.

For instance, the *number of link changes f*or a pair of nodes i and j reflects the number of transitions between "on" and 'off' status and is defined as [40]:

$$LC(i,j) = \sum_{t=1}^{T} C(i,j,t)$$

Here *C(i,j,t)* is an indicator random variable such that *C(i,j,t)=1* if *X(i,j,t-1)*= 0 and *X(i,j,t)*= 1, reflecting the situation when the link between nodes i and j is "off" at time *t-1* and becomes "on" at time t. The *average number of link changes* is calculated with averaging over node pairs satisfying certain conditions, i.e.,:

$$\overline{LC} = \frac{\Sigma_{i=1}^{N} \Sigma_{j=i+1}^{N} LC(i,j)}{P}$$

Similar formulations are used for the other mentioned metrics. So, the *link duration* expresses the average duration of the link existing between nodes i and j, while *averaged link duration* is obtained with averaging over node pairs satisfying certain conditions. The *path availability* expresses the fraction of time during which a path is available between two nodes i and j. More details about these metrics definition can be found in [17, 40].

Node degree (ND) is defined as the number of neighbor nodes averaged over the total number of nodes (N) and in every time instant [17]. Two nodes are neighbors if they are in transmission range R. The node degree is given by:

$$ND = \frac{\Sigma_{t=1}^{T} \Sigma_{i=1}^{N} N(i,t)}{T \cdot N}$$

where T is the total simulation time.

Another approach to adapt to changes in network topology is proposed in [52]. A translation of the mobility model to the resulting topology dynamics is achieved through *topology change rate (TCR)*, defined as the number of link changes per time unit as observed by a single node. Calculated analytical value for TCR for different mobility models gives a possibility to compare different scenarios or to select the parameter of mobility model in a way that a specific TCR is obtained. It gives the possibility to generate a sequence of mobility

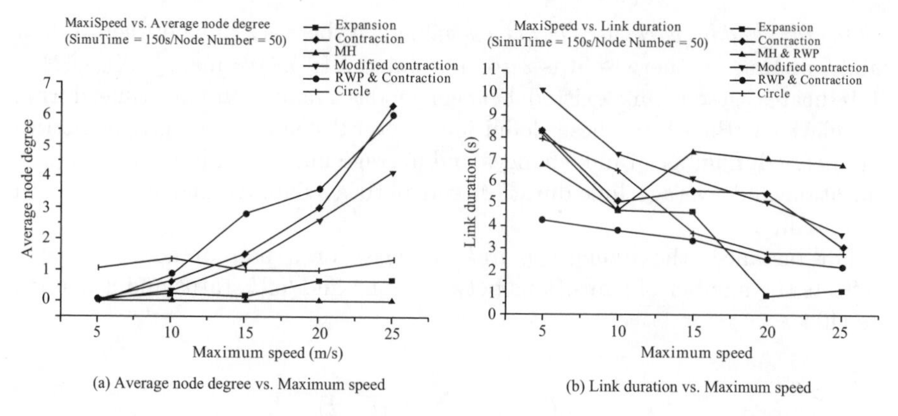

Fig. 7.16. (a) Average node degree and (b) Average link duration vs. mobility

patterns with a pre-described network dynamics and can contribute to building blocks towards a systematic understanding of the impact of mobility on ad hoc network performance [40].

The efficiency and analysis of proposed metrics is usually completed through appropriate simulations [6, 17–19, 40, 53]. As an illustration, the dependence of average node degree and average link duration on nodes mobility is depicted on Fig. 7.16(a) and Fig. 7.16(b), respectively [17].

It is obvious how the mentioned metrics differentiate between the different mobility models involved.

7.4.3 Protocol Performance Metrics

These metrics are protocol dependent. For instance, the performance of routing protocols is evaluated using metrics such as *throughput* and *routing overhead* [1, 9, 17]. Fig. 7.17(a) and (b), [1], show the throughput as a function of mobility for two different routing protocols: DSDV and HSR (see Chapter 3). Several mobility models are considered. It is obvious that the performance metrics is strongly dependent on the actual routing protocol and behaves differently for different mobility models.

Similar evaluations can be done for other performance metrics. Different performance metrics (e.g., control overhead [1, 9], average end-to-end delay and average hop count [9], packet delivery fraction [54]) depend differently on involved networking protocols. Intensive on-going research activities evaluate various routing protocols across different mobility models in order to develop further insight and better understanding of realistic scenarios. The development of trace mobility models, based on actual collected mobility traces, can further contribute to that process [1].

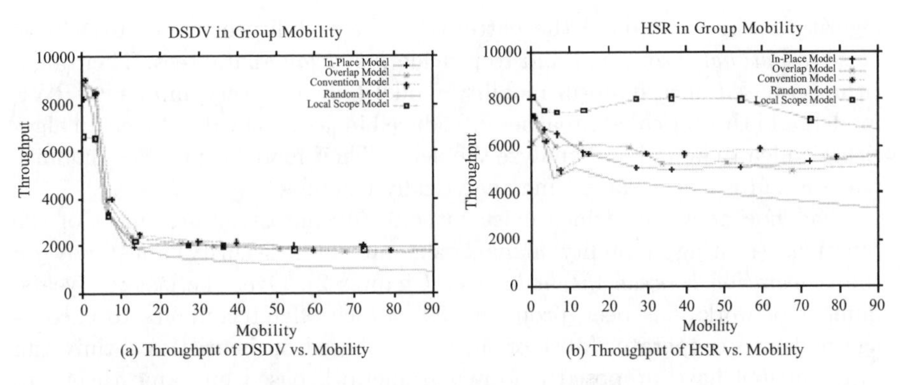

Fig. 7.17. (a) Throughput of DSDV and (b) Throughput of HSR vs. mobility

7.5 Impact of Mobility

The nodes' mobility has impact on various aspects of ad hoc networking. These influences are widely investigated through numerous theoretical and simulation studies [55–77]. The previous subsection shows the impact that mobility has on performance metrics and how it affects different routing protocols. Routing is the most eminent area in ad hoc networking. Performances of routing protocols are very sensitive to underlying mobility pattern [2]. Mobility environment also has a strong impact on QoS in ad hoc networks [55]. There are separate chapters in this book dedicated to routing protocols in ad hoc networks (see Chapter 3) and QoS (see Chapter 5). The following subsection presents some examples where mobility affects particular networking features such as capacity and delay, cooperativeness and scalability, multicasting and clustering, energy conservation, security and query resolution.

7.5.1 Impact on Capacity and Cooperation

Understanding the capacity limits in ad hoc networks took different direction since Gupta's and Kumar's work [56] who showed that the average available throughput per node decreases as the square root of the number of nodes n in a static ad hoc network (see Chapter 2). Equivalently, the total network capacity increases at most as $\sqrt{n}$. Their results hold quite generally, irrespective of the network topology, power control policy or any transmission scheduling strategy. In a network where nodes move randomly in a circular disk (with uniform steady state distribution), Grossglauser and Tse [57], showed that it is possible for each sender-receiver node pair to obtain a constant fraction of the total available bandwidth (independent of the number of node pairs). However, their scheme does not provide the delay guarantee. Bansal and Liu in [58] present a theoretical framework and propose a routing algorithm which exploited the patterns in the mobility of nodes to achieve *close-to-optimal throughput* and to provide *low delay* guarantees. Their algorithm is based on a uniform mobility model (which is very similar to RWP model) and theoretical approaches introduced in [57]. They developed an algorithmic idea to bound and analyze the delay. Their model captures a mobility pattern and is at the same time analytically tractable.

In ad hoc networks, which exist without any infrastructure, most of the functions (routing, mobility management and even security) must rely on the *cooperation between the nodes* (see Chapter 2). Over the last few years, number of works has been proposed [59], which offer techniques to encourage nodes to cooperate. Most of them were based on heuristics. Srinivasan et al. in [60] have proposed a formal framework based on game theory to study cooperation concerning energy-efficiency. They identified the conditions under which cooperation is a Nash equilibrium (meaning that none of the nodes can increase its utility by unilaterally changing its strategy). Felegyhazi, Hubaux and Buttyan investigated the conditions for the existence of cooperation, showing that cooperative Nash equilibrium is much more likely to happen with mobile than with static nodes [59]. They investigated the effect of mobility by introducing a novel mobility metric called *duration of a step* and applied the RWP model in ad hoc environment. The node cooperation with the network was measured through *generosity* of the node. They have concluded that the absence of mobility is a major hurdle for spontaneous cooperation.

Papadopouli and Schulzrinne in [61] were focused on how user mobility patterns affect the *spread of information* in specific situations in a dense urban area. They have investigated pedestrian on the streets (using the RWP model) and a subway environment. The information propagated according to the epidemic model. They have described and evaluated a system, 7DS (seven degrees of separation, a variation of six degrees of information, see Chapter 2), that facilitates the information exchange for mutual benefits of the users. 7DS allows mobile users to quickly obtain popular data items, even if most users are disconnected from wide-area network infrastructure. They have estimated how long a mobile user has to wait before obtaining a desired piece of information. This delay depends strongly on popularity of the data and the frequency with which mobiles query for the data.

7.5.2 Impact on Scalability and Clustering

Scalability issues in ad hoc networks (MANETs) have been typically addressed using hybrid routing schemes operating in the hierarchical network architecture [35, 62]. The impact of mobility on temporal stability of the clusters is obtained using a mobility-aware clustering framework [63]. The authors come to conclusions about how the mobility (e.g., random motion, group mobility) affects the temporal stability of the identified clusters. They also propose a predictive clustering scheme which is able to adapt to varying network conditions by dynamically adjusting the cluster size.

McDonald and Znati in [64] present a novel framework for dynamically organizing mobile nodes in wireless ad hoc networks into clusters in which the probability of *path availability* can be bounded. The expression for path availability, as a function of time, is derived for a particular mobility model. Since the criteria for cluster organization depends directly upon path availability, the structure of the cluster topology is adaptive with respect to node mobility (single and group). So, this framework supports an adaptive hybrid routing architecture that can be more restrictive and effective when mobility rates are low and more efficient when mobility rates are high.

Large ad hoc networks mostly consist of nodes which differ in resources and transmission ranges. They are not equally involved in network operations such as packet forwarding, flooding, etc. [31] proposes a backbone protocol (B-protocol), in which a small subset of nodes are selected to form a backbone based on their status. The simulation result shows the efficiency of B-protocol in case of node mobility and node/link failure. The number of selected backbone nodes and the links between them in the backbone are considerably smaller. The link stability under different mobility patterns is investigated in [65].

7.5.3 Impact on Multicasting

Multicasting is an important communication paradigm in ad hoc networking [66]. Effect of mobility on multicast routing is investigated in [47]. Pandey and Zappala observe flooding [67], ODMPR [68] and ADMR [69] routing protocols under different mobility models (RWP, uniform, Manhattan, Exhibition and Battlefields) using techniques that have been demonstrated to be effective for unicast routing protocols (e.g., IMPORTANT framework). To better differentiate between mobility models, two new metrics, *reachability* and *node density*, were introduced. Results show that lower node density and low spatial dependence degrade the throughput. Higher node density, which is typically exhibited by group-based mobility patterns, can lead to lower delay and lower transmission overhead. However, the group mobility pattern can cause the partitioning [1, 70]. ASMR can provide good throughput even at high speeds, under low level of traffic. Under high level of traffic, ADSM experiences congestion collapse. It is challenging to design a multicast protocol that can achieve good throughput during period of high mobility.

7.5.4 Impact on Energy Conservation

Providing routing mechanisms is the fundamental issue in ad hoc networks. The performance of an ad hoc routing protocol is highly dependent on the mobility model beneath. Moreover, the mobile nodes are often power constrained. Most of the studies done for energy conservation are focused on metrics such as packet delivery ratio, delay and route optimality, interference. The impact of mobility on energy conservation of ad hoc routing protocols (AODV, DSR, TORA and DSDV, see Chapter 3) under three different mobility models (RWP, RPGM and Manhattan) is investigated in [71–74]. The conclusions could be useful when deploying ad hoc networks with power constraints. Results show that reactive protocols are more speed sensitive. In highly dynamic environment, proactive protocols save more power. For group motion, on-demand protocols perform better in terms of energy conservation. DSR performs best among the evaluated protocols, except for high speed cases in Manhattan model. Flooding is very expensive in energy cost and reducing of flooding approach is an important issue in designing of energy conserving protocols for ad hoc networks.

7.5.5 Impact on Security

Contrary to belief that mobility makes security more difficult, authors in [75] show that it can be useful to provide security in ad hoc networks (see Chapter 8) [76]. The security association established between the nodes in vicinity can help exchange appropriate cryptographic material. They propose a generic technique, which can be applied to either self-organized ad hoc network or to ad hoc networks placed under an (off-line) authority. Their security mechanisms can work in any network configuration. However, the time necessary to set up the security associations is strongly influenced by the size of deployment area, the mobility patterns and the number of "friends". The new mechanisms (friends), which quantify the benefit of mobility, support the building of security associations, even between nodes that do not meet physically. The security aspect in ad hoc networks is observed in more details in Chapter 8.

7.5.6 Impact on Query Resolution

Query resolution and resource discovery is an important issue in large scale ad hoc networks (e.g., sensor networks). Helmy in [77] proposes an architecture which is based on the concept of small world (see Chapter 2) and introduces concept of *contact*. Contacts are initially chosen from nearby neighbors. As they move away, they discover new neighbors and hence become more effective in query resolution. Helmy develops mechanisms for efficient, mobility assisted, contact selection which is totally distributed and self-organized. Results show drastic improvement obtained by using contacts, especially in high mobility scenarios. In fact, queries utilize mobility.

7.6 Projects and Implementations

Mobility in ad hoc environment is extremely important issue that strongly influenced the system performances. It was a target research area in several projects. Mobility must be taken in consideration in order to achieve realistic results or even extract some benefits out of nodes movement patterns. The following text lists some of these projects, pointing on their major research considerations.

Realistic Mobility Modelling for Mobile Ad Hoc Networks [78]. The objective of this project is to provide better mobility modelling for practical research in mobile ad hoc networking. The project provides a realistic mobility model based on a vehicular traffic. It makes use of the publicly available TIGER (Topologically Integrated Geographic Encoding and Referencing) database from the U.S. Census Bureau, giving detailed street maps for the entire United States of America, and model automobile traffic on these maps.

Obstacle Mobility Model [79]. This project proposes a creation of more realistic movement models through the incorporation of obstacles. These obstacles are utilized to both restrict node movement as well as wireless transmissions. In addition to the inclusion of obstacles, movement paths using the Voronoi diagram of obstacle vertices are constructed. Nodes can then be randomly distributed across the paths, and can use shortest path route computations to destinations at randomly chosen obstacles.

Mobility Models for Mobile Ad Hoc Network Simulations [80]. This project has two related goals. The first is to develop realistic mobility models from movement data collected on wireless users and the second is to develop stationary distributions of mobility models used in mobile ad hoc network simulations.

STRAW [81]. Unlike many other mobile ad-hoc environments where node movement occurs in an open field (such as conference rooms and cafés), vehicular nodes are constrained to streets often separated by buildings, trees or other objects. Street layouts and different obstructions increase the average distance between nodes and, in most cases, reduce the overall signal strength received at each node. It is argued that a more realistic mobility model with the appropriate level of detail for vehicular networks is critical for accurate network simulation results. The STRAW (STreet RAndom Waypoint) mobility model addresses this issue by constraining node movement to streets defined by map data for real US cities and limit their mobility according to vehicular congestion and simplified traffic control mechanisms.

Mobility Management and Resource Allocation in Wireless Networks [82]. This project deals with two central issues in Multimedia Mobile Computing, namely Mobility Management and Connection Admission. More specifically, the project asks the questions of how to efficiently locate users within massively accessed internet.

The approach is based on the idea of "User Mobility Profile" (UMP). As a user roams within an internet, the system continuously collects information on the users' mobility and connection patterns. This information is used to anticipate the future location of the active users, thus, reduces the search space and, consequently, the cost of locating the user.

DAVIS Smart Mobility Modeling Project: Initial Scoping and Planning Project [83]. The goal of the Davis Smart Mobility Model project is to optimize individual mobility options on the University of California (Davis) campus through improved connectivity among nodes, enhanced techniques to link land-use planning and transportation system design, advanced information technologies and clean-fuel vehicles. The project learns how innovative mobility services and technologies (such as carsharing and smart parking management) might help to alleviate the transportation impacts of a campus expansion, expected to result in the arrival of more than 9,000 additional students, staff and faculty in the coming decade.

AGILE: Architectures for Mobility [84]. The AGILE project provides means for addressing the novel level of mobility introduced complexity in communication and distribution by developing an architectural approach in which mobility aspects can be modelled explicitly and mapped on the distribution and communication topology made available at physical levels. The whole approach is developed over a uniform mathematical framework based on graph-oriented techniques that support sound methodological principles, formal analysis and refinement.

7.7 Concluding Remarks

The previous sections support the conclusions on the importance of mobility modelling highlighting the impact of mobility on different network protocols and parameters in ad hoc networks. The search for more realistic movement patterns results in variety of mobility models ranging from simple Random Walk Mobility Model [10] to sophisticated Obstacle Model [21, 22] and hybrid models [17], and from entity [9] to group mobility models [1, 70]. Number of combinations occurred in order to combine the benefits of different approaches (e.g., two-tiered models [24]). Existing and novel metrics are used to compare analytical or simulation modes and achieving performances. Mobility models for ad hoc networks are still in focus of worldwide research community paving the path towards better understanding and more realistic approach to network design towards seamless communications.

References

[1] Hong, X., Gerla, M., Pei, G., and Chiang, C. -C., "A Group Mobility Model for Ad Hoc Wireless Networks," *ACM MSWiM'99*, August 1999.

[2] Bettstetter, C., "Mobility Modeling in Wireless Networks: Categorization, Smooth Movement and Border Effects," *ACM Mobile Computing and Communications Review*, 5(3), 2001, pp. 55–69.

[3] Groenevelt, R. B., Altman, E., and Nain, P., "Relaying in Mobile Ad Hoc Networks: The Brownian Motion Mobility Model," *INRIA Technical Report RR-5311*, September 2004.

[4] Lei, Z. and Rose, C., "Wireless Subscriber Mobility Management Using Adaptive Individual Location Areas for PCS Systems," *IEEE ICC'98*, Atlanta, USA, June 1998.

[5] Guerin, R. A., "Channel Occupancy Time Distribution in a Cellular Radio System," *IEEE Transaction on Vehicular Technology*, 36, August 1987, pp. 89–99.

[6] Le Boudec, J. -Y. and Vojnovic, M., "Perfect Simulation and Stationarity of a Class of Mobility Models," *IEEE INFOCOM 2005*, Miami, USA, March 2005.

[7] ETSI, "Selection procedures for the choice of radio transmission technologies of the UMTS (UMTS 30.03, version 3.2.0)," *Technical Report*, European Telecommunication Standards Institute, April 1998.

[8] Boche, H. and Jugl, E., "Extension of ETSI's Mobility Models for UMTS In Order to Get More Realistic Results," *UMTS Workshop*, Gunzburg, Germany, November 1998.

[9] Camp, T., Boleng, J., and Davies, V., "A Survey of Mobility Models for Ad Hoc Network Research," *Wireless Communication & Mobile Computing (WCMC): Special issue on Mobile Ad Hoc Networking: Research, Trends and Applications*, 2 (5), 2002, pp. 483–502.

[10] Sun, Y., Belding-Royer, E. M., and Perkins, C. E., "Internet Connectivity for Ad Hoc Mobile Networks," *International Journal of Wireless Information Networks*, 9, April 2002.

[11] Chiang, C., "Wireless Network Multicasting," Ph.D. Thesis, University of California, 1998.

[12] Lassila, P., Hyytia, E., and Koskinen, H., "Connectivity Problems of Random Waypoint Mobility Model for Ad Hoc Networks," *4th Annual Med-Hoc-Net Workshop*, Ile de Porquerolles, France, June 2005.

[13] Haas, Z. J., "A New Routing Protocol for the Reconfigurable Wireless Networks," *IEEE 6th International Conference on Universal Personal Communications (ICUPC'97)*, 1997.

[14] Sanchez, M., *Mobility Models*, 2001, Information available at: http://www.disca.upv.es/misan/mobmodel.htm

[15] Sanchez, M. and Manzoni, P., "A Java Based Simulator for Ad Hoc Networks," 2001, Information available at: http://www.scs.org/confernc/wmc99/errata/websim/w408/w408.html

[16] Zhou, B., Xu, K., and Gerla, M., "Group and Swarm Mobility Models for Ad Hoc Network Scenarios Using Virtual Tracks," *IEEE MILCOM 2004*, Monterey, USA, 2004.

[17] Lu, Y., Lin, H., Gu, Y., and Helmy, A.,"Towards Mobility-Rich Analysis in Ad Hoc Networks: Using Contraction, Expansion and Hybrid Models," *IEEE ICC'04*, June 2004.

[18] Hsu, W. -j., Merchant, K., Shu, H. -w., Hsu, C. -h., and Helmy, A., "Weighted Waypoint Mobility Model and Its Impact on Ad Hoc Networks," *ACM MobiCom'04*, Philadelphia, USA, 2004.

[19] Hsu, W.-j., Merchant, K., Shu, H. -w., Hsu, C. -h., and Helmy, A., "Preference-based Mobility Model and the Case for Congestion Relief in WLANs using Ad hoc Networks," *IEEE VTC 2004 Fall*, Los Angeles, USA, September 2004.
[20] Vieira, P., Vieira, M., Queluz, M. P., and Rodrigues, A., "A New Realistic Vehicular Mobility Model for Wireless Networks," *8th International Symposium on Wireless Personal Multimedia Communications — WPMC 2005*, Aalborg, Denmark, September 18–22, 2005.
[21] Jardosh, A. P., Belding-Royer, E. M., Almeroth, K. C., and Suri, S., "Real-World Environment Models For Mobile Network Evaluation," *Journal on Selected Areas in Communications special issue on Wireless Ad hoc Networks*, March 2005, pp. 622–632.
[22] Jardosh, A., Belding-Royer, E. M., Almeroth, K. C., and Suri, S., "Towards Realistic Mobility Models For Mobile Ad hoc Networks," *ACM MobiCom'03*, San Diego, USA, September 2003.
[23] Le Boudec, J. -Y. and Vojnovic, M., "Perfect Simulation and Stationarity of a Class of Mobility Models," *IEEE INFOCOM 2005*, Miami, USA, March 2005.
[24] Zaidi, Z. R., Mark, B. L., and Thomas, R. K., "A Two-Tier Representation of Node Mobility in Ad Hoc Networks," *IEEE SECON 2004*, Santa Clara, USA, October 2004.
[25] Lam, D., Cox, D. C., and Widom, J., "Teletraffic Modeling for Personal Communication Services," *IEEE Communications Magazine*, 35, August 1997, pp. 79–87.
[26] Bar-Noy, A., Kessler, I., and Sidi, M., "Mobile Users: To Update or not to Update?," *ACM/Baltzer Wireless Networks*, 1 (2), 1995.
[27] Akyildiz, I. F., Lin, Y. -B., Lai, W. -R., and Chen, R. -J., "A New Random Walk Model for PCS Networks," *IEEE Journal on Selected Areas in Communications*, 18 (7), 2000, pp. 1254–1260.
[28] Kim, T. S., Chung, M. Y., and Sung, D. K., "Mobility and Traffic Analyses in Three-Dimensional PCS Environments," *IEEE Transactions on Vehicular Technology*, 47, May 1998, pp. 537–545.
[29] Kim, T. S., Kwon, J. K., and Sung, D. K., "Mobility and Traffic Analysis in Three-Dimensional High-Rise Building Environments," *IEEE Transactions on Vehicular Technology*, 49, May 2000, pp. 1633–1640.
[30] Hong, X., Kwon, T. J., Gerla, M., Gu, D. L., and Pei, G., "A Mobility Framework for Ad Hoc Wireless Networks," *Lecture Notes in Computer Science*, Vol. 1987, pp. 185–196.
[31] Saha, A. K. and Johnson, D. B., "Modeling Mobility for Vehicular Ad Hoc Networks," *ACM VANET 2004*, Philadelphia, USA, October 2004.
[32] Davies, V. A., "Evaluating Mobility Models within an Ad Hoc Network," Master Thesis, Colorado School of Mines, 2000.
[33] Shukla, D., "Mobility Models in Ad Hoc Networks," Master Thesis, KReSIT-IIT Bombay, India, 2001.
[34] Nain, P., Towsley, D., Liu, B., and Liu, Z., "Properties of Random Direction Models," *IEEE INFOCOM 2005*, Miami, USA, March 2005.
[35] Groenevelt, R. B., Altman, E., and Nain, P., "Relaying in Mobile Ad Hoc Networks: The Brownian Motion Mobility Model," *INRIA Technical Report RR-5311*, September 2004.
[36] Basagni, S., Chlamtac, I., Syrotiuk, V. R., and Woodward, B. A., "A Distance Routing Effect Algorithm for Mobility (DREAM)," *ACM/IEEE MobiCom'98*, 1998.

[37] Yoon, J., Liu, M., and Noble, B., "Random Waypoint Considered Harmful," *IEEE INFOCOM 2003*, San Francisco, USA, 2003.

[38] Lin, G., Noubir, G., and Rajamaran, R., "Mobility Models for Ad hoc Network Simulation," *IEEE INFOCOM 2004*, Hong Kong, March 2004.

[39] Schindelhauer, C., Lukovszki, T., Ruhrup, S., and Volbert, K., "Worst Case Mobility in Ad Hoc Networks," *ACM SPAA'03*, San Diego, USA, June 2003.

[40] Bai, F., Sadagopan, N., and Helmy, A., "IMPORTANT: A Framework to Systematically Analyze the Impact of Mobility on Performance of RouTing Protocols for Adhoc NeTworks," *IEEE INFOCOM 2003*, San Francisco, USA, 2003.

[41] Dijkstra, E. W., "A Note on Two Problems in Connection with Graphs," *Numerische Mathematik*, 1959, pp. 269–271.

[42] Tugcu, T. and Ersoy, C., "How a New Realistic Mobility Model Can Affect the Relative Performance of a Mobile Networking Scheme," *Wireless Communications and Mobile Computing*, 4, 2004, pp. 383–394.

[43] Markoulidakis, J. G., Lyberopoulos, G. L., Tsirkas, D. F., and Sycas, E. D., "Mobility Modeling in 3rd Generation Mobile Telecommunication Systems," *IEEE Personal Communications* 4, 1997, pp. 41–56.

[44] Awerbuch, B., Berenbrink, P., Brinkmann, A., and Schneideler, C., "Simple Routing Strategies for Adversarial Systems," *42nd Annual IEEE Symposium on Foundations of Computer Science*, October 2001.

[45] Basch, J., Guibas, L. J., and Hershberger, J., "Data Structures for Mobile Data," *Journal of Algorithms*, 31, 1999.

[46] Czumaj, A. and Sohler, C., "Soft Kinetic Data Structures," *SODA'01*, 2001.

[47] Pandey, M. and Zappala, D., "The Effects of Mobility on Multicast Routing in Mobile Ad Hoc Networks," *Computer and Information Science Technical Report CIS-TR-2004-2*, University of Oregon, March 2004.

[48] Johansson, P., Larsson, T., Hedman, N., Mielczarek, B., and Degermark, M., "Scenario-Based Performance Analysis of Routing Protocols for Mobile Ad Hoc Networks," *ACM/IEEE MobiCom'99*, 1999.

[49] Kwak, B. -J., Song, N. -O., and Miller, L. E., "A Standard Measure of Mobility for Evaluating Mobile Ad Hoc Network Performance," *IEICE Transactions on Communications*, E86-B (11), November 2003, pp. 3236–3243.

[50] Kwak, B. -J., Song, N. -O., and Miller, L. E., "A Mobility Measure for Mobile Ad-Hoc Networks," *IEEE Communications Letters*, 7, August 2003, pp. 379–381.

[51] Kwak, B. -J., Song, N. -O., and Miller, L. E., "A Canonical Measure of Mobility for Mobile Ad Hoc Networks," *IEEE MILCOM 2003*, Boston, USA, October 2003.

[52] Perez-Costa, X., Bettstetter, C., and Hartenstein, H., "Towards a Mobility Metric for Reproducible and Comparable Results in Ad Hoc Networks Research," *ACM MobiCom'03*, San Diego, USA, September 2003.

[53] Yoon, J., Liu, M., and Noble, B., "Sound Mobility Models," *ACM MobiCom'03*, San Diego, California, USA, September 2003.

[54] Hofmann, P., Bettstetter, C., Wehren, J., and Prehofer, C., "Performance Impact of Mobility in an Emulated IP-based Multihop Radio Access Network," Submitted to *ACM Mobile Computing and Communications Review*, 2005.

[55] Boumerdassi, S., Renault, E., and Wei, A., "Impact of Mobility Environment on the QoS in Ad-Hoc Networks," *Technical Report*, CEDRIC Laboratory, France, 2003.

[56] Gupta, P. and Kumar, P. R., "The Capacity of Wireless Networks," *IEEE Transactions on Information Theory*, 46, 2000, pp. 388–404.
[57] Grossglauser, M. and Tse, D. N. C., "Mobility Increases the Capacity of Ad Hoc Wireless Networks," *IEEE/ACM Transactions on Networking*, 10 (4), August 2002, pp. 477–486.
[58] Bansal, N. and Liu, Z., "Capacity, Delay and Mobility in Wireless Ad-Hoc Networks," *IEEE INFOCOM 2003*, San Francisco, USA, 2003.
[59] Felegyhazi, M., Hubaux, J. -P., and Buttyan, L., "The Effect of Mobility on Cooperation in Ad Hoc Networks," *WiOpt'04*, Cambridge, UK, March 2004.
[60] Srinivasan, V., Nuggehalli, P., Chiasserini, C. F., and Rao, R. R., "Cooperation in Wireless Ad Hoc Networks," *IEEE INFOCOM'03*, San Francisco, USA, 2003.
[61] Papadopouli, M. and Schulzrinne, H., "Seven Degrees of Separation in Mobile Ad Hoc Networks," *IEEE GLOBECOM 2000*, San Francisco, USA, November 2000.
[62] Yu, J. Y. and Chong, P. H. J., "A Survey of Clustering Schemes for Mobile Ad Hoc Networks," *IEEE Communications Survey*, 7, 2005, pp. 32–48.
[63] Venkateswaran, A., Sarangan, V., Gautam, N., and Acharya, R., "Impact of Mobility Prediction on the Temporal Stability of MANET Clustering Algorithms," *ACM PE-WASUN'05*, Montreal, Canada, October 2005.
[64] McDonald, A. B. and Znati, T., "A Mobility Based Framework for Adaptive Clustering in Wireless Ad-Hoc Networks," *IEEE Journal on Selected Areas in Communications*, (17) (8), August 1999, pp. 1466–1487.
[65] Lin, T. and Midkiff, S. F., "Mobility versus Link Stability in the Simulation of Mobile Ad Hoc Networks," *Communication Networks and Distributed Systems Modeling and Simulation Conference (CNDS)*, Orlando, USA, January 2003.
[66] Helmy, A., Jaseemuddin, M., and Bhaskara, G., "Multicast-Based Mobility: A Novel Architecture for Efficient Micro-Mobility," *IEEE Journal on Selected Areas in Communications: Special Issue on All-IP Wireless Networks*, 22 (4), May 2004, pp. 677–690.
[67] Ho, C., Obraczka, K., Tsudik, G., and Viswanath, K., "Flooding for Reliable Multicast in Multi-Hop Ad Hoc Networks," *International Workshop on Discrete Algorithms and Methods for Mobile Computing and Communications*, 1999.
[68] Lee, S., Su, W., and Gerla, M., "On-Demand Multicast Routing Protocol in Multihop Wireless Mobile Networks," *ACM/Baltzer Mobile Networks and Applications, Special Issue on Multipoint Communication in Wireless Mobile Networks*, 2000, pp. 441–453.
[69] Jetcheva, J. and Johnson, D. B., "Adaptive Demand-Driven Multicast Routing in Multi-Hop Wireless Ad Hoc Networks," *ACM MobiHoc'01*, October 2001.
[70] Wang, K. H. and Li, B., "Group Mobility and Partition Prediction in Wireless Ad-Hoc Networks," *IEEE ICC 2001*, Helsinki, Finland, June 2001.
[71] Chen, B. -r. and Chang, C. H., "Mobility Impact on Energy Conservation of Ad Hoc Routing Protocols," *International Conference on Advances in Infrastructure for Electronic Business, Education, Science, Medicine, and Mobile Technologies on the Internet SSGRR 2003*, L'Aquila, Italy, 2003.
[72] Feeney, L. M., "An Energy-Consumption Model for Performance Analysis of Routing Protocols for Mobile Ad Hoc Networks," *Mobile Networks and Applications*, 3 (6), June 2001, pp. 239–249.

[73] Feeney, L. M. and Nilsson, M., "Investigating the Energy Consumption of a Wireless Network Interface in an Ad Hoc Networking Environment," *IEEE INFOCOM'01*, Anchorage, USA, April 2001.

[74] Cano, J. -C. and Manzoni, P., "A Performance Comparison of Energy Consumption for Mobile Ad Hoc Network Routing Protocols," *IEEE/ACM MASCOTS 2000*, San Francisco, USA, August 2000.

[75] Capkun, S., Hubaux, J. -P., and Buttyan, L., "Mobility Helps Security in Ad Hoc Networks," *ACM MobiHoc'03*, Annapolis, USA, June 2003.

[76] Buttyan, L. and Hubaux, J. -P., "Report on a Working Session on Security in Wireless Ad Hoc Networks," *Mobile Computing and Communications Review*, 6(4), 2002, pp. 74–94.

[77] Helmy, A., "Mobility-Assisted Resolution of Queries in Large-Scale Mobile Sensor Networks (*MARQ*)," *Computer Networks Journal - Elsevier (Special Issue on Wireless Sensor Networks)*, 43(4), November 2003, pp. 437–458.

[78] Realistic Mobility Modelling for Mobile Ad Hoc Networks Project, Rice University, Information available at: http://www.cs.rice.edu/~amsaha/Research/MobilityModel

[79] Obstacle Mobility Model Project, Information available at: http://moment.cs.ucsb.edu/mobility/index.html

[80] Mobility Models for Mobile Ad Hoc Network Simulations Project, Center for Automatics, Robotics and Distributed Intelligence, Colorado School of Mines, Golden, USA, Information available at: http://egweb.mines.edu/cardi/MANET.html

[81] C3 (Car-to-car cooperation) Project, Information available at: http://aqualab.cs.northwestern.edu/projects/C3.html

[82] Mobility Management and Resource Allocation in Wireless Networks Project, Wireless Networks Lab, Cornell University, Information available at: http://people.ece.cornell.edu/~haas/wnl/wnlprojects.html#Publications_MM

[83] Shaheen, S. A., "DAVIS Smart Mobility Modeling Project: Initial Scoping and Planning Project," *Final Report*, University of California, Davis, USA, March 2003.

[84] AGILE: Architectures for Mobility (IST-2001–32747), Information available at: www.pst.informatik.uni-muenchen.de/projekte/agile

8

Security in Ad Hoc Networks

8.1 Introduction

Security in wireless ad hoc networks has recently gain a momentum and became a primary concern in attempt to provide secure communication in a hostile wireless ad hoc environment. Numerous proposals were suggested without deriving a general solution. Securing wireless ad hoc networks is particularly difficult for many reasons including the:

- *Vulnerability of channels*: message can be eavesdropped and fake messages can be injected into the network, with no necessity of physical access;
- *Vulnerability of nodes*: nodes can be easily captured or stolen and can fall under the control of the attacker;
- *Absence of infrastructure*: ad hoc networks operate independently of any infrastructure, which makes inapplicable any classical solutions based on certification authorities and on-line servers;
- *Dynamically changing topology*: sophisticated routing protocols designed to follow the permanent changes in topology can be attacked by incorrect routing information generated by compromised nodes, which is difficult to distinguish.

The unique characteristics of mobile ad hoc networks pose a number of nontrivial challenges to security design. Moreover, different applications have different security requirements. The complexity and diversity of the field has led to a multitude of proposals, which focus on different parts of the problem domain. They vary between trust and key management, secure routing and intrusion detection, availability and cryptographic protocols [1]. The ultimate goal of the security solutions is to provide security services, such as authentication, confidentiality, integrity, anonymity and availability to mobile users. In order to achieve this goal, the implemented security mechanisms should provide complete protection spanning the entire protocol stack [2]. Table 8.1

Table 8.1. Security solutions for MANETs

Layer	Security issues
Application layer	Detecting and preventing viruses, worms, malicious codes and application abuses
Transport layer	Authenticating and securing end-to-end communications through data encryption
Network layer	Protecting the ad hoc routing and forwarding protocols
Link layer	Protecting the wireless MAC protocol and providing link-layer security support
Physical layer	Preventing signal jamming denial-of-service attacks

depicts relevant security issues corresponding to different networking layers in a mobile ad hoc network (MANET).

Security design in ad hoc mobile networks has to face the lack of *clear line of defense.* Each node in an ad hoc network may function as a router and forward packets for other peer nodes. Unlike in wired networks (e.g., routers), there is no well defined place where the traffic monitoring or access control mechanisms can be deployed. This makes the separation of inside from outside network domain obscure.

Protection in wireless ad hoc networks can be *proactive* and *reactive.* The proactive approach attempts to prevent the attacker from malicious actions. This is usually achieved through various cryptographic techniques. The reactive approach attempts to detect security threats a posteriori and reacts accordingly. To be effective, the security in ad hoc networks must integrate both approaches and contrive there components: *prevention, detection* and *reaction.* Introducing enhanced security mechanisms and vigor cryptographic solutions can produce increasing computation, communication and management overhead. It can affect the resource-constrained ad hoc network performances such as scalability, service availability, robustness, etc. So, one of the fundamental challenges in security design becomes achieving a good balance between the security strength and network performances.

This chapter focuses in the security aspect of communication in ad hoc wireless networks and gives a general overview of security issues. It explains the security goals and challenges in achieving secure communication in ad hoc environment. Most usual attacks that can threat functioning of ad hoc networks are reviewed. The cryptographic mechanisms implemented in security algorithms and protocols are further listed. Key management is briefly disscused. Special attention is reserved to security routing in ad hoc wireless networks and different secure architectures. Concluding remarks highlight the open challenges and roadmaps in this challenging and demanding area.

8.1.1 Security Goals

Security is the possibility of a system to withstand an attack [3]. The requirement of system security is to have controlled access to resources. Different security mechanisms (preventive and detective) are involved in order to achieve this requirement. Security is an important issue for ad hoc networks, especially because of the vulnerable transmission media, specifics of ad hoc networking and possible secure sensitive applications [4]. The major security goals in ad hoc networks consist of providing the following security attributes: *availability, confidentiality, integrity, authentication and non-repudiation.*

Availability ensures the survivability of network services despite denial of service attacks. The adversary can attack the service at any layer of an ad hoc network. For instance, at physical and media control layer it can employ jamming to interfere with communication on physical channels; on network layer it could disrupt the routing protocol and disconnect the network; or on higher layers it could bring down some high-level services (e.g., the key management service).

Confidentiality ensures that certain information is never disclosed to unauthorized entities. It protects the network transmission of sensitive information such as military, routing, personal information, etc.

Integrity guarantees that the transferred message is never corrupted. A corruption can occur as a result of transmission disturbances or because of malicious attacks on the network.

Authentication enables a node to ensure the identity of the peer node with whom it is communicating. It allows manipulation-safe identification of entities (e.g., enables the node to ensure the identity of the peer node), and protects against an adversary gaining unauthorized access to resources and sensitive information, and interfering with the operation of other nodes.

Non-repudiation ensures that the origin of a message cannot later deny sending the message and the receiver cannot deny the reception. It enables a unique identification of the initiator of certain actions (e.g., sending of a message) so that these completed actions can not be disputed after the fact.

There are other security goals that are of concern to certain applications such as *authorization, anonymity, self-stabilization, Byzantine robustness, location privacy,* etc. The security of mobile ad hoc networks also has to consider features such as privacy, correctness, reliability and fault tolerance. More details are elaborated in [3, 5, 6].

8.1.2 Security Challenges

Achieving security performances in wireless ad hoc environment is a challenging task [4]. Unlike the wire-line networks, the unique characteristics of ad hoc networks pose a number of nontrivial challenges to security design, such as open peer-to-peer architecture, insecure operational environment and shared broadcast radio channel, stringent resource constraints, roaming of nodes, highly dynamic network topology combined with lack of central authority and association, scalability and physical vulnerability.

Vulnerable wireless links can be exposed to link attacks ranging from passive eavesdropping to active impersonation, message reply and message distortion.

Eavesdropping may violate confidentiality and expose secret information to an adversary. Active attacks can delete the messages, inject erroneous messages, modify message and impersonate a node. These actions may violate availability, integrity, authentication and non-repudiation. A threat is enlarged because of shared broadcast channel and insecure operational environment (e.g., insecure enemy territory in battlefields).

Roaming nodes with relatively poor physical protection can be exposed to malicious attacks by compromised nodes. To reduce the vulnerability, which may be caused by compromised centralized entity, and to achieve high survivability, ad hoc network should have distributed architecture.

Dynamic topology and changeable nodes membership may disturb the trust relationship among the nodes. The trust may also be disturbed if some nodes are detected as compromised. Nodes in wireless ad hoc networks may be dynamically affiliated to different administrative domains. This dynamism could be better protected with distributed and adaptive security mechanisms [4].

Scalability is an important issue concerning security. Security mechanisms should be capable of handling a large network as well as small ones [2, 7].

Resource availability (bandwidth, battery and computational power) in ad hoc networking is a scarce feature. Providing secure communication in such changing and dynamic environment, as well as protection against specific threats and attacks, leads to development of various security schemes and architectures. Collaborative ad hoc environments also allow implementation of self-organized security mechanisms. The next section describes specific attacks in wireless ad hoc networking.

8.2 Attacks in Wireless Ad Hoc Networks

A *threat* in a communication network is a potential event or series of events that could result in the violation of one or more security goals [5]. The actual implementation of a threat is called *attack*. Wireless ad hoc networks are target for all the threats that occur in fixed networks, i.e., masqueraded identities, authorization violations, eavesdropping, data loss, modified and falsified data units, repudiation of communication processes and sabotage. Moreover, the existence of the wireless transmission links and dynamic network topology contributes considerably towards increasing the threat potential.

8.2.1 Types of Attacks

Attacks against ad hoc networks are generally divided into two groups:

- *Passive attacks* (typically involve only one eavesdropping of data);
- *Active attacks* (involve actions performed by adversaries, such as replication, modification and deletion of exchanged data).

Attacks are considered as *external attacks* if they are targeted to cause congestion, propagate incorrect routing information, prevent services of working properly or shut down them completely. External attacks can be active or passive.

External active attacks that can be usually easily performed against ad hoc network are: black hole, routing table overflow, sleeps deprivation and location disclosure [3]. External active attacks can usually be prevented by using standard security mechanism such as firewalls, encryption, etc.

Internal attacks are more severe attacks, since malicious nodes have already been authorized and are thus protected with the security mechanisms the network and its services offer. These kind of malicious parties are called *compromised nodes*. They may operate as a group using standard security protection to protect their attacks, compromising the security of the whole ad hoc network.

Attacks can also be classified as malicious or rational. *Malicious attacks* aim to harm the members or the functionality of the network, involving any means disregarding corresponding costs and consequences [8]. *Rational attacks* are more predictable in terms of the attack means and the attack target, since they involve personal profit.

According to the affected mechanisms, two levels of attacks can be distinguished:

- *Attacks on basic mechanisms* of ad hoc network (e.g., the routing)
- *Attacks on security mechanisms* (e.g., the key management).

Vulnerabilities of the basic mechanisms, besides already mentioned features, include cooperativeness of the nodes (attempt to work according to the rules in order to have fair allocation of resources), node selfishness (deny to relay packets for other nodes in order to save battery), neighbor discovery (as in Bluetooth [9]), etc.

Vulnerability of security mechanisms points out the importance of a good cryptographic design with proper management and safe keeping of a small number of cryptographic keys [10]. With node mobility and variable connectivity these objectives are not trivial.

Nodes that perform active attacks with the aim of damaging other nodes by causing network outage are considered as *malicious*, while nodes that perform passive attacks aiming to save battery life for their own communications are considered *selfish*.

Functionalities of different layers can be threat of different security attacks. Fig. 8.1 gives a classification of attacks possible in wireless ad hoc networks.

The following subsections present different types of attacks and the way they compromise security.

8.2.1.1 Denial of Service

The *denial of service (DoS)* can produce a severe security risk in any distributed system. This threat is produced either by unintentional failure or by

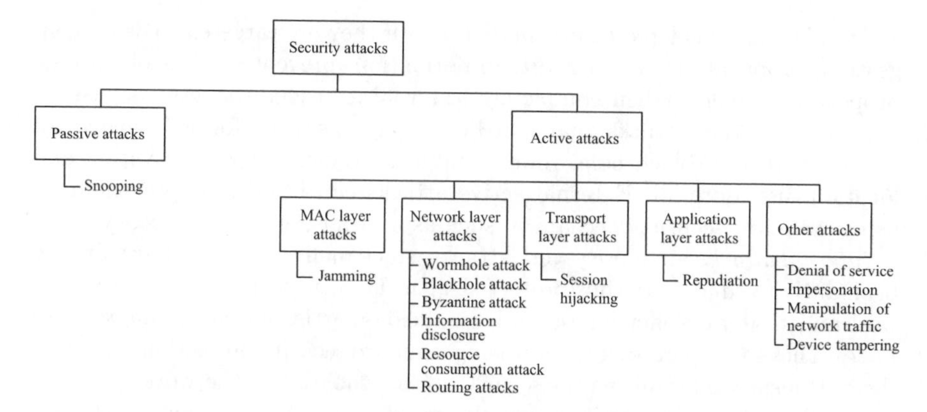

Fig. 8.1. Classification of attacks in wireless ad hoc networks

malicious actions. The consequences strongly depend on particular application. For instance, in the battlefields scenario, the consequences of shutting down the network can be catastrophic, while during the conference connection in the conference room, it may only cause some disturbances.

The denial of service attack has many forms. The classical way is to flood any centralized resource, which can disturb its correct operation or completely crash it. Ad hoc networks which exist without centralized infrastructure are more sensitive to distributed denial of service attacks. So, if the attackers have enough computing power and bandwidth for their operation, they can easily crash and congest the smaller ad hoc networks.

Compromised nodes can initiate severe threats to ad hoc networks if they are able to modify the routing protocol (or part of it) and send the routing information very frequently. It can cause congestion or even prevent nodes to gain new routing information about the changed network topology. If the compromised nodes and if the changes of the routing protocol are not detected, the consequences are severe (even the network seems to operate normally). This invalid operation initiated by malicious nodes in ad hoc networks is called a *Byzantine failure*.

8.2.1.2 Impersonation

Impersonation attacks can be serious security risk in ad hoc networking, concerning critical operations in all levels. Compromised nodes can masquerade itself as trusted nodes and initiate false or even dangerous behavior such as sending false routing information, gaining access to configuration system as a super-user, certify public key without proper credentials, give false status information to other nodes, etc.

Impersonation threats are mitigated by applying strong *authentication mechanisms*. Authentication provides a party to be able to trust the origin of data it receives or stores. It usually is performed in every layer by application of digital signatures or keyed fingerprints over routing messages, different information (configuration or status) or exchanged payload data of the used services. Digital signatures and public-key cryptography requires relatively significant computation power and secure key management, which is inappropriate for wireless ad hoc network capabilities. Lighter solutions include keyed hash functions or a priori negotiated and certified keys and session identifiers.

8.2.1.3 Disclosure

Exchanging confidential information must be protected from eavesdropping and unauthorized access. In ad hoc networks, confidential information can concern specific status details of a node, location of nodes, private or secret keys, passwords, etc. The disclosure of the exchanged or stored information can be especially critical in military applications (strategic routing attacks and tactical attacks).

8.2.1.4 Trust Attacks

Trust is a privilege associated between the identity of the user with particular trust level. So, a *trust hierarchy* is an explicit representation of *trust levels.* Attacks on the trust hierarchy can be initiated by inside or outside nodes, if they try to impersonate anyone else and obtain higher level privileges. Different mechanisms can be used to protect against trust attacks, such as strong access control mechanisms (Authentication, Authorization and Accounting or AAA) and cryptographic techniques (encryption, public key certificates, and shared secrets). Some techniques to prevent insider attacks include secure transient association and tamper proof and tamper resistant nodes [3].

8.2.1.5 Attacks on Information in Transit

Compromised or enemy nodes can utilize the information carried in the routing protocol packets to launch attacks. These attacks can cause corruption of the information, disclosure of sensitive information, misusing of the legitimate service from other protocol entities or even denial of service. Threats to information in transit include: interruption, intersection and subversion, modification of the information integrity, and fabrication, i.e., insertion of false routing information or metrics.

8.2.1.6 Attacks Against Secure Routing

Attacks against secure routing are severe threat to ad hoc networks. Internal attacks are difficult to differentiate, because the network topology dynamically changes. Malicious nodes can disrupt the correct functioning of a routing protocol by *modifying* routing information, *fabricating* false routing information and by *impersonating* other nodes [11].

Malicious nodes can easily perpetrate *integrity* attack, by simply altering protocol fields in order to subvert the traffic. A special case of integrity attack is *spoofing.* A malicious node impersonates a legitimate node due to the lack of authentication in the current ad hoc routing protocols. It can result in miss-presentation of network topology and undesirable network loops or network portioning.

Another type of active attack is the creation of a tunnel or *wormhole* attack [12]. During this attack, two colluding malicious nodes link through a private connection (i.e., tunnel) bypassing the network. This allows a node to short-circuit the normal flow of routing messages and create a virtual vertex cut in the network. This is controlled by the two colluding attackers.

Lack of cooperation among ad hoc nodes due to node selfishness can result in denying participation in routing protocol or forwarding packets. The selfishness problem may cause so *called black hole attack.*

Some security mechanisms implemented in wireless ad hoc networking are described in more details in the following sections.

8.3 Security Mechanisms

A comprehensive solution to communication security includes *protocols, algorithms* and *key management.* The breakdown of any of these components compromises security. Traditional security mechanisms include authentication, digital signature and encryption/decryption. This section gives an overview of basic concepts concerning security mechanisms.

8.3.1 Cryptographic Issues

Preventive security controls are often protocols that utilize *cryptography.* Cryptography analyzes and develops methods for transforming of unsecured *plaintext* into *ciphertext* that can not be read by unauthorized entities. There are two main applications of cryptography: *data encryption* and *data signing.* Data encryption (and inverse operation — *decryption*) implements some cryptographic algorithms to achieve data transformation. Data signing performs manipulation-safe calculations of the checksum, based on *signature check*, in order to determine whether data was modified after it was created and signed.

8.3.2 Cryptographic Primitives

Three types of *cryptographic primitives* are used in order to authenticate the content of messages exchanged among nodes: *message authentication code (HMAC), digital signature*, and *one-way HMAC key chain* [2]. HMAC refers to keyed hashing for message authentication [13]. It is applicable in case when nodes share a secret symmetric key which allow them to generate and verify a *message authenticator* $h_k(.)$, using a cryptographic one-way function. The computation is light and can be applied even on small sensor nodes. Only the nodes that share a secret key may use it, which makes HMAC inapplicable for broadcast messages. SRP and DSR (see Chapter 3) take this approach with pair-wise shared keys [14].

Digital signature is based on asymmetric key cryptography and involves much more computational overhead in signing/decrypting and verifying/encrypting operations. It is sensitive to DoS attacks because of possibility of bogus signatures. Each node needs to keep a certificate revocation list (CRL) of revoked certificates. Any node that knows the public key of the signing nodes may use it. Digital signature approach is implemented in S-AODV [15] and ARAN [16].

One-way HMAC key chain provides a cryptographic one-way function $f(x)$, designed to make the input x invisible. When applied repeatedly on the input, a chain of outputs $f^i(x)$ is obtained. The reverse order of generation is used to authenticate messages. A message with an HMAC using $f^i(x)$ as the key is proven to be authentic when the sender reveals $f^{i-1}(x)$. The computation is lightweight, even the storage of the hush chain is nontrivial in case of long chains. This approach is used in TESLA, a hash-chain-based protocol, to authenticate broadcast messages [17].

8.3.3 Cryptographic Algorithms

Cryptographic algorithms perform a mathematical transformation of input data (e.g., data, keys) to output data to conceal it [3, 18]. They may use one or both mentioned security applications and have to be embedded into a semantic context which usually occurs as a part of a cryptographic protocol [5, 8]. A *cryptographic protocol* is a procedural instruction for a series of processing steps and message exchanges between multiple entities, aiming to achieve specific security objectives.

Cryptographic algorithms can be classified according to the number of used different keys into:

- *Hash algorithms*, use *no key*;
- *Secret-key cryptography*, use *one key (symmetric algorithms)*;
- *Public-key cryptography*, use *two different keys* for encryption and decryption or signing and signature check (*asymmetric algorithms*).

A hash algorithm is a one way function that maps a message of any size into a fixed size *digest*, considering it as a *fingerprint of a message* [19]. Hash algorithms do not use keys (see Fig. 8.2) and are mainly used to create cryptographic check values and to generate pseudo-random numbers.

Secret-key cryptography uses one key and involves a pair of functions: *encryption* and *decryption* (see Fig. 8.3). A *key* is a small amount of information used by cryptographic algorithms. The *secret key* is a shared secret

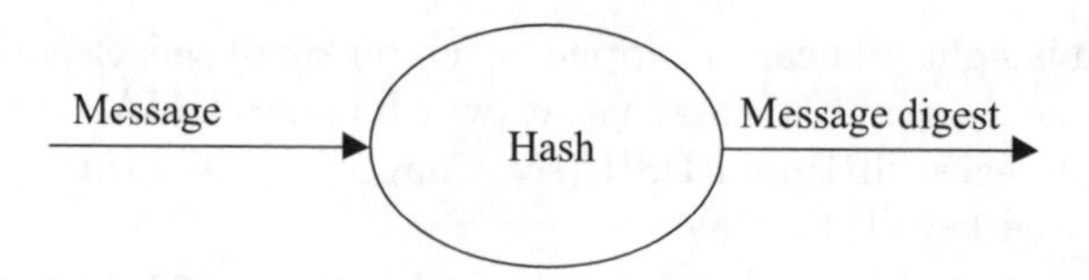

Fig. 8.2. Hash function cryptography

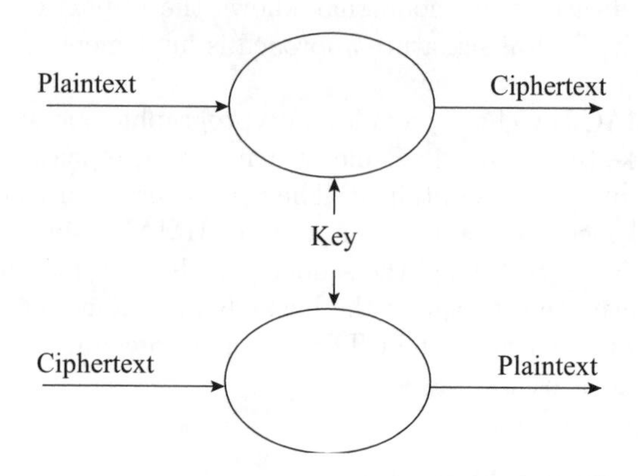

Fig. 8.3. Secret-key cryptography

between the communicating entities. The same key is used to transform the message (*plaintext*) into *ciphertext,* and vice versa, from ciphertext back to plaintext (*symmetric key* algorithm). Secret-key encryption provides confidentiality. Only the communicating entities that know the secret key are eligible to participate and uncover the plaintext message. The only bottleneck can be the ability to securely send the secret key to the receiver. The secret key approach is non-scalable, since each communicating pair must share the secret key.

Public-key cryptography uses a pair of keys, a *public key* and a *private key.* These pairs are uniquely associated with each other. Each entity has a key pair, $<K_p, K_s>$, where K_p is a public key (used for encryption) and K_s is a private (secret) key (used for decryption) of an entity (see Fig. 8.4). For instance, if the message is mentioned to particular receiving entity, then its key pair is used in encryption/decryption processes (i.e., K_{pr} and K_{sr} keys). Public-key encryption provides confidentiality, because the private key is only known to the key user. Public-key encryption uses *asymmetric key* algorithms and is primarily based on difficulties in solving the underlying mathematical principles. A very popular example of public key cryptography is RSA system developed by Rivest, Shamir and Adleman [18].

Cryptography algorithms can also involve digital signatures, which are based on public key encryption. Digital signatures are verified by an arbitrator (see Fig. 8.5). The digital signature binds an entity with a message.

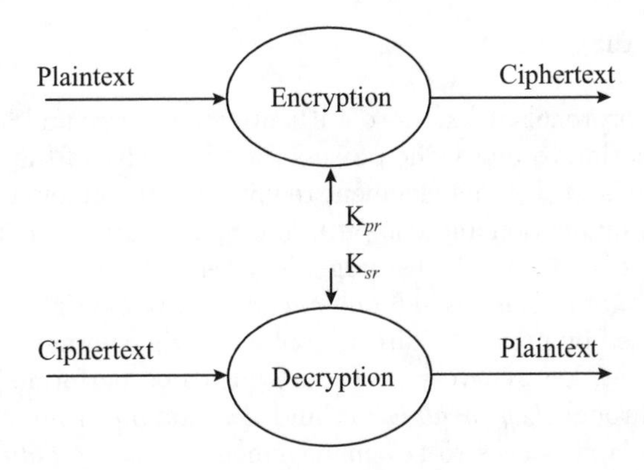

Fig. 8.4. Public-key encryption

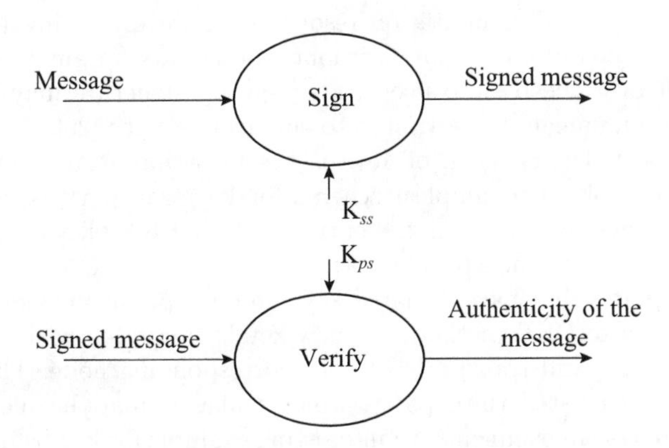

Fig. 8.5. Public-key digital signature

A sender signs the message using his private key K_{ss}. The signature is verified by arbitrator which uses the sender's public key K_{ps}. Public key signature is more often used to sign digest instead of message itself. In practice, a trusted third party (TTP) is agreed in advance upon who is responsible for issuing digital signatures and for solving any disputes regarding these signatures [6]. These provide authentication, integrity and non-repudiation since they bind a message, a signature and the private key used to generate the signature.

The cryptographic algorithms are of central importance to network security. However, their detailed observation overcomes the scope of this book. The secure administration of cryptographic keys is called *key management*. The following section is dedicated to key management, since the keys required in these algorithms are essential for efficient cryptographic protection.

8.3.4 Key Management

Cryptographic protocols use keys to authenticate entities and allow access to guarded information to those who possess the knowledge of the keys [5, 8, 18]. Different methods for key management require varying amount of initial configuration, communication and computation. Special attention is dedicated to key generation, distribution and storage. Key generation is the creation of the keys in a *random* or at least *pseudorandom-controlled* way. The generated keys can be distributed by personal (direct) contact or via communication channel. In case of lost key, key recovery procedure has to be performed. Key control is performed through *key invalidation* and *destruction* of no longer required keys. The main approaches to key management are *key pre-distribution, key transport, key arbitration* and *key agreement.*

Key pre-distribution involves distributing keys to all interested parties before the start of the communications [6]. This method is not demanding in communication and computation resources. However, all involved parties and a key must be known *a priori*, without a possibility for any change. As an improvement over the basic concept, a sub-group structure may be formed, with some communications restricted to sub-group nodes [20].

Key transport. Distribution of keys is the most important issue of cryptographic protocols. The simplest scheme for key transport assumes that a shared key already exists among the participants (KEK — key encrypting key method). It is used to encrypt a new key and transmit it to all corresponding nodes. Only nodes that have a shared key can decrypt the new key. In public key infrastructure (PKI) method, the new key is encrypted with each participant's public key and transported to the corresponding node. This assumes the existence of trusted third party, which makes it inapplicable to ad hoc infrastructureless environment. An interesting example for key transport without prior shared keys is *three-pass protocol* that is based on special type of encryption [21].

Key arbitration. Key arbitration schemes use a central arbitrator in order to create and distribute keys among all participants and can be considered as a class of key distribution schemes. There is a difference in distribution of public keys which belong to a public knowledge, and private (secret) keys which are shared by multiple entities. Private keys can be distributed through a pre-established secure channel or an open channel [18]. Public keys are usually distributed through certificates. A *certificate* binds a public key with an entity. Certificates are certified, stored, and distributed by one or more trusted parties in a *centralized* or *decentralized* approach. In a centralized approach, there is only one trusted third party, called *Certificate Authority* (*CA*). Decentralized public-key distribution is performed either via a *decentralized key distribution center*, or through *individual nodes* that comprise the network. Since the ad hoc networks exist without predefined infrastructure, the decentralized approach that involves comprising nodes is a preferable approach.

Key agreement. Most of key agreement schemes are based on asymmetric key algorithms. Key agreement is used when it is necessary for several parties to agree upon a secret key and its exchanges, used in later communications. In case of group key agreement, each participant contributes a part to the secret key. This involves high computational complexity. The most popular key agreement scheme is Diffie-Hellman exchange [18].

There are different types of keys used in different technologies. Bluetooth devices, e.g., implement a link layer security, using *link keys* (mostly for authentication) and *encryption keys* [22].

8.3.4.1 Key Management in Ad Hoc Wireless Networks

Key management in ad hoc environments shares the disadvantages of infrastructureless nature concerning the *network infrastructure* (missing dedicated routers and stable links), *services infrastructure* (missing naming resolution, directories and trusted third parties — TTPs) and *administrative infrastructure* (missing the administrative support of certifying authorities). There are different approaches in the attempt to enhance key management in wireless ad hoc networks.

Group Key Management Homogeneous ad hoc wireless networks (e.g., MANETs) consider nodes with similar transmission capabilities. In such networks, the bandwidth available to each node rapidly decreases as network size grows (see Chapter 2). Recently, heterogeneous network concepts involve large number of nodes with different communication ranges and capabilities and hierarchical topology. They prove efficiency regarding capacity (see Chapter 2), while posing new security concerns.

Secure group communication requires scalable and efficient group membership management with access control capable to protect data and cope with potential compromises [4]. Several group management schemes are proposed to secure communications via different group key management architectures, ranging from key distributing scheme for large scale single sender multicast to key agreement schemes for small any-to-any peer groups [4, 20, 23].

Group management must be performed securely with relevant keying material delivered via secure channels. It must adjust to group secrets usually triggered by either timeouts or membership changes in the underlying group communication system. Group key management must provide key *independence* property [24] and scalability. Proposed group key management types can be classified as:

- *Centralized*, where the key distribution is managed by a key center (e.g., Key Distributed Center — KDC in [4]). Centralized key trees have one point of failure and reduce the cost of re-keying from $O(n)$ to $O(log\ n)$, (where n is the group size). Examples are key graph [25], One-Way Function three (OFT) [26], etc.

- *Collaborative*, where the group collectively recomputes the new key whenever a membership change occurs [10]. It is applicable for larger groups and, due to multiple communication rounds, does not have a single point of failure. Examples are GDH [27], Tree-based Group Diffie-Hellman (TGDH) scheme [28] and STR [29].
- *Hybrid*, as in [20, 23], where Rhee, Park and Tsudik propose combination of centralized OFT group key management (for cell groups) and collaborative TGDH key group management (for control groups). Another example can be found in [30].

Authors in [20, 23] for UAV-MBN network propose three level architecture composed of three kinds of networking units with heterogeneous communications capabilities and computation power: the *regular ground mobile nodes* (*cell group*), the *ground mobile backbone* (*MBN*) *nodes* (*control group*) and the *UAV nodes*. Each UAV (unmanned aerial vehicle) leads a *single-area theater* (see Fig. 8.6). When the third level is absent (e.g., without backbone structure), the network operates in an infrastructureless mode. For the reason of scalability, group key management within a cell is performed by the cell group manager (an MBN node) in a centralized fashion. At the same time, key management within the control group is done in a collaborative fashion by all MBN nodes that are members of the control group. In order to avoid exchanging and certification of public key certificates between all nodes, they also propose using *implicitly certified public key* (*ICPK*).

An interesting modification towards adaptive security solutions in order to adapt to dynamic infrastructure changes is proposed in [31].

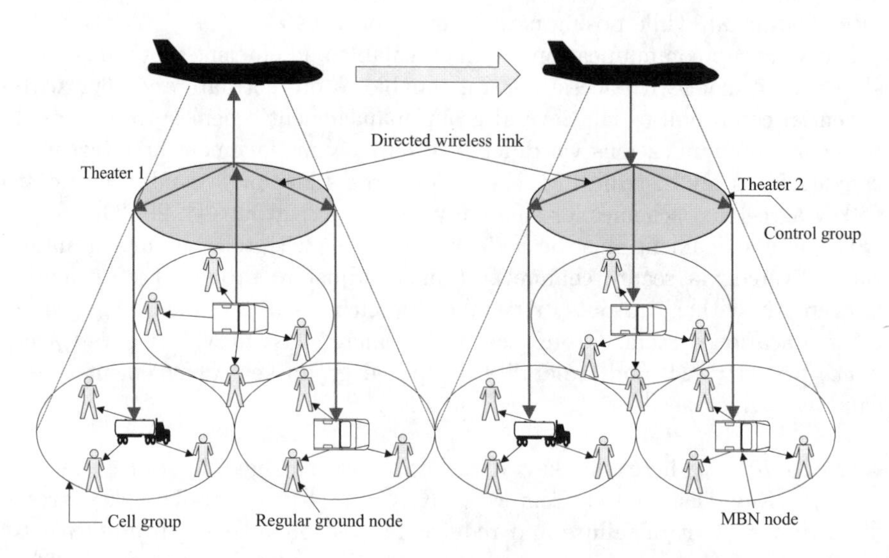

Fig. 8.6. Group communication and key management model in UAV-MBN network

Threshold Cryptography Ad hoc networks exhibit a need for some kind of authentication. Identity based cryptosystems are dealing with binding public keys to identities. In these systems, the entities' public identification information plays the role of its public key. The drawback of this method is that a centralized authority (CA) is needed for establishment of private keys of the users. A centralized certification authority in mobile ad hoc networks can be avoided with:

- emulating a conventional certification authority by distributing it on several nodes,
- distributed solution where all nodes have to authenticate each other by setting up an appropriate context,
- self-organized public-key infrastructure [10].

One way to deal with low physical security and availability constraints is the *distribution of trust*. The trust can be distributed to a collection of nodes. Since it is unlikely that all $t+1$ nodes will be compromised, it is acceptable that a consensus of $t+1$ nodes is trustworthy. Zhou and Haas in [4] propose a key management *emulating* service distributed over a certain number of nodes called *servers*. The whole service is based on public/private K/k pairs. The public key K is known to all nodes, while the private key k is divided into n shares $(s_1, s_2, \ldots, s_n)$ with one share per server. Each server also has a public/private key K_i/k_i and knows the public keys of all nodes. The n servers are chosen arbitrarily among the nodes.

The system uses *threshold cryptography* to protect itself against potential compromise of one of the n servers. An $(n,\ t+1)$ threshold cryptography scheme allows n parties to share ability to perform a cryptographic operation (e.g., creating a digital signature), so that $t+1$ parties can perform this operation jointly, whereas it is infeasible for at most t parties to do so (even by collision). The $(n,\ t+1)$ threshold cryptography is used to make the service tolerate compromised nodes. In this $(n,\ t+1)$ model there are more than one CAs. To sign a certificate, each server signs the certificate with its share and transmits the certificate to a combiner. The combiner is able to sign the certificate if it gets $t+1$ correct partial signatures. Compromised servers are not able to sign certificates, because there is at most t of them at any time. In this scheme $n \geq (3t+1)$ meaning that at least n nodes are needed to enable the scheme. After building this key management system, a public key cryptography can be used to do the authentication. The concepts of threshold secret sharing and secret share updates were combined with localized certification service [32] to provide robust and ubiquitous security support for mobile ad hoc networks.

In order to deal with mobile attackers which temporarily compromise the servers, *share refreshing* is proposed. It means that servers create a new independent set of shares periodically. This method is confidential and robust.

An establishment of secure session with share of a priori *context* can be arranged among small group of people at a conference in a same room, when they want to establish ad hoc connections during the meeting. It is assumed that they do not have access to third party key management service. The present person can share a fresh password among them. However, the password-authenticated key exchange method [33] is proposed as a protection against vulnerable dictionary attacks.

Self-Organized Public Key Management The authors in [10] have proposed a new public key description system suitable for self-organized ad hoc networks. It is similar to *pretty good privacy* (*PGP*) system in the sense that the public used certificates are issued by the users. However, the proposed key management system does not rely on certificate directories for the distribution of certificates and certificates are stored and distributed by the users only based on personal acquaintance. Certificates are issued only for a specified period of time and contain their time of expiry. Before it expires, the certificate is updated by the user who issued it.

Each user maintains a *local certificate repository* that contains a limited number of certificates selected by the user according to some algorithm. Each repository has two parts. In the first part each user stores the certificate that it issues. In the other part, it stores a set of selected certificates issued by other users in the system, which are periodically updated. When user A wants to obtain the public key of user B, they merge their local certificate repositories and A tries to find an appropriate certificate chain from A to B in the merged repository. If any of certificates are conflicting (e.g., the same public key issued to different users or the same user having different public key), it is possible that some malicious nodes have issued a false certificate. The node labels such certificates as *conflicting*.

The authors define a *certificate graph* as a graph whose vertices are public keys of some nodes and whose edges are public-key certificates issued by users. The local certificate repositories are represented by sub-graphs and certificate chains are represented by paths. When a user A wants to obtain the public key of another user B, it finds a chain of valid public key certificates leading to B. In this chain, the first hop is an edge from A (certificate issued by A) and the last hop leads into B (this is certificate issued by B) [10]. All intermediate nodes are trusted (through previous certificates in the path). The trust path that may occur in self-organized systems exhibits small world [34] properties (highly clustered with a small average diameter, see Chapter 2). Shortcuts can reduce average diameter of small world graph. The *shortcut hunter algorithm* takes into account shortcuts when building a sub-graph.

Different authentication models for ad hoc networks are listed in Table 8.2. A comprehensive overview of existing authentication models is presented in [35].

The following section discusses some security schemes implemented in ad hoc networks.

Table 8.2. Authentication models for ad hoc networks

Encryption scheme	Model	Implementation
Symmetric	IEEE 802.11b Model	No authentication or keys exchanged over secure side-channel outside the IT systems [36]
	Bluetooth Model	PIN manually entered in all devices [37]
	Resurrecting Duckling Model	Key exchanged by physical contact [38]
	Pairwise Key Pred-Distribution Model	Sensors are initialized with subset of key pool before deployed; random subset [39], subset based on expected location [40]
Hybrid	Password Model	Shared password is used for authentication and securely establishing a session key [41]
	Key Chain Model	Anchor x_0 of hash chain serves as private key and x_n as the public key [42, 43]
Asymmetric	Distributed CA Model	CA represented by n special server nodes using a threshold scheme [4],
		Any K nodes represent the CA using a threshold scheme [44, 45]
	Identity-Based Model	Identity used as public key and no certificates are required; CA distributed by using threshold scheme [46]
	Self-Certified Public Key Model	Certificate is embedded in public key [47]
	Self-Organization Model	Trusted path between 2 nodes, idea similar to PGP [10], or more advanced [21]
	Trusted Subgroup Model	Small groups of nodes that trust each other a priori, building trusted path by joining subgroups [48]
	Public Key without Certificate Model	Public key distributed over short range (visual or physical contact) [49]

8.4 Security Schemes

It is difficult to differentiate between malicious network activity and specific problems associated with an ad hoc networking environment. In an ad hoc network, malicious nodes may enter and leave the immediate radio transmission range at random intervals or may collude with other malicious nodes to disrupt network activity and avoid detection. Malicious nodes may behave maliciously only intermittently, further complicating their detection. Dynamic topologies make it difficult to obtain a global view of the network and any approximation can become quickly outdated. Many security schemes have been proposed to deal with security aspects in wireless ad hoc networks although neither of them has succeed to completely achieve all security goals.

Link-layer security schemes protect the one-hop connectivity between two direct neighbors that are within communication range of each other through secure MAC (medium access control) protocol. The most prevalent security solutions among wearable wireless devices are presented in data-link-layer security schemes implemented in IEEE 802.11 [36] and Bluetooth [37] standards. IEEE 802.11, primarily using vulnerable WEP (wired equivalent privacy) [50], has been enhanced into IEEE 802.11i/WPA version [2, 51]. The Target Group TGi proposed long-term architecture based on IEEE 802.1x that supports various authentication modes. Bluetooth specification includes a set of security profiles for service-level and for data-link-layer security [22]. Each Bluetooth device uses a set of security components (unique device address, authentication key, symmetric data encryption key and random number RAND) to provide management, authentication and confidentiality. Data link-layer security is justified by necessity to establish a trusted infrastructure and practically serves as an access control and privacy enhancement. Although extensively used, these data-link-layer security schemes can not meet the end-to-end security requirements and are usually combined with some higher layer security schemes.

Particular attention is dedicated to *authentication schemes* that deal with problems from simple authentication approaches, such as in resurrecting duckling [35, 38, 52, 53], zero configuration (where nodes must be able to authenticate each other without any infrastructure [10]) and face-to-face authentication over short-range link in small spontaneous networks [7], up to the large scale self-organized distribution concepts (for key-based and cryptography-based schemes) and trust graphs which behave like small-world graph (see Chapter 2) and appropriate certificate chains necessary to verify any public key. Scalable distributed authentication is achieved in self-securing ad hoc wireless networks combined with localized trust model [45].

Cooperative security schemes try to cope with node selfishness in a different way. Authors in [54] propose introducing a virtual currency (*nuglets*) for any transaction in order to motivate node participation (see Chapter 5). Neighbor's verification of well behaved nodes through *token* release is implemented in [55]. CONFIDANT [56] relies on isolating misbehaving nodes, CORE [1, 57] collects observation of request packets, etc. S*ecure packet forwarding schemes* [2] provide protection against node's failure to correctly forward data packets and consists of detection technique and reaction scheme. An example of localized detection and end-host reaction are *watchdog* and *pathrather* (see Chapter 5) proposed in [57].

Intrusion detection and *secure routing* play important role in providing secure environment and communications in wireless ad hoc networks and they will be described in more details in the following text.

8.4.1 Intrusion Detection

Intrusion detection includes capturing audit data and providing evidence if the system is under attack [58]. Based on the type of the used audit data, the *intrusion detection system* (*IDS*) can be categorized as *network-based* and *host-based*. The former usually runs on the gateway of a network and exams the packets that go through the network hardware interface, while the latter monitors and analyzes the events generated by programs or users on the hosts [59]. The techniques used in intrusion detection systems can be classified as: *misuse detection* (use patterns of known attacks) and *anomaly detection* (flag deviation of known attacks). Both techniques rely on sniffing packets and using those packets for analysis [60].

Zhang and Lee in [58] propose architecture for intrusion detection and response where every node in the wireless ad hoc network participates in intrusion detection and response through individual *IDS agents*. Since there are no fixed "concentration points", where real-time traffic monitoring can be done, audit collection is limited by the radio-range of the devices. Anomalies are not easily distinguishable from localized, incomplete and possibly outdated information which makes anomaly detection schemes not directly applicable in wireless ad hoc networks. So, the authors in [58] propose a new architecture for IDS, based on IDS agents.

Intrusion detection *complements* intrusion prevention techniques such as encryption, authentication, secure MAC, secure routing and so on, improving the network security. To be efficient, it should have distributed and cooperative architecture and preferably implement anomaly detection (as in [58]). Further improvement can be achieved if it is incorporated into all networking layers and in an integrated cross-layer manner (see Chapter 4).

Similar proposal to Zhang and Lee deploys IDS monitors on individual nodes for intrusion detection within radio range [61]. Authors in [61] propose a platform for enabling collaboration for dissemination of such IDS data (i.e., propagation of information on misbehaving nodes) and offer a collective response to misbehaving or intrusive nodes. Architecture Technology Corporation [62] (one of the leading organizations in the world in the field of network security) has proposed TIARA (techniques for intrusion-resistant ad hoc routing algorithms) [63] as a set of collective design techniques which provide distributed intrusion detection mechanisms and a comprehensive protection against intruders.

SQUAWB (Secure QoS-enhanced UltrA WideBand) protocol is another protocol developed to support the efficient use of the frequency spectrum among multiple simultaneous users [62]. The middleware layer handles the security and QoS details that most application programs currently do not.

SQUAWB will dramatically improve the usability of UWB technology. Commercial users benefit from technology that can be used for wireless personal area networks (WPANs). Without SQUAWB, current UWB technology is very limited in its commercial usefulness.

8.4.2 Secure Routing in Wireless Ad Hoc Networks

The routing in wireless ad hoc networks can not rely on dedicated routers as in wireline networks. This functionality is spread over all nodes which act as regular terminals as well as routers for other nodes. Providing secure routing in such environment faces many problems specific for ad hoc networking and requirements to resist to possible security attacks [3, 6, 64, 65]. Most of the well known routing protocols for ad hoc networks (see Chapter 3) do not include security aspects. The protection against vulnerability of wireless ad hoc networks from different security attacks and especially attacks at the networking layer must fulfill certain requirements [66, 67]. Current efforts are mainly oriented to *reactive* (on-demand) routing protocols such as DSR [68] and AODV [69] (see Chapter 3).

The secure routing protocols must take into account *active* attacks performed by the malicious nodes that aim to intentionally tamper the execution of the routing protocols [70]. They should be able to detect the presence of malicious nodes in the network and avoid them from the routing and route discovery processes. Also, they should guarantee the correct route discovery and preserve the confidentiality of the network topology. The routing protocol must be self-stable and return to its normal operating state within some time limit. It should support permanent functionality of the routing process and ensure Byzantine robustness (to work properly even if some of the nodes become malicious [71, 72]). Since ad hoc networks deploy multi hop routing protocols, where each node provides routing capabilities, selfish behavior may present a significant saving in battery power and reserving more bandwidth for its own traffic. This may result in complete denial of service misbehavior (dropping of packets, injecting incorrect routing information or distorting routing information in order to partition the network). However, most of the current routing security solutions do not take into account the lack of cooperation (i.e., node selfishness) and do not include cooperation enforcement mechanisms [11].

Severe threat against ad hoc routing is the wormhole attack which can completely disable the routing and disrupt the communication. Number of proposals for detection of a wormhole use approach based on packet leashes (e.g., temporal leashes, geographical leashes) [12].

A lot of different approaches are proposed in order to achieve security-aware routing in wireless ad hoc networks. They implement different mechanisms such as strong encryption, digital signatures, timestamps, secret-key cryptography, hashing function, MACs, etc. Table 8.3 presents the most important security-aware routing properties and appropriate resolving techniques [3].

Table 8.3. Secure aware routing properties and techniques

Authenticity	Password, certificate
Authorization	Credentials
Integrity	Digest, digital signature
Confidentiality	Encryption
Non-repudiation	Changing of digital signatures
Timeliness	Timestamp
Ordering	Sequence number

A comprehensive overview of existing routing protocols is given in [3, 11, 64–84]. Several security routing protocols are briefly discussed in the following subsections.

SRP. The *secure routing protocol* (*SRP*) can be applied to a multitude of existing *reactive* routing protocols to protect against attacks that disrupt the route discovery process and guarantee the acquisition of correct topological information [14]. This protocol guarantees that fabricated, compromised or replayed route replies would either be rejected or never reach back to the querying node.

The trusted relationship could be initiated by knowledge of the public key. There exists a *security association* (*SA*) between the source and the destination node which can verify the trusted node using the shared secret key. SRP incorporates the security features into the packet forwarding mechanism (e.g., based on the dynamic routing protocol (DSR) packets), (see Chapter 3). SRP requires additional six-word header containing unique identifiers that tag the discovery process and a message authentication code (MAC). The source node initiates a route request generating a MAC and using the entire IP packet as an input. It uses a keyed hash algorithm, shared secret key and the basic protocol route request (RREQ) packet. The intermediate nodes measure the frequencies of received queries from the neighbors. They provide priority ranking as inversely proportional to the queries rate. A malicious node will be served last or completely ignored due to low priority rank. Based on reply packets which contain the route priority information, the source nodes choose one or more routes to forward their data. The route replies contain accurate connectivity information and safeguard the network functionality. The scheme is robust in presence of a number of non-colliding nodes and provides accurate routing information in a timely manner. It is also immune to IP spoofing. However, SRP is not immune to the wormhole attacks.

SAR. The *security aware ad hoc routing protocol* (*SAR*) defines *level of trust* as a metric for routing and as one of the attributes for security which is taken into consideration while routing [73]. Different privileged levels and levels of trust can be defined among the nodes following the desired trust hierarchy. A node which initiates a route discovery sets the required minimal trust level for the nodes participating in the query/reply propagation. Nodes at each trust level share symmetric keys for encryption/decryption distributing a common key among themselves and with nodes with higher level of trust. The SAR security mechanisms can be incorporated into the traditional routing protocols for ad hoc wireless networks, both into on-demand and table-driven. However, the protocol requires different keys for different level of security, which increases the total number of keys in the network. Also, the fixed assignment of trust levels can be inconvenient for dynamic ad hoc wireless environments.

SEAD. The *secure efficient ad hoc distance vector (SEAD) routing protocol* is based on the destination-sequenced distance vector (DSDV) routing protocol [74] (see Chapter 3) designed to overcome the DoS and resource computation attacks. Proactive routing protocols provide periodic exchange of routing information among the nodes in order to update the current routing information from each node to all destinations. SEAD was inspired by DSDV-SQ [75] routing protocol (a beneficial version of DSDV) and deals with the attackers that modifiy the sequence number and the metric field of a routing table update message. To secure the DSDV-SQ routing protocol, SEAD implements the *one-way hash chain* and does not rely on expensive asymmetric cryptography. Security mechanisms implemented in SEAD are authentication of sequence number and metric of a routing table update message using hash chain elements. The receiver also authenticates the sender of SEAD routing information in attempt to eliminate malicious nodes (either by implementation of a broadcast authentication mechanism, e.g., TESLA, or by using the message authentication codes (MACs)). SEAD is sensitive to wormhole attacks.

ARAN. The *authenticated routing for ad hoc networks (ARAN) routing protocol* is based on cryptographic certificates [16] and provides protection against malicious actions carried by third parties and peers in the ad hoc environment. The implemented minimum security approach introduces *authentication, message integrity* and *non-repudiation* and consists of a *preliminary certification* process followed by a mandatory *end-to-end authentication* (and an optional second stage that provides *secure shortest path*).

ARAN is robust against unauthorized participation, spoofed route signaling, fabricated routing messages, alternation of routing messages, securing shortest path and reply attacks. The use of asymmetric cryptography makes it costly in terms of computation and power consumption. ARAN is not immune to wormhole attacks.

ARIADNE. On-demand secure routing protocol (ARIADNE) is based on DSR (see Chapter 3) and relies only on highly efficient *symmetric cryptography* [76]. It needs some mechanism to distribute the authentic keys required by the protocol. Each node needs a shared secret key (between a source and a node), an authentic key for each node in the network and an authentic route discovery chain element for each node. ARIADNE provides point-to-point *authentication* of a routing message using a message authentication code (MAC) and a shared key. In case of broadcast packets, the authentication is performed by TESLA [17] protocol. Route discovery chain helps in authentication of route discovery and protects against a cache poisoning attacks (caused by flooding of RREQ packets). ARIADNE is efficient against anomalies in routing traffic flow in the network (modification, fabrication and impersonation). However, it does not protect against wormhole attacks, except in its advanced version.

S-AODV. Security-aware AODV (S-AODV) protocol is an efficient solution to eliminate a *black-hole attack* caused by a single malicious node. A malicious intermediate node could advertise that it is the shortest path to the destination resulting in the black-hole problem. Proposed solutions deal with limitations in generating a route reply packet or are realized through checking the neighbors of the malicious intermediate node [77].

The S-AODV protocol assumes that each intermediate node can validate all transit routing packets. The originator of a control message appends a RSA signature and the last element of a hash chain. A message traverses the network and the intermediate nodes cryptographically validate the signature and the hash value. S-AODV requires considerable control overhead and is incapable of dealing with malicious nodes that work in a group.

Sec-AODV. Sec-AODV is a secure routing protocol, based on AODV over IPv6, and further reinforced by a routing protocol-independent Intrusion Detection and Response system for ad-hoc networks [61]. Security features in the routing protocol include mechanisms for non-repudiation, authentication using statistically unique and cryptographically verifiable (SUCV) identifiers, without relying on the availability of a Certificate Authority (CA) or a Key Distribution Center (KDC). SUCVs associate a host's IPv6 address with its public key that provides verifiable proof of ownership of that IPv6 address to other nodes (similar as cryptographically generated address (CGA)). Sec-AODV is based on the protocol proposed in BSAR [78] and SBRP [79] for DSR and does not require: prior trust relations between pairs of nodes (e.g., a trusted third party or a distributed trust establishment); time synchronization between nodes or prior shared keys or any other form of secure association. It provides on-demand trust establishment among the nodes collaborating in detection of malicious activities and establishes a secure communication channel based on the concept of SUVC identifiers.

General comments. The previous list of security routing protocols is not exclusive. Variety of different security features continue to enhance routing protocols aiming to achieve more secure communications and even better QoS provisioning (see Chapter 5) in wireless ad hoc networks. Each security protocol provides protection against one or more security attacks and targets different layer in the protocol stack, as depicted in Table 8.4.

Comparison of some existing security-aware protocols is given in Table 8.5.

8.5 Security Architectures for Ad Hoc Networks

Most of the security schemes concentrate on intrusion detection, secure routing, authentication and key management aiming to provide protection against

Table 8.4. Defense against attacks

Attack	Targeted layer in the protocol stack	Proposed solutions
Jamming	Physical and MAC layers	FHSS, DSSS
Wormhole attack	Network layer	Packet Leashes [80]
Blackhole attack	Network layer	[81]
Byzantine attack	Network layer	[82]
Resource consumption attack	Network layer	SEAD [74]
Information disclosure	Network layer	SMT [83]
Location disclosure	Network layer	SRP [14], NDM [84]
Routing attacks	Network layer	SEAD [74], ARAN [16], ARIADNE [76]
Repudiation	Application layer	ARAN [16]
Denial of Service	Multi-layer	SEAD [74], ARIADNE [76]
Impersonation	Multi-layer	ARAN [16]

particular security attack or attacks. There are several attempts to provide generic approach defining *generic security architectures* for wireless ad hoc networks which usually jointly considers more than one security schemes [85–97]. Examples of such approaches are: group collaboration architectures [85], a general intrusion detection architecture [86], a subscriptionless service architecture [87], Archipelago security architecture [88], layered architecture [89], integrated architecture [90], etc. Following sections give a short description of several such architectures.

Group collaboration security architectures present security models and mechanisms capable of supporting group collaboration [91] in wireless ad hoc networks [85]. Short term or long term collaboration and sharing of resources require effective security models that allow such grouping to work. The basic of these models are *context sensitive security* and developing of object-centered group interaction models for collaboration in wireless ad hoc networks. For this security approach many open questions are still under consideration.

Hierarchical hybrid networks present security architecture capable of defending against link attacks [92]. Wireless nodes are organized into groups. The security schemes utilize encryption/decryption and public key based authentication techniques.

Authors in [93] propose security architecture for access control in wireless ad hoc networks which is fully distributed and based on securely creating groups and managing group membership. The design comprises key-oriented certificates and is survivable in situations where nodes fail and communication is only occasional. It introduces a best-effort solution in membership revocation.

Table 8.5. Security features in some of the routing protocols in ad hoc networks

Protocol	Security positives	Security negatives
SRP	Fabricated, compromised or replay route replies rejected; no online CA; guaranteed acquisition of correct topological information in a timely manner; no complete knowledge of keys by all nodes	Security association as a requirement; possible attack when nodes collude during the two phases of a single route discovery; each SRP query can only discover one route, while diverse routes should be set up to ensure robustness
SAR	Can be easily incorporated on different routing protocols; defines different trust levels	Requirement for different keys for different level of trust (large number of keys); dynamic key assignment
SEAD	Implements one-way hash chain which is a cheaper solution; uses access node authentication; overcomes the DoS attacks	Sensitive to wormhole attacks
ARAN	Uses cryptographic certificates and is robust against modification, fabrication and impersonation	Requires preliminary certification process; costly protocol due to assymmetric cryptography, not immune to wormhole attacks
ARIADNE	Uses symmetric cryptography and is based on authentication (shared key, MAC and authentic route discovery chain); guarantees that the target node of a route discovery authenticates the source	Needs mechanisms to bootstrep authentication keys; only the enhanced version protects against a wormhole attack
S-AODV	Public key cryptography used	High overhead; possible route discovery corruption; compromise of IP portion
Sec-AODV	Uses SUCV, provides on-demand trust establishment	Sensitive to DoS attacks
SMT	Guarantees integrity, replay protection and origin authentication; interoperability with accepted procedures such as source routing; symmetric key cryptography used	Limited protection against compromised topological information
OSPF	Flooding and information least dependency; hierarchy routing and information hiding; two authentication methods: a simple password scheme and a cryptographic message digest; a digital signature scheme to protect the OSPF routing protocol	Age field not protected by digital signature; internal routers can generate incorrect routing information; public key cryptography very expensive and will slow performance of the router; Area Border Routers and Autonomous System Boundary Routers can generate false routing information

Bechler et al. in [94] propose *cluster-based security architecture* for securing communication in mobile ad hoc networks which is highly *adaptable* to their characteristics. They divided the network into clusters and implemented a decentralized certification authority. Decentralization is achieved by using threshold cryptography and a network secret distributed over a number of nodes. Different types of keys (symmetric cluster key, asymmetric public key) and certificates can be used in communication. Nodes decide adaptively about the security level and appropriate encryption (no encryption, secret cluster key for intra-cluster only, public node keys directly exchanged or public node keys certified by the cluster head node). Allowing adaptable complexity is a novelty and advantage which must be tradeoff between required security and availability, acceptable overhead and expected performance. Multi-party communications have recently received particular attention (e.g., group applications like video-conferencing) focusing on security as inevitable networking challenge. A new *framework for multicast (tree-based) security* proposed in [95] concerns with the security in large dynamic multicast groups, involving a one-to-many communication pattern, with a dynamic set of recipients. This framework addresses the conflicting requirements (scalability and security) and defines basic properties of set of cryptographic functions that assure confidentiality (either for encryption of bulk data or only for encryption of short messages as required by key distribution) including intermediate components.

A General Intrusion Detection Architecture [96] defines architecture which is well adapted to the specifics of mobile ad hoc networks. It couples simple trust-based mechanisms with a mobile agent-based intrusion detection system. They define dynamic transient trust relations between the members of dynamic *communities* and complement those mechanisms with distributed hierarchical IDS architecture. The distribution of the IDS mechanisms is achieved by implementing a *local intrusion detection system* (*LIDS*) which is responsible for exchanging security data and intrusion alerts (see Fig. 8.7). The LIDS consists of a local LIDS agent (intrusion detection and response), mobile agents (collect and process data on remote hosts), local MIB (management information base) agents (usually an interface to SNMP). The authors propose to use *SNMP* (*simple network management protocol*) to audit source for LIDS. This architecture is well adapted to the specifics of MANETs. The introduction of a mobile agent makes the global architecture evolutive.

A subscriptionless service architecture combines a *security association management* with a dedicated identity management in a service manager architecture located on a service layer (see Fig. 8.8) with additional units (for personalization) on the network and MAC layers. A service manager controls services (starts, stops and personalizes them) according to a defined policy. It also controls authentication of remote services which want to communicate with some local services. A valid *security association* (long term, event or transaction association) is a precondition for establishment of any communication session. The association contains the user identity, security relevant

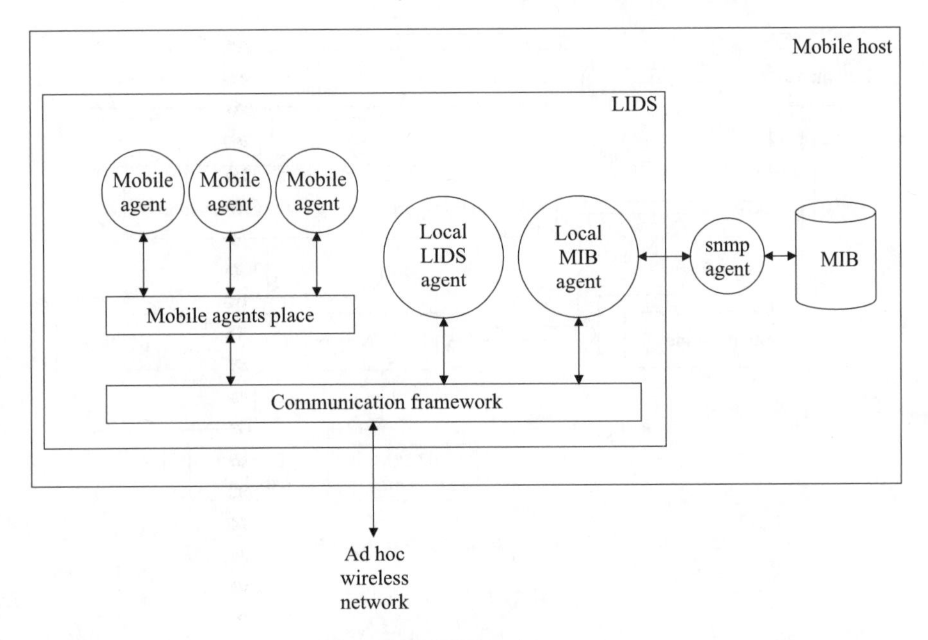

Fig. 8.7. LIDS architecture

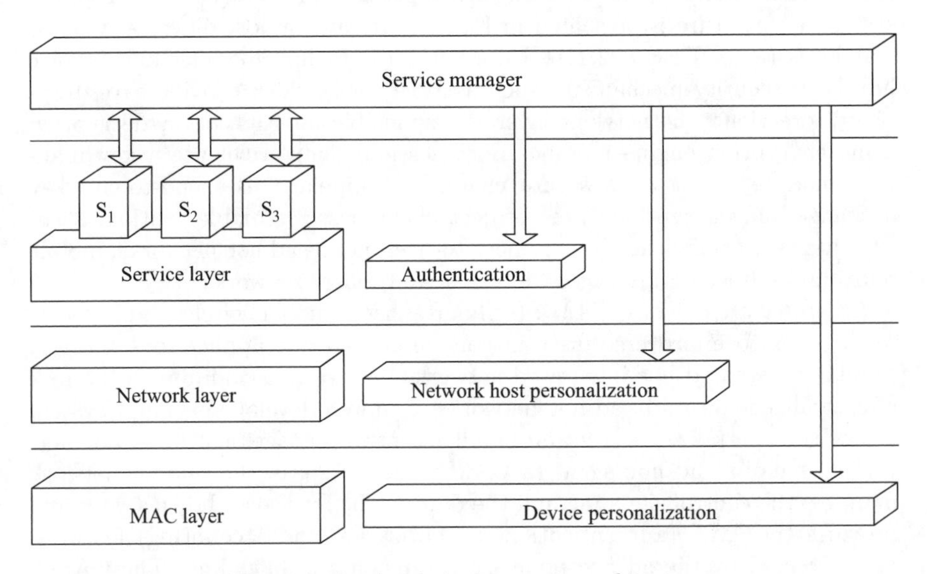

Fig. 8.8. A subscriptionless service architecture

information (e.g., minimum key, length requirements for confidentiality and integrity protection) and personal profile data of the session.

The *Archipelago project* aims to form a secure extended ad hoc network of different wireless capable devices (e.g., laptops, PDAs) and bridge them to

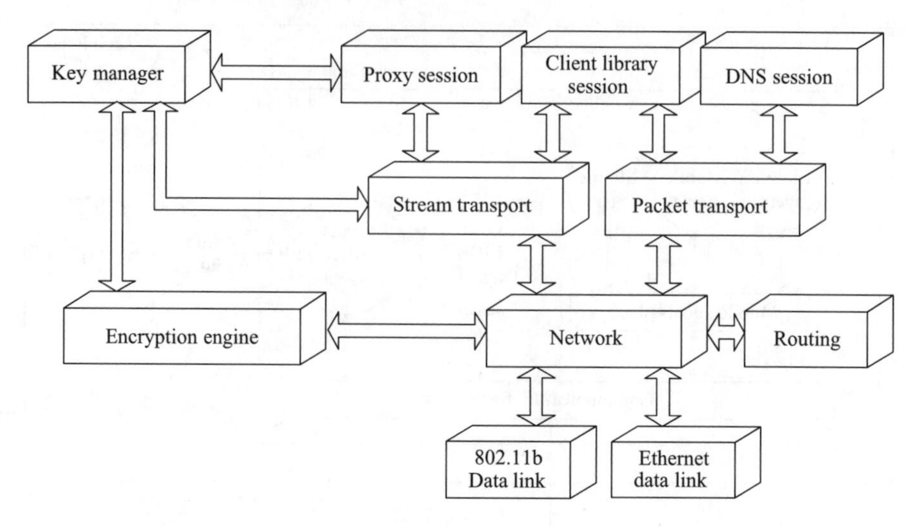

Fig. 8.9. Archipelago system architecture

the Internet via archipelago nodes (which act as a proxy) [88]. The proposed system architecture is presented in Fig. 8.9. It can support different wireless standards (e.g., IEEE 802.11b, Ethernet) in data link modules and involve link layer security mechanisms such as WEP [50], which provide eavesdropping protection. The network layer is responsible for encryption/decryption using encryption engine module. Each session module can use a key manager module to set up a secure channel. Arhipelago uses end-to-end key exchange and encryption. This project encompasses scalable routing, security, and energy efficiency in pure peer-to-peer mobile ad hoc networks, mobile multi-hop infrastructure access networks and sensor networks.

Integrated architecture CAMA (cellular-assisted mobile wireless ad hoc network) aims to improve security in commercial ad hoc applications through cellular-assisted ad hoc integrated networks [90]. It aims to improve the lack of centralized control in ad hoc networks. The well developed cellular system works as a centralized server and handles control management in ad hoc networks. A mobile ad hoc agent (a CAMA agent) [90] in the cellular network manages the control signaling and the data traffic for the ad hoc network and provides the AAA (authentication, authorization and accounting) features. It is accessed by the ad hoc node with a one-hop cellular link. The CAMA agent is a centralized control point and works as a key certificate center and an authentication center. Different keys are used for different communication patterns ranging from secret key, general CAMA key to session and broadcast keys (see Fig. 8.10). Since integrating different networks is gaining momentum (e.g., cellular/WLAN network, cellular/satellite network) CAMA has potential to be widely used.

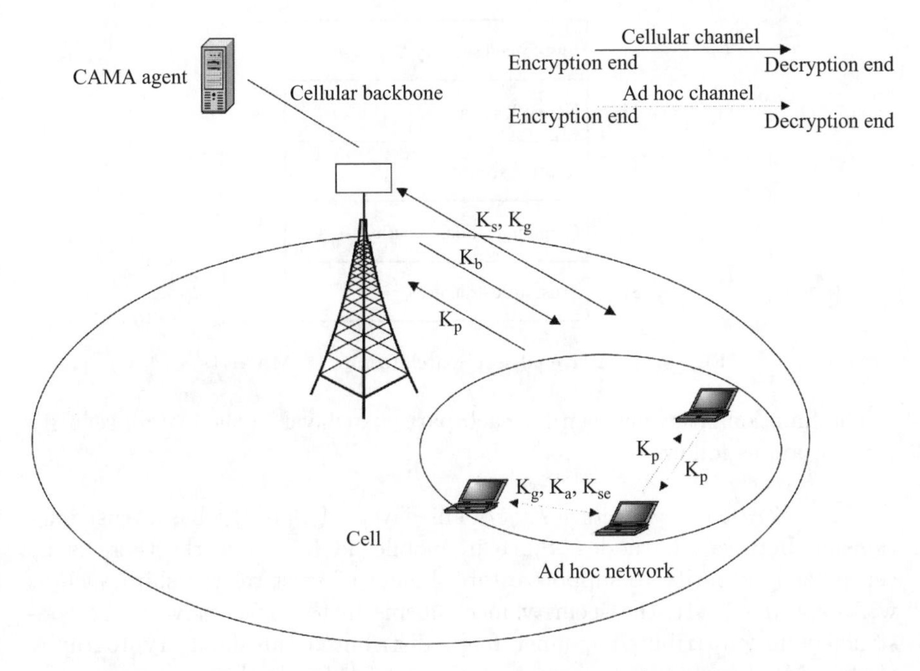

Fig. 8.10. An illustration of key application in CAMA

Security architecture for wireless ad hoc networks should address a number of issues that threaten the overall security of a network and not only some specific needs [97]. The following section presents an example of such generic approach which accepts layering philosophy.

Layered approach. Security is usually considered isolated, after the whole infrastructure of the network has been designed. However, it is desirable that the security is considered as an inseparable part together with the development of the network and not as mechanisms added after-through. Moreover, it cannot be considered separately from the layer view. Most of the existing security protocols are designed to fulfill specific security requirements. Their overlapping functionalities can make the whole system inefficient and complex.

The lack of methodology to manage the complexity of security requirements in various situations will lead to misplacement of security mechanisms and overlapping of security functionalities. The relevant proposals usually deal with establishment of the trusted infrastructure alone, or just concern with security routing protocol based on certain assumptions that security associations are already established. A layered approach to security design can provide such advantages as modularity, simplicity, flexibility and standardization of protocols [89]. A layer secured architecture for mobile ad hoc networks (MANETs) proposed in [89] considers five-layer security, as depicted in Fig. 8.11.

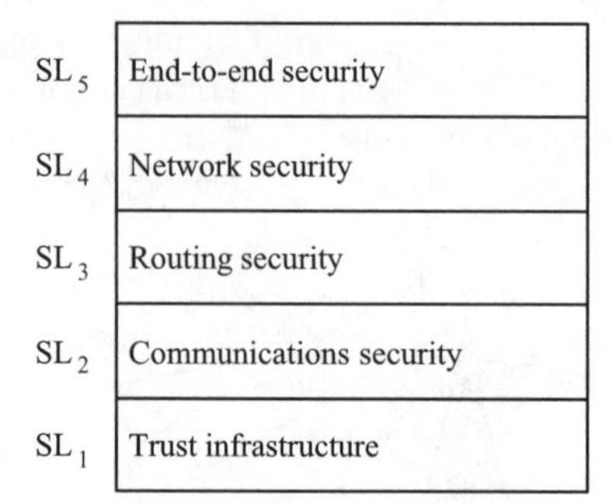

Fig. 8.11. A layered secure architecture for MANETs

The functionalities foreseen for each particular layer in the overall security concept are as follows:

SL1 — Trust Infrastructure Layer. This layer refers to the basic trust relationship between the nodes. Since in mobile ad hoc networks there is no centralized authority to support establishment of trust relationship (such as well deployed PKI), the security mechanisms in this layer have to be constructed in a distributed manner (e.g., distributed threshold cryptography [4, 44]). Moreover, the security association established in this layer must serve for the upper layer security mechanisms.

SL2 — Communication Security Layer. This layer refers to the security mechanisms applied in transmitting data frames in a node-to-node manner. SL2 is a layer which provides hop-to-hop communication security and is related to data link security and physical layer security (according to the OSI model) in the wireless communication channel. Examples are WEP (Wired Equivalent Privacy) or 802.1x working in data link layer and physical protection mechanisms like frequency hopping security mechanisms, which protect the data frame from eavesdropping, intersection, alteration or dropping from the unauthorized party along the route.

SL3 — Routing Security Layer. This layer refers to security mechanisms applied to routing protocols. The security mechanisms in this layer are highly related to the network topology and are always designed with respect to the specific routing protocols (as already mentioned in previous sections). The routing security layer involves two aspects: *secure routing* and *secure data forwarding*. In secure routing aspects, nodes are required to cooperate to share correct routing information, in order to keep the network connected efficiently. In secure data forwarding, the transmitted packets should be protected against tampering, dropping and altering by any unauthorized party.

SL4 — Network Security Layer. This layer refers to the security mechanisms used by the network protocols, which perform sub-network operations from end system to end system. This layer deals with the network access control and network layer data protection. It works at the end of the network fabric and deploys the mechanisms that address the security problems which can not be solved satisfactorily in the underlying routing protocols, implementing the security services like peer entity authentication, confidentiality and integrity. For instance, in SMT (Secure Message Transmission protocol), security efforts are implemented in the end systems as enhancements to the unreliable routing protocol [83].

SL5 — End-to-end Security Layer. This layer refers to the end system security and any application specific security protocol (e.g., SSL-Secure Socket Layer, SSH-Secure SHell program). The security protocols in this layer are independent of the underlying networking technology. However, the related security mechanisms are restricted to only intend parties and the provision of any security services highly depends on specific applications.

The functionalities of each layer are usually designed according to the security requirements of particular applications and they may differ for military applications or for commercial or personal scenarios.

8.6 Projects

Security in wireless ad hoc networks becomes a challenging issue observated in many ongoing research projects. Although an important networking aspect, it usually accompanies other issues, making it difficult to list projects dedicated exclusively to security in wireless ad hoc networks. The following section presents some of the existing projects which dedicate major portion of their research to security aspects in wireless ad hoc networking.

Secure Mobile Networking [98]. This project aims at integration of security and network-layer mobility into real systems that tackle the issues of secure enclaves. The undertaken work results in the development of a high performance Secure Mobile Network and insights into its use as part of the National Information Infrastructure in the USA. The goals of the project have included tight integration of Mobile-IP and IPSEC so that all packets originating from a Mobile-IP wireless node can be protected under an IPSEC umbrella, be the mobile node at home or away. The project also investigated other security, redundancy and wireless network reliability issues, with the overall goal of developing a system with multiple security defense and redundancy mechanisms.

Security of Vehicular Networks [99]. To become a real technology that can guarantee public safety on the roads, vehicular networks need an appropriate security architecture that will protect them from different types of security attacks. This project explores different security aspects of vehicular networks, including threat model, authentication and key management, privacy and secure positioning.

SPOT [100]. The goal of the SPOT (Secure PosiTioning) project is to investigate techniques for secure localization of wireless devices. Several mechanisms have been proposed in the framework of SPOT which inlcude mechanisms for secure position, distance and encounter verification of wireless devices.

Mobile Ad Hoc Network Security [101]. The objective of this research project is to develop security mechanisms that support secure routing, communication and intrusion detection within small-scale wireless mobile ad-hoc networks (MANET). The project team is working to implement a secure bootstrapping and routing protocol for MANETs that does not rely on pre-existing trust associations between nodes or the availability of an on-line service to establish these trust associations. Additional areas of research include: secure ad hoc communications, secure distributed storage management, distributed trust management and ad hoc wireless testing tools.

IPonAir [102]. The goal of this project is the seamless interconnection of different wireless and fixed network areas, which might be ordered hierarchically. This makes an efficient, flexible and secure communication based on the Internet protocol suite possible. Mobile communication with multiple domains in a wireless, self-configuring and secure heterogenous environment is seen as a long-term challenge in the IPonAir project.

Moby Dick [103]. The EU IST project Moby Dick provides a solution that integrates QoS, mobility and AAA into a heterogeneous access environment, using IPv6 and Mobile IPv6 as base technologies. As examples of heterogeneous access, Ethernet, Wireless LAN and TD-CDMA are used. The goal is to provide seamless services using the flexibility of the Internet, while providing authentication, *security* and QoS as known from circuit-switched networks.

8.7 Concluding Remarks

This chapter gives an insight in the security problems and solutions of wireless ad hoc networks, area of increasing importance in the design of future wireless communication systems (e.g., 4G). Wireless ad hoc networks impose additional challenges in front of the designers, due to the lack of infrastructure and the dynamic and ephemeral character of the relationship between the network nodes. They require more sophisticated, efficient and well designed security mechanisms to achieve security goals [104]. Increasing number of ad hoc networking applications (including sensor networks, ubiquitous computing and peer-to-peer applications) emphasize a need for strong privacy protection and security mechanisms. Security in ad hoc networking is a huge topic and this chapter gives an overview of the most relevant issues. It defines security goals and possible threats and attacks, explains major security mechanisms and schemes and presents several security architectures and existing projects.

There are still many challenges ahead. Ad hoc networks rely on cooperation of involved nodes which can be threatened by node selfishness and result in denial of service, network break down and depriving all users of cooperation. So, demands for efficient resolvement of secure grouping, membership management and trust management are still under investigation. Accountability is another open issue. Concerning trust metrics and repudiation mechanisms, these are partially solved with multi-agent design. Cooperation approach, similar to game theory, may be used in designing ad hoc communication systems. Compromise between anonymity and accountability is applied through micro-payment enforcements. Micro-payments are enforcing security and enhancing QoS provisioning at the same time. Secure service provisioning [105, 106] incorporates security mechanisms into service discovery procedures. Security features intend to be embedded in ad hoc devices providing secure link layer functionalities. Efficient use of computation resources and guarding against parasitic computation is another challenge.

The research in the area of authentication and key management concentrates on designing cryptographic algorithms that should be efficient in sense of computational and message overhead. Variety of broadcast and multicast scenarios are still waiting to be resolved. Designing self-enforcing privacy policies and enhancing privacy mechanisms are challenging issues for ubiquitous computing environments.

References

[1] Buttyan, L. and Hubaux, J. -P., "Report on a Working Session on Security in Wireless Ad Hoc Networks," *Mobile Computing and Communications Review*, 6(4), 2002, pp. 74–94.
[2] Yang, H., Luo, H., Ye, F., Lu, S., and Zhang, L., "Security in Mobile Ad Hoc Networks: Challenges and Solutions," *IEEE Wireless Communications Magazine*, February 2004, pp. 38–47.
[3] Ilyas, M., *The Handbook of Ad Hoc Wireless Networks*, CRC Press, 2003.
[4] Zhou, L. and Haas, Z. J., "Securing Ad Hoc Networks," *IEEE Network Magazine*, 13(6), 1999, pp. 24–30.
[5] Gunter Schafer, *Security in Fixed and Wireless Networks*, John Wiley and Sons, 2003.
[6] Siva Ram Murthy, C. and Manoj, B. S., *Ad Hoc Wireless Networks: Architectures and Protocols*, Prentice Hall Communications Engineering and Emerging Technologies Series, 2004.
[7] Feeney, L. M., Ahlgren, B., Westerlund, A., and Dunkels, A., "Spontnet: Experiences in Configuring and Securing Small Ad Hoc Networks," *Fifth International Workshop on Networked Applicances* (*IWNA5*), Liverpool, UK, October 2002.
[8] Karpijoki, V., "Security in Ad Hoc Networks," *Tik-110.501 Seminar on Network Security*, 2000.
[9] Marti, S., Giuli, T., Lai, K., and Baker, M., "Mitigating Routing Misbehavior in Mobile Ad Hoc Networks," *Mobicom'00*, Boston, MA, USA, August 2000.
[10] Hubaux, J. -P., Buttyan, L., and Capkun, S., "The Quest for Security in Mobile Ad Hoc Networks," *ACM MobiHoc 2001*, Long Beach, California, USA, October 2001.
[11] Molva, R. and Michiardi, P., "Security in Ad hoc Networks," *Personal Wireless Communications* (*PWC 2003*), Venice, Italy, September 2003.

[12] Perrig, A., Hu, Y. -C., and Johnson, D. B., "Wormhole Protection in Wireless Ad Hoc Networks," *Technical Report TR01-384*, Department of Computer Science, Rice University.

[13] *The Keyed-Hash Message Authentication Code (HMAC)*, No. FIPS 198, National Institute for Standards and Technology (NIST), 2002. http://csrc.nist.gov/publications/fips/index.html.

[14] Papadimitratos, P. and Haas, Z., "Secure Routing for Mobile Ad Hoc Networks," *CNDS*, 2002.

[15] Zapata, M. and Asokan, N., "Securing Ad Hoc Routing Protocols," *ACM WiSe*, 2002.

[16] Dahill, B., et al., "A Secure Protocol for Ad Hoc Networks," *IEEE ICNP*, 2002.

[17] Perrig, A., et al., "The TESLA Broadcast Authentication Protocol," *RSA CryptoBytes*, 5(2), 2002.

[18] Summers, R. C., *Secure Computing: Threats and Safeguards*, McGraw-Hill, New York, 1996.

[19] Sterjev, M. and Gavrilovska, L., "WPS: A Security Protocol for WPAN Networks," *2003 IEEE Sarnoff Symposium (2003 IEEE SSAWWC)*, New Jersey, March 2003.

[20] Rhee, K. H., Park, Y. H., and Tsudik, G., "A Group Key Management Architecture for Mobile Ad-hoc Wireless Networks," *Journal of Information Science and Engineering*, 21, 2005, pp. 415–428.

[21] Capkun, S., Hubaux, J. -P., and Buttyan, L., "Self-Organized Public-Key Management for Mobile Ad Hoc networks," *IEEE Transactions on Mobile Computing*, 2(1), 2003, pp. 52–64.

[22] Vanhala, A., "Security in Ad-hoc Networks," *Research Seminar on Security in Distributed Systems*, Department of Computer Science, University of Helsinki, 2000.

[23] Rhee, K. H., Park, Y. H., and Tsudik, G., "An Architecture for Key Management in Hierarchical Mobile Ad-hoc Networks," *Journal of Communications and Networks*, 6(2), June 2004, pp. 156–162.

[24] Steiner, M., Tsudik, G., and Waidner, M., "CLIQUES: A New Approach to Group Key Agreement," *International Conference on Distributed Computing Systems*, 1998.

[25] Wong, C., Gouda, M., and Lam, S., "Secure Group Communications using Key Graphs," *ACM SIGCOMM '98*, 1998.

[26] Balenson, D., McGrew, D., and Sherman, A., "Key Management for Large Dynamic Groups: One-Way Function Trees and Amortized Initialization," *IETF Internet Draft:draft-balensongroupkeymgmt-oft-00.txt*, 1999.

[27] Marti, S., Giuli, T., Lai, K., and Baker, M., "Mitigating Routing Misbehavior in Mobile Ad Hoc Networks," *MOBICOM*, 2000.

[28] Diffie, W. and Hellman, M., "New Directions in Cryptography," *IEEE Transactions on Information Theory*, 1976, pp. 644–654.

[29] Buttyan, L. and Hubaux, J. -P., "Nuglets: A Virtual Currency to Stimulate Cooperation in Self-Organized Ad Hoc Networks," *Technical Report DSC/2001/001*, Swiss Federal Institute of Technology, Lausanne, 2001.

[30] Xu, G. and Iftode, L., "Locality Driven Key Management Architecture for Mobile Ad-hoc Networks," *IEEE MASS 2004*, Florida, USA, October 2004.

[31] Kong, J., Luo, H., Xu, K., Gu, D. L., Gerla, M., and Lu, S., "Adaptive Security for Multi-level Ad-hoc Networks," *Journal of Wireless Communications and Mobile Computing* (*WCMC*), 2, 2002, pp. 533–547.

[32] Kong, J., Zerfos, P., Luo, H., Lu, S., and Zhang, L., "Providing Robust and Ubiquitous Security Support for Mobile Ad-Hoc Networks," *ICNP'01*, Riverside, California, USA, November 2004.

[33] Perkins, C. E., ed., *Ad Hoc Networking*, Addison-Wesley, 2001.

[34] Watts, D., "Small Worlds," *Princeton University Press*, 1999.

[35] Hoeper, K. and Gong, G., "Models of Authentications in Ad Hoc Networks and Their Related Network Properties," *Technical Report CACR 2004-2003*, Centre for Applied Cryptographic Research, University of Waterloo, 2004.

[36] IEEE 802.11, Standard Specifications for Wireless Local Area Networks, http://standards.ieee.org/wireless

[37] Bluetooth SIG, Specification of the Bluetooth system, Version 1.1; February 22, 2001, available at https://www.bluetooth.com

[38] Stajano, F. and Anderson, R., "The Resurrecting Duckling: Security Issues for Ad-Hoc Wireless Networks," *Proceedings of 7th International Workshop on Security Protocols*, B. Christianson, B. Crispo, J. A. Malcolm, and M. Roe (eds), LNCS 1796, Springer-Verlag, pp. 172–194, 1999.

[39] Eschenauer, L. and Gligor, V. D., "A Key-Management Scheme for Distributed Sensor Networks," *9th ACM conference on Computer and Communications Security*, 2002.

[40] Liu, D. and Ning, P., "Location-Based Pairwise Key Establishments for Static Sensor Networks," *1st ACM Workshop Security of Ad Hoc and Sensor Networks (SASN) '03*, 2003.

[41] Asokan, N. and Ginzboorg, P., "Key Agreement in Ad Hoc Networks," *Computer Communications*, 23(17), 2000, pp. 1627–1637.

[42] Weimerskirch, A. and Westhoff, D., "Zero Common-Knowledge Authentication for Pervasive Networks," *Tenth Annual International Workshop on Selected Areas in Cryptography* (*SAC 2003*), 2003.

[43] Weimerskirch, A. and Westhoff, D., "Identity Certified Authentication for Ad-Hoc Networks," *1st ACM workshop on Security of ad hoc and sensor networks* (*SASN*), 2003.

[44] Kong, J., Zerfos, P., Luo, H., Lu, S., and Zhang, L., "Providing Robust and Ubiquitous Security Support for Mobile Ad-Hoc Networks," *International Conference on Network Protocols (ICNP) 2001*, 2001.

[45] Luo, H., Zerfos, P., Kong, J., Lu, S., and Zhang, L., "Self-Securing Ad Hoc Wireless Networks," *Seventh IEEE Symposium on Computers and Communications* (*ISCC '02*), 2002.

[46] Khalili, A., Katz, J., and Arbaugh, W., "Toward Secure Key Distribution in Truly Ad-Hoc Networks," *2003 Symposium on Applications and the Internet Workshops* (*SAINT 2003*), IEEE Computer Society, 2003.

[47] Girault, M., "Self-Certified Public Keys," *Advances in Cryptology — EUROCRYPT '91*, D. W. Davies (ed.), LNCS 547, Springer-Verlag, 1991, pp. 490–497.

[48] Gokhale, S. and Dasgupta, P., "Distributed Authentication for Peer-to-Peer Networks," *Symposium on Applications and the Internet Workshops 2003* (*SAINT'03 Workshops*), 2003.

[49] Balfanz, D., Smetters, D. K., Stewart, P., and Chi Wong, H., "Talking to Strangers: Authentication in Ad-Hoc Wireless Networks," *Network and Distributed System Security Symposium 2002 (NDSS '02)*, 2002.

[50] ANSI/IEEE Std 802.11, "Wireless LAN Medium Access Control (MAC) and Physical Layer (PHY) Specifications: Authentication and Privacy," 1999.

[51] IEEE Std. 802.11i/D30, "Wireless Medium Access Control (MAC) and Physical Layer (PHY) Specifications: Specification for Enhanced Security," 2002.

[52] Anderson, R. and Kuhn, M., "Tamper resistance — a cautionary note," *2nd USENIX Workshop on Electronic Commerce*, 1996.

[53] Anderson, R. and Kuhn, M., "Low Cost Attacks on Tamper Resistant Devices," in *Mark Lomas et al., editor, Security Protocols, 5th International Workshop Proceedings, Vol. 1361 of Lecture Notes in Computer Science*, Springer-Verlag, 1997, pp. 125–136.

[54] Buttyan, L. and Hubaux, J. -P., "Nuglets: A Virtual Currency to Stimulate Cooperation in Self-Organized Ad Hoc Networks," *Technical Report DSC/2001/001*, Swiss Federal Institute of Technology, Lausanne, 2001.

[55] Yang, H., Meng, X., and Lu, S., "Self-Organized Network-Layer Security in Mobile Ad Hoc Networks," *WiSe '02*, 2002.

[56] Buchegger, S. and Le Boudec, J. -Y., "Performance Analysis of the CONFIDANT Protocol," *MobiHoc '02*, 2002.

[57] Michiardi, P. and Molva, R., "Core: A COllaborative REputation mechanism to Enforce Node Cooperation in Mobile Ad Hoc Networks," *IFIP Communication and Multimedia Security Conference*, 2002.

[58] Zhang, Y. and Lee, W., "Intrusion Detection in Wireless Ad Hoc Networks," *Mobicom'00*, Boston, MA, USA, 2000.

[59] Wai, F. H., Aye, Y. N., and James, N. H., "Intrusion Detection in Wireless Ad-Hoc Networks," *CS4274 Introduction to Mobile Computing, term paper*, Fall 2005, School of Computing, National University of Singapore.

[60] Anjum, F., Subhadrabandhu, D., and Sarkar, S., "Intrusion Detection for Wireless Adhoc Networks," *Vehicular Technology Conference, Wireless Security Symposium*, Orlando, Florida, October 2003.

[61] Patwardhan, A., Parker, J., Joshi, A., Karygiannis, A., and Iorga, M., "Secure Routing and Intrusion Detection in Ad Hoc Networks," *3rd IEEE International Conference on Pervasive Computing and Communications*, Kauaii Island, Hawaii, March 2005.

[62] Architecture Technology Corporation. Information available at: http://www.atcorp.com/atcny/

[63] Ramanujan, R., Ahamad, A., Bonney, J., Hagelstrom, R., and Thurber, K., "Techniques for Intrusion-Resistant Ad Hoc Routing Algorithms (TIARA)," *MILCOM 2000*, Los Angeles, CA, USA, 2000.

[64] Brandao, P., Sargento, S., Crisostomo, S., and Prior, R., "Secure Routing in Ad Hoc Networks," *Ad Hoc & Sensor Wireless Networks*, 1, 2005.

[65] Hu, Y. -C., Johnson, D. B., and Perrig, A., "SEAD: Secure Efficient Distance Vector Routing for Mobile Wireless Ad Hoc Networks," *Ad Hoc Networks*, 1, Elsevier, 2003, pp. 3–13.

[66] Dennis, L. S. E. and Xianhe, E., "Study of Secure Reactive Routing Protocols in Mobile Ad Hoc Networks," *CS4274 Term Paper*, School of Computing, National University of Singapore.

[67] Kotzanikolaou, P., Mavropodi, R., and Douligeris, C., "Secure Multipath Routing for Mobile Ad Hoc Networks," *Second Annual Conference on Wireless On-demand Network Systems and Services (WONS'05)*, St. Moritz, Switzerland, January 2005.
[68] Johnson, D. B. and Maltz, D. A., "Dynamic Source Routing in Ad Hoc Wireless Networks," in T. Imielinski and H. Korth (eds), *Mobile Computing*, pp. 153–181. Kluwer Academic Publishers, 1996.
[69] Perkins, C. E. and Royer, E. M., "The Ad Hoc On-Demand Distance Vector Protocol," in C. E. Perkins (ed.), *Ad Hoc Networking*, pp. 173–219. Addison-Wesley, 2000.
[70] Eichler, S., Dotzer, F., Schwingenschlogl, C., Caro, F. J. F., and Eberspacher, J., "Secure Routing in a Vehicular Ad Hoc Network," *IEEE VTC 2004 Fall*, Los Angeles, USA, September 2004.
[71] Awerbuch, B., Curtmola, R., Holmer, D., Nita-Rotaru, C., and Rubens, H., "ODSBR: An On-Demand Secure Byzantine Routing Protocol," *Technical Report Version 1*, The Archipelago Project, October 2003.
[72] Burmester, M., Le, T. V., and Weir, M., "Tracing Byzantine Faults in Ad Hoc Networks," *Communication, Network and Information Security — 2003*, New York, USA, 2003.
[73] Yi, S., Naldurg, P., and Kravets, R., "A Security-Aware Routing Protocol for Wireless Ad Hoc Networks," *6th World Multi-Conference on Systemics, Cybernetics and Informatics (SCI 2002)*, 2002.
[74] Perkins, C. E. and Bhagwat, P., "Highly Dynamic Destination-Sequenced Distance-Vector Routing (DSDV) for Mobile Computers," *Proceedings of SIGCOMM 1994*, 1994.
[75] Broch, J., Maltz, D. A., Johnson, D. B., Hu, Y. -C., and Jetcheva, J. G., "A Performance Comparison of Multi-Hop Wireless Ad Hoc Network Routing Protocols," *Proceedings of MOBICOM 1998*, 1998.
[76] Hu, Y. -C., Perrig, A., and Johnson, D. B., "Ariadne: A secure On-Demand Routing Protocol for Ad Hoc Networks," *Proceedings of MOBICOM 2002*, 2002.
[77] Deng, H., Li, W., and Agrawal, D. P., "Routing Security in Wireless Ad Hoc Networks," *IEEE Communications Magazine*, 40(10), October 2002, pp. 70–75.
[78] Bobba, R., Eschenauer, L., Gligor, V., and Arbaugh, W., "Bootstrapping Security Associations for Routing in Mobile Ad-Hoc Networks," *Technical Report TR 2002-44*, University of Maryland, May 2002.
[79] Tseng, Y. -C., Jiang, J. -R., and Lee, J. -H., "Secure Bootstrapping and Routing in an IPv6-Based Ad Hoc Network," *ICPP Workshop on Wireless Security and Privacy*, 2003.
[80] Hu, Y. -C., Perrig, A., and Johnson, D. B., "Packet Leashes: A Defense against Wormhole Attacks in Wireless Ad Hoc Networks," *INFOCOM 2003*, San Francisco, CA, USA, April 2003.
[81] Ramaswamy, S., Fu, H., Sreekantaradhya, M., Dixon, J., and Nygard, K. E., "Prevention of Cooperative Black Hole Attack in Wireless Ad Hoc Networks," *International Conference on Wireless Networks*, Las Vegas, USA, June 2003.
[82] Awerbuch, B., Curtmola, R., Holmer, D., Nita-Rotaru, C., and Rubens, H., "Mitigating Byzantine Attacks in Ad Hoc Wireless Networks," *Technical Report Version 1*, Archipelago project, March 2004. Information available at: http://www.cnds.jhu.edu/research/networks/archipelago

[83] Papadimitratos, P. and Haas, Z., "Secure Data Transmission in Mobile Ad Hoc Networks," *ACM Workshop on Wireless Security*, 2003.

[84] Fasbender, A., Kesdogan, D., and Kubitz, O., "Variable and Scalable Security: Protection of Location Information in Mobile IP," *46th IEEE Vehicular Technology Society Conference*, Atlanta, USA, March 1996.

[85] Danesh, A. and Inkpen, K., "Collaborating on Ad Hoc Wireless Networks," Available at: http://www.parc.xerox.com/csl/projects/ubicomp-workshop/positionpapers/danesh.pdf

[86] Albers, P., Camp, O., Percher, J. -M., Jouga, B., Me, L., and Puttini, R., "Security in Ad Hoc Networks: A General Intrusion Detection Architecture Enhancing Trust Based Approaches," *1st International Workshop on Wireless Information Systems (WIS-2002), ICEIS 2002, 4th International Conference on Enterprise Information Systems*, Ciudad Real, April 2002.

[87] Schmidt, M., "Subsciptionless Mobile Networking: Anonymity and Privacy Aspects Within Personal Area Networks," *IEEE WCNC 2002*, 2002.

[88] The Archipelago Project, Information available at: http://www.cnds.jhu.edu/research/networks/archipelago

[89] Yu, S., Zhang, Y., Song, C., and Chen, K., "A Security Architecture for Mobile Ad Hoc Networks," *Proceedings of Asia-Pacific Advanced Network (APAN)*, 2004.

[90] Wu, X. and Bhargava, B., "Improving Security in Ad Hoc Networks through Integrated Architecture," *1st NSF/NSA/AFRL Workshop on Secure Knowledge Management*, Buffalo, New York, USA, September 2004.

[91] Berket, K. and Agarwal, D., "Enabling Secure Ad-hoc Collaboration," *WACE03*, Seattle, Washington, USA, June 2003.

[92] Lu, Y., Bhargava, B., and Hefeeda, M., "An Architecture for Secure Wireless Networking," Available at: http://www.cs.purdue.edu/homes/yilu/papers/wireless-sec.pdf

[93] Aura, T. and Maki, S., "Towards a Survivable Security Architecture for Ad-Hoc Networks," *Lecture Notes in Computer Science*, Springer-Verlag, 2001, pp. 63–73.

[94] Bechler, M., Hof, H. -J., Kraft, D., Pahlke, F., and Wolf, L., "A Cluster-Based Security Architecture for Ad Hoc Networks," *IEEE INFOCOM 2004*, Hong Kong, March 2004.

[95] Molva, R. and Pannetrat, A., "Scalable Multicast Security with Dynamic Recipient Groups," *ACM Transactions on Information and System Security*, 3, August 2000, pp. 136–160.

[96] Albers, P., Camp, O., Percher, J. -M., Jouga, B., Me, L., and Puttini, R., "Security in Ad Hoc Networks: A General Intrusion Detection Architecture Enhancing Trust Based Approaches," *1st International Workshop on Wireless Information Systems (WIS-2002), ICEIS 2002, 4th International Conference on Enterprise Information Systems*, Ciudad Real, April 2002.

[97] Seys, S., "Security Architecture for Wireless Ad hoc Networks," *Revue HF Tijdschrift* 2005(1), 2005.

[98] Secure Mobile Networking Project, Portland State University, Information available at: http://www.cs.pdx.edu/research/SMN

[99] EPFL Vehicular Networks Security Project, Information available at: http://ivc.epfl.ch

[100] SPOT Project, Information available at: http://lcawww.epfl.ch/capkun/spot

[101] NIST Mobile Ad Hoc Network Security Project, Information available at: http://csrc.nist.gov/manet

[102] IPonAir Project, Information available at: http://www.ccrle.nec.de/Projects/IPonAir.htm

[103] Moby Dick Project, IST-2000-25394, Information available at: http://www.ist-mobydick.org

[104] Wrona, K., "Distributed Security: Ad Hoc Networks & Beyond," *PAMPAS Workshop*, London, September 2002.

[105] Handorean, R. and Roman, G. -C., "Secure Service Provision in Ad Hoc Networks," *Technical Report WUCSE-03-47*, Washington University, Department of Computer Science, St. Louis, Missouri, USA, 2003.

[106] Eronen, P., Gehrmann, C., and Nikander, P., "Securing Ad Hoc Jini Services," *NordSec2000*, Reykjavik, Island, October 2000.

9

Towards Seamless Communications

9.1 Introduction

This concluding chapter highlights the importance of ad hoc networking as an inevitable paradigm towards future 4G networks enabling intelligent, adaptive, robust and user-sensitive communications. It presents the major characteristics of the 4G concept and explains the role of ad hoc networks in achieving seamless communication and pervasive computing.

Ad hoc networks, as an access technology, will play important role within the 4G in providing high bandwidth and obtaining Internet connectivity to end mobile users. They enable mobile devices to connect to each other (within the transmission range and through automatic configuration) and to extend the Internet connectivity through Internet gateway nodes [1]. In addition, the revealing interest of operators towards licence-free frequencies also focuses attention on IP-compatible, multi-hop ad hoc networks.

This chapter explains the 4G fundamental issues and challenges, where the ad hoc networks are foreseen as potential cornerstones. It gives a brief overview of the problems connected to ad hoc implementation towards achievement of seamless communications focusing on the role of ad hoc networks in building the all-IP concept, resource management, end-to-end QoS, reference model and technology roadmap.

9.2 Market Trends and Growth

The abundance of mobile computing and communication devices (e.g., laptops, handheld digital devices, PDAs, cell phones and wearable computers) generates a movement from the *Personal Computer* age (i.e., a one computing device per person) to the *Ubiquitous Computing* age (a user utilizes, at the same time, several electronic platforms through which he can access all required information whenever and wherever needed). The huge number of personal wireless devices, which are constantly getting smaller, cheaper, more convenient and more powerful and the numerous demanding multimedia applications result in persistent enlargement of mobile multimedia traffic. The future mobile applications should be context aware, reconfigurable and personalized [2].

Wireless networks are becoming the easiest solution for interconnections. They have experienced an exponential growth in the past decade. Market analyses [3] foresee impressive market growth for the challenging wireless multimedia communications in the upcoming years (see Fig. 9.1).

For the end mobile users, 4G [4] should be seamless technology which is not costly, is simple and works, and is personalized according to their needs. It should support the paradigm shift from technology centric to user centric concepts and should provide "anytime, anywhere, anyhow and always-on" connectivity in a seamless manner. It is expected that the number of mobile subscribers will outperform the number of fixed subscribers in the forthcoming years. The number of subscribers to wireless data services will grow rapidly from 170 million worldwide in 2000 to more than 2.3 billion in 2010 (according to "World Mobile Subscriber Markets 2005"). The same conclusions go for the number of mobile vs. number of fixed Internet subscribers (see Fig. 9.2).

The continuous evolution of wireless networks and the emerging variety of different heterogeneous wireless network platforms with different properties require integration of these heterogeneous networks into a single platform. The platform should be capable of supporting user roaming and transport of Internet traffic, while not interrupting active communications. This process is followed by the development of new mobile devices designed to deal with these various network platforms and protocols. The evolving 4G wireless technology is a common umbrella that covers and integrates all these requirements.

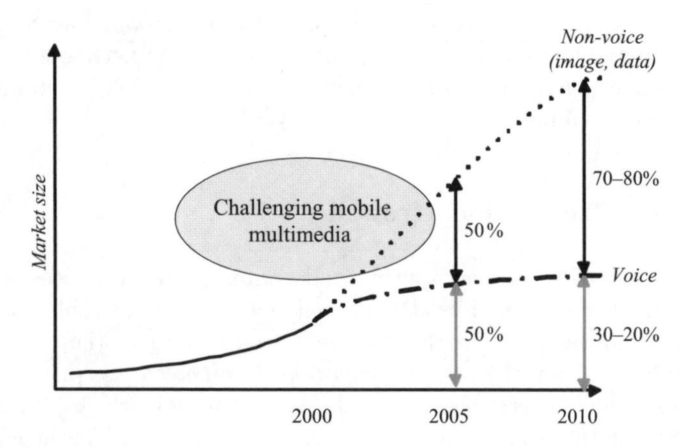

Fig. 9.1. Voice to non-voice communications (DoCoMo vision for 2010) [3]

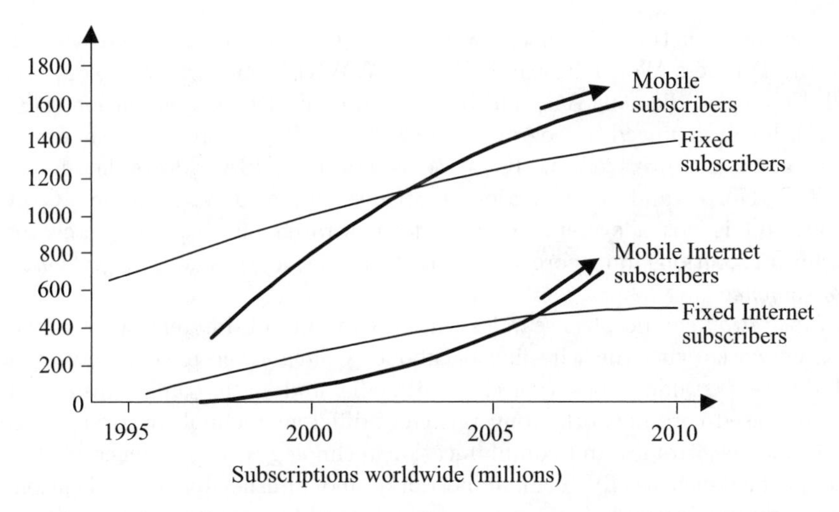

Fig. 9.2. The growth of the number of mobile subscribers [5]

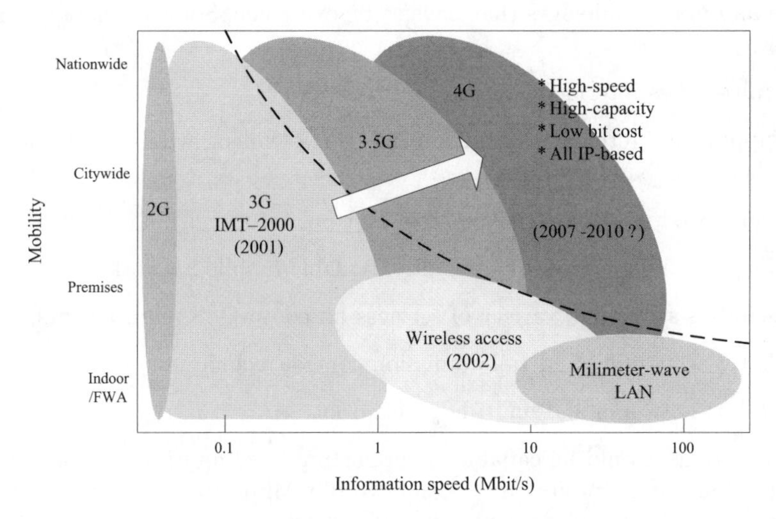

Fig. 9.3. Mobile communications systems

9.3 4G: What Is It all About?

4G is all about a global wireless system presenting an open system approach and an all-IP based seamless connectivity, which is expected to step on the scene around 2010, after the 3G and B3G system solutions (see Fig. 9.3). 4G is foreseen to encompass all existing, planned and future mobile and fixed wireless networks, both terrestrial and satellite.

The work on the 4G concept was initiated in the industry and academia by the Wireless World Research Forum (WWRF) (includes Ericsson, Alcatel, Nokia and Siemens AG) and by the National Science Foundation (NSF), which has announced a program in 1999 that calls for proposals that would look at issues involved in 4G systems. In the USA, Motorola, Lucent, AT&T, Nortel and other major companies are also working on 4G systems. 4G is not a system designed form scratch offering completely new technical solutions, but more a concept whose major goals are: *integration* and *convergence.*

Integration should offer seamless interoperability of different types of wireless networks with the wire-line backbone. *Convergence* means convergence of different traffic types (voice, multimedia and data traffic) over a single IP-based core network, convergence of different technologies (computers, consumer electronics and communication technology), convergence of different media (such as TV, cellular networks and internet-based applications), convergence of services (e.g., television and satellite communications), etc. [6].

4G is still in its formative stages with many open issues. In order to achieve the defined goals, there are demanding tasks that have to be fulfilled and many challenges that have to be overcome. Some of them are listed below:

- Multi-access interfaces, timing and recovery;
- Support for interactive multimedia, voice, video, wireless internet and other broadband services;
- High speed, high capacity and low cost per bit;
- Global mobility, service portability, scalable mobile networks;
- Seamless switching, variety of services based on QoS requirements;
- Better scheduling and call admission control techniques;
- Ad hoc networks and multi-hop networks.

4G systems should be capable of supporting huge multimedia traffic (from several tens of megabits per second to 100 Mbps for outdoor and up to 1Gbps for indoor environments) [6]. 4G systems should be secure, offering end-to-end QoS and quickly deployable services which users may require (anytime, anywhere and from any device), in a cost-effective manner, under one billing mechanisms and fully personalized. 4G is likely to incorporate global positioning services (GPS). Moreover, it should be accessible via diversified radio access (e.g., cellular, WLANs, ad hoc networks), should provide advanced mobility management and support vertical handovers, seamless services, integration of cellular and ad hoc networks etc.

4G is going to be all digital and packet switched network. 4G systems are expected to provide *real-time* and *internet-like* services. The real-time services can be classified into two classes: *guaranteed* and *better-than-best effort* [7]. For the guaranteed services, a pre-computed delay bound is required (e.g., voice). The better-than-best effort services can be *predictive* (need upper bound on end-to-end delay), *controlled delay* (allow dynamically variable delay) and *controlled load* (services which need significant resources for bandwidth and packet processing). Among them, the *guaranteed* and *controlled load* services are proposed to appear in 4G.

The mobility management includes *location registration*, *paging* and *handover*. The mobile terminal should be able to access the services at any possible place. The *global roaming* can be achieved with the help of *multi-hop* networks that can include the WLANs or the satellite coverage in remote areas in a *seamless* manner. The handover techniques should be designed to efficiently use the networks and minimize the number of handoffs. Location management techniques may be also implemented.

Congestion control in the high performance 4G networks is another important issue towards improving QoS parameters. Different approaches implement either *avoidance/prevention* or *detection/recovery* techniques. The avoidance schemes consider admission control based on measurements model or pre-computed model, while detection/recovery schemes implement flow control and feedback traffic management mechanisms.

Voice over multi-hop networks will be an interesting problem, because of the strict delay requirements for voice. The network protocols should be adaptive to dynamically changing channel conditions. The digital to analog conversions at high data rates, as well as multi-user detection and estimation at base stations, smart antennas, complex error control techniques and dynamic routing will require sophisticated signal processing [8].

Ad hoc networking is an important piece within this networking puzzle. The following section reveals the integration and contribution of wireless ad hoc networking within the 4G paradigm.

9.4 Ad Hoc Networks Contribution to 4G

The 4G platform is considered as an "evolution of opportunity to radically change the architecture of the converging fixed and mobile networks towards an evolved Internet core with cellular and ad hoc wireless technology at the edge" [9]. 4G integrates different network topologies and platforms (see Fig. 9.4). Network integration can occur either as an integration of heterogeneous wireless networks with different characteristics (e.g., WLAN, PAN, WAN and mobile ad hoc networks) or as an integration of wireless networks and fixed backbone network infrastructure, the Internet and the PSTN. Wireless ad hoc networks contribute to the 4G ideas offering cost-effective and flexible solutions [10].

Ad hoc networks can *enlarge the wireless coverage* of existing infrastructure wireless network [11]. However, achieving seamless integration (e.g., extend IP to support mobile devices) is still a challenging issue. For instance, seamless

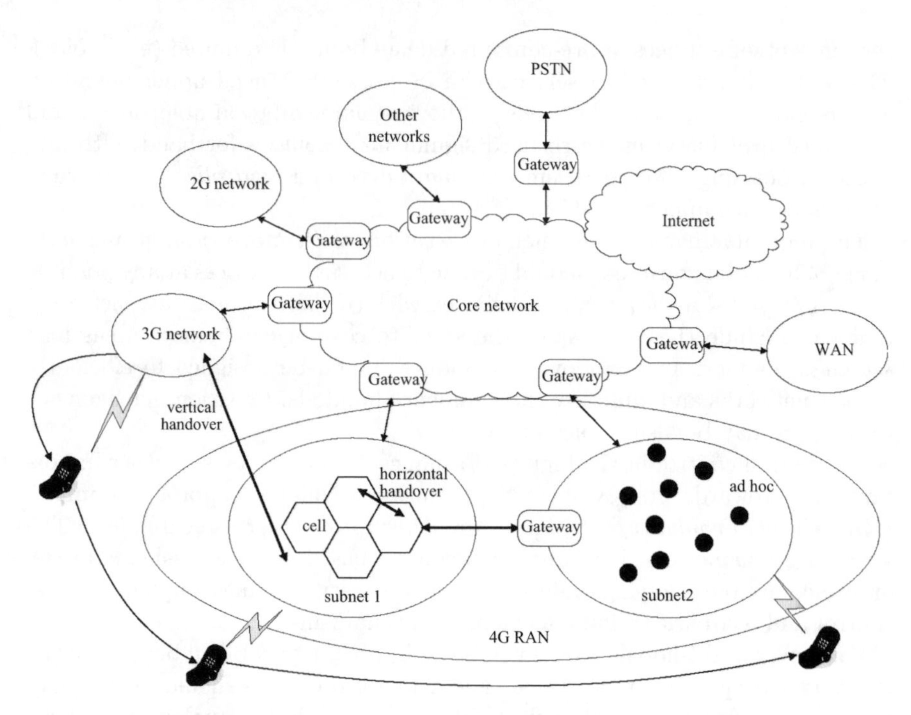

Fig. 9.4. 4G network

combination of cellular and ad hoc networks assumes that applications in cellular and ad hoc will merge providing seamless services. The gateway between ad hoc and cellular networks should support the peer-to-peer applications in combined networks. Developed scenarios should support handover and local reconfiguration, TCP session performance and AAA mechanisms [3].

The 4G IP-based systems are cost-effective solutions comparing to 2G and 3G due to the lower equipment price, cheaper, open converged, IP wireless environment than regular network installation, efficient spectrum reuse and operation in the ISM bands of many ad hoc networks (e.g., WLAN, WPAN).

Ad hoc networks are at the center of the evolution towards the 4G wireless technology. They became a prime candidate for achieving personal pervasive communications. They do not require preinstalled infrastructure, are easy and flexible to maintain and act in a self-organized and auto-configured manner [11]. Combination of self-organized network capabilities, legacy and content oriented services and applications could become a future "killer application" (such as SMS) [1]. In addition, the development of application and system solution specific to ad hoc networking can open opportunities for the new operators offering new services. Ad hoc networks can extend the capacity/ coverage of Wi-Fi spots exploiting multi-hop wireless network connectivity (e.g., 802.11a) [12].

9.5 Towards Seamless Connectivity

Wireless ad hoc networks complete the 4G picture towards seamless communications and ubiquitous computing. They aim to achieve the 4G goals in providing all-IP based platform, dynamic resource management, end-to-end QoS and security, global mobility and shift towards the user-centric paradigm. Intensive research and standardization activities are going on in order to resolve critical issues and prepare wireless ad hoc networks for the global scene.

Some solutions and ongoing research activities were observated in more details in the previous chapters dedicated to particular topics. The following subsections highlight the general ideas that stand behind the wireless ad hoc networking paradigm.

9.5.1 All-IP Solutions

4G systems have recognized the Internet protocol and its extensions as a technology that allows integration of heterogeneous networks into a single, *all-IP based*, integrated network platform [13–15]. The advantages of this approach are: availability for seamless global roaming between all technologies that support IP services, integration of telecommunication services and transparent selection of the underlying technology with respect to the requirements.

The idea of creation of an all-IP environment was supported by the fact that most of the wireless broadband multimedia application in the future will be IP based and that all heterogeneous networks will also use Internet protocol [16]. IP provides independence of the underlying networks in sense of transparency. Moreover, the new mobile terminals will be capable of operating with different network platforms (e.g., UMTS and WLAN). In such circumstances, IP compatible, multi-hop ad hoc networks play important role in realization of alternate dynamic network structures that will operate in a license-free domain with a variety of wireless, non-expensive, devices [16].

Several features characterize the seamless all-IP architecture. The architecture should support new mobile Internet services (real-time, quality-assured conventional and streaming services) available through open interfaces to the globally roaming users. Then, the separation of the access networks from the core network (done via these open interfaces) allows evolution of wireless access networks to be independent of the core part. And finally, the all-IP architecture supports the employment of the IP technology on the transport layer (e.g., SCTP) in various parts of the network. It should provide smooth interconnection with IP-based networks of different RANs (radio access networks) and the core network.

The *end-to-end IP transparency* assumes that the IP packets are transmitted to the end user over the air interface with optimized spectrum efficiency (minimizing redundant and header bits). For true IP transparency, access

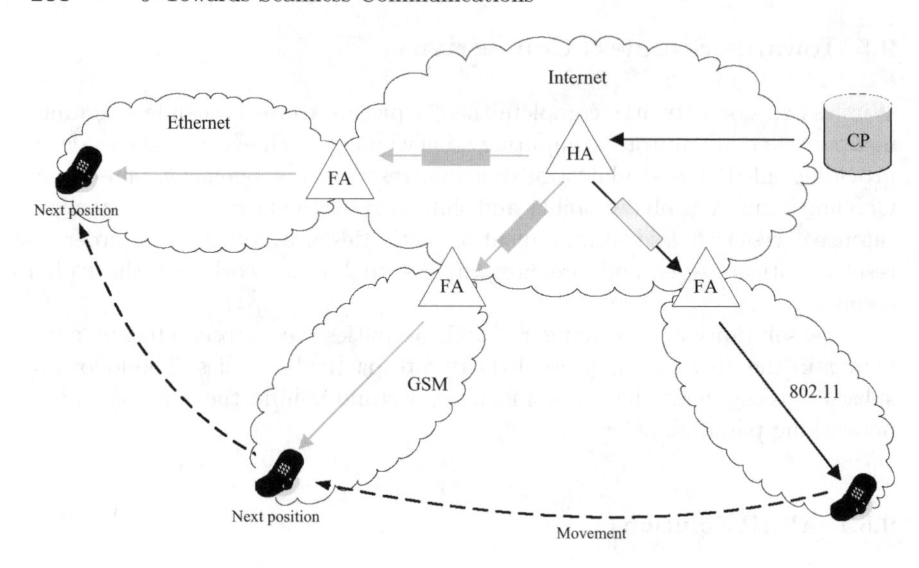

Fig. 9.5. The mobile internet protocol (MobileIP)

technologies are independent. It may require introduction of *IP convergence layer* [13]. In that case, wireless specific signaling and resource reservation mechanisms are not necessary to achieve specific level of QoS for some applications.

The most relevant extensions are the *Mobile IP protocol* [17], considered as the major player in integration between all wireless as well as wire-lined network platforms, and the up-coming standard from the MANET working group [18]. Mobile IP protocol introduces three new network entities (HA — home agent, FA — foreign agent and MN — mobile node). HA acts as a proxy to every registered node, redirecting incoming traffic from the content provider (CA) through encapsulation to the most recent registered location of the MN, see Fig. 9.5. For the Mobile IP protocol, the underlying network platform that provides service is transparent. Its extensions can support regional registration management (in Gateway Foreign Agent) and fast handovers [12, 15, 19]. Mobile IP allows seamless mobility, even if it occurs between domains with previously incompatible routing models. Also, IPv6 and its mobile counterparts provide advantages regarding routing advertisements and address auto-configuration.

A mobile ad hoc network is an infrastructure-free wireless network that is built on the fly. Dynamic address assignment is a desirable feature for deploying mobile ad hoc networks. *IP address auto-configuration* in mobile ad hoc networks is still an unresolved issue [20]. The centralized approach used in traditional networks (centralized server such as DHCP — Dynamic Host Configuration Protocol) is not acceptable because of its distributed and dynamic nature. Possible solutions are: detection of duplicated addresses (DAD proposed by IETF Zenoconf group), and a solution based on IPv6 stateless auto-configuration [21]. An ad hoc mobile node performs the auto-configuration picking two addresses: a temporary address and the actual address. For IPv4 the messages are ICMP (internet control message protocol) packets [22], while for IPv6 the NDP (network discovery protocol) is used [23]. Methods for address aggregation, based on efficient hierarchical address space (similar to hierarchical topology solutions), lead to significant reduction of the routing overhead information [24]. Ad hoc networks, either as self-organized or relay-based, are enablers of high bandwidth and Internet connectivity to the end users [25]. They are desirable and cost-effective solutions for remote and for highly concentrated areas.

Realization of the all-IP concept is strongly related to achievement of end-to-end QoS and distributed resource management [26]. Mobility management (macro and micro mobility), location and context awareness, and security are another issues connected to IP solutions. The IP-based core networks enable using available VoIP set of protocols (MEGACOP, MGCP, SIP, H.232, SCTP, etc.) and support packetized voice and multimedia on top of data transport [1, 27].

9.5.2 Radio Resource Management (RRM)

Radio resource management (RRM) has significant impact on the fulfilment of QoS requirements and on obtaining higher spectral efficiency [28]. RRM encompasses several functions such as handover, power control, admission control, congestion control and packet scheduling [29]. Efficient algorithms for RRM are spread over each node or mobile station in a distributed RRM architecture in case of self-configurable ad hoc networks. Potentials of the wireless ad hoc networks and the advent of the all-IP wireless platforms require new perspectives on RRM architecture and the way it integrates in the overall network resource control [7]. RRM is positioned between the underlying wireless technologies and higher layer and services (see Fig. 9.6).

There are numerous challenges yet to be addressed. Novel schemes for distributed dynamic channel allocation and call/connection admission control are searching for near-optimal computationally inexpensive distributed solutions [30]. New, adaptive and hybrid schemes should optimize the control of the radio access bearers for real time and non-real time traffic. Sofisticated traffic engineering should offer new approaches to handle bursty multimedia traffic, random arrival and reading times and random number of packets per session, without jeopardizing network coverage and capacity [29, 31].

RRM should fit in the overall picture of network management for next generation networks [32] that include techniques such as mobility management, security management, power management [33], location management, etc. Different areas of management interfere with particular techniques to improve the performance. For instance, the cross-layer optimizations will be introduced as the key feature of future communication networks. They

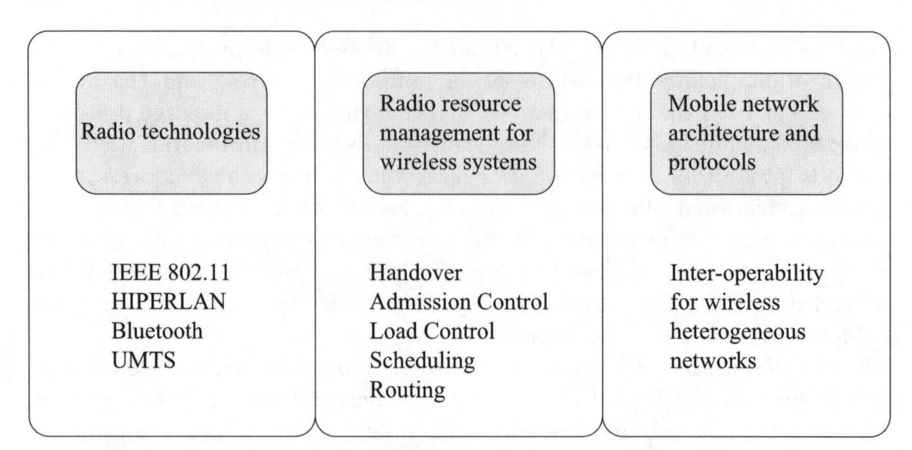

Fig. 9.6. RRM for wireless systems

can influence the resource management and the achieved QoS performance of the overall network. Other examples can be found in service discovery techniques, security mechanisms, mobility modelling, location and context awareness.

9.5.3 End-to-End QoS

Providing end-to-end QoS in ad hoc networks requires close cooperation with the network management (i.e., mobility management and resource management mechanisms). This is a demanding task due to the dynamic nature of the network and the nonnegotiable QoS between users and networks (which may have different QoS support). QoS provisioning in ad hoc networks combines pure IP solutions with underlying radio technologies [28, 34]. Specifics of ad hoc and all-IP approaches include mobility, aggregation, channel sharing, stateless and routing. The IP QoS consists of two areas: service quality management and QoS enforcement techniques (adopted and balanced through RRM control). They both are standardized by IETF. Quality enforcement mechanisms, such as IntServ and DiffServ techniques can be used to control the IP flows through queuing, marking and dropping the packets [28]. Adopted enforcement techniques should conform to customer expectations, network type and traffic characteristics.

QoS management interacts with users, in the form of Service Level Agreements (SLAs) [35] and concerns with the propagation of those expectations through the network in the form of network-level and element-level policies. Different QoS metrics are defined as appropriate for ad hoc networks [36]. They may take the form of *metapolicies* which combine different metrics such as mobility, characteristics of radio channel, cell size and orientation, and handover acceptance metrics [28].

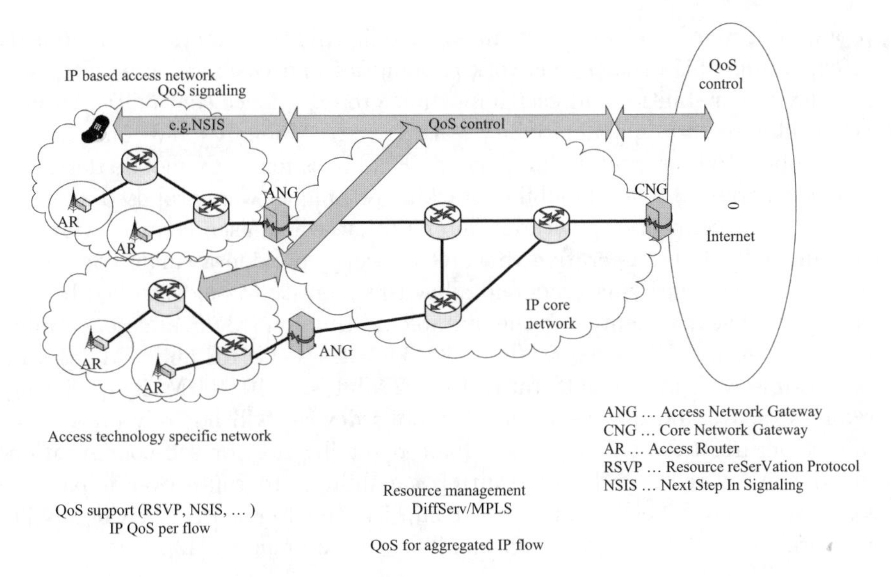

Fig. 9.7. End-to-end QoS architecture

Fig. 9.7 depicts the end-to-end QoS architecture based on DiffServ enforcement technique. Its role is to police a particular bottleneck in the system.

9.5.4 Towards the User-Centric Paradigm

Wireless ad hoc networks play important role towards pervasive computing environment and fulfilling the concept of *ambient intelligence* [37, 38]. Ambient intelligence places the user at the center of the information society and supports the shift towards the *user-centric* paradigm. The user shoud be connected in an *always best connected (ABC)* [39] principle, meaning that he or she is not only always connected, but also connected through the best available device and access technology at any time. The definition of *best* depends on the number of different aspects such as personal preferences, size and capabilities of the device, application requirements, security, operator or corporate policies, available network resources and network coverage. Ad hoc networks play important role enabling ubiquitous computing with enhanced system mobility and portability support [1]. They can cope with future, bandwidth demanding, multimedia applications.

9.5.5 Global Mobility

4G is a system in motion, assuming that nodes are mobile and may need to change their point of attachment to the network [40, 41]. There are numerous proposals towards achieving seamless mobility. Three factors are dominant in expanding this concept: network capabilities and coverage, user interface and device capabilities and user importance to experience continuity. Ad hoc networking requires global mobility features (e.g., MobileIP) and connectivity to a global IPv6 network to support an IP address for every mobile device.

The generalized N-level mobility tracking techniques were proposed in order to keep the connections without interruption when user changes wireless domains [42]. It is imperative that 4G networks provide a single consistent mechanism of identifying users and allow this identity to be controlled by the user and efficiently mapped to the mutable destination [13]. Seamless roaming in heterogeneous IP networks (e.g., 802.11 WLAN, GPRS and UMTS) will be possible with higher data rates, from 2 Mbit/s to 10–100 Mbit/s, offering reduced delays and new services. As mobile devices will not rely on a fixed infrastructure, they will require enhanced intelligence for self-configuration in ad hoc networks and have routing capabilities to route over a packet-switched network. Seamless mobility comprises the user centric view of mobile networking. A global view on 4G mobility can be found in [42].

9.6 Conclusion

This chapter concludes the book where many presented details and specifics of ad hoc networks should help the reader understand this tempting networking paradigm. Ad hoc networks are major enablers of the future person-centric networking philosophy which integrates into the 4G picture [43]. Development of ad hoc networks towards seamless communications and 4G remains open, regarding technology solutions, services and application and business plans. As Niels Bohr had formulated in his famous words:

"Prediction is always difficult, above all of the future"

However, mixed with hope and knowledge, it becomes a joy of our technical minds.

References

[1] Chlamtac, I., Conti, M., and Liu, J. J. -N., "Mobile Ad Hoc Networking: Imperatives and Challenges," *Ad Hoc Networks 1*, Elsevier Publishing, 2003, pp. 13–64.

[2] Raatikainen, K., "A New Look At Mobile Computing," *IWCT2005*, Oulu, Finland, June 2005.

[3] Berndt, H., "Moving towards 4G — Mobile Adventure," *International Forum on 4th Generation Mobile Communications*, London, UK, May 2002.

[4] WWRF — Wireless World Research Forum: http://www.wireless-world-research.org/.

[5] Uusitalo, M.A, "The Wireless World Research Forum — Global Visions of a Wireless World," *IWCT2005*, Oulu, Finland, June 2005.

[6] Prasad, R., "Convergence Paving A Path Towards 4G," *IWCT2005*, Oulu, Finland, June 2005.
[7] Maltz, D. A., "Resource Management in Multi-hop Ad Hoc Networks," *CMU-CS-00*-150, School of Computer Science, Carnegie Mellon University, November 1999.
[8] Polydoros, A., "Towards Ubiquitous Wireless Communications: A Signal Processing Perspective," *10th National Symposium of Radio Science*, Poznan, March 2002.
[9] O'Mahony, D., "Prospects for 4G Telecommunications Systems," *URSI Symposium on Radio Science*, Royal Irish Academy, Dublin, December 2002.
[10] Gavrilovska, L. and Atanasovski, V., "Ad Hoc Networking Towards 4G: Challenges and QoS Solutions," *TELSIKS2005*, Nis, Serbia and Montenegro, September 2005, pp. 71–80.
[11] Raychaudhury, D., "Topics in 4G Wireless Networks: Ad Hoc Nets, Adaptive Services and QoS," *ANWIRE*, Glasgow, UK, April 2003.
[12] Bayer, N., Sivchenko, D., Xu, B., Hischke, S., Rakocevic, V., and Habermann, J., "Integration of Heterogeneous Ad hoc Networks with the Internet," *IWWAN2005*, London, UK, May 2005.
[13] Dixit, S. and Prasad, R., *Wireless IP and Building the Mobile Internet*, Artech House, 2002.
[14] O'Mahony, D. and Doyle, L., "Beyond 3G: 4G IP-Based Mobile Networks," Chapter 6, in Sudhir Dixit and Ramjee Prasad (Eds), *Wireless IP and Building the Mobile Internet*, Artech House, 2002.
[15] Fikouras, N. A., Gorg, C., and Fikouras, I., "Achieving Integrated Network Platforms through IP," *WWRF — Wireless World Research Forum*, Munich, Germany, March 2001.
[16] Perkins, C. E., Malinen, J. T., Wakikawa, R., Belding-Royer, E. M., and Sun, Y., "IP Address Autoconfiguration for Ad Hoc Networks," *IETF MANET Internet Draft*, November 2001.
[17] Perkins, C., "IP Mobility Support," *IETF RFC 2002*, October 1996.
[18] IETF Mobile Ad-Hoc NETworks (MANETs) Charter, http://www.ietf.org/html.charters/manet-charter.html
[19] Aust, S., Proetel, D., Gorg, C., and Pampu, C., "Mobile Internet Router for Multihop Ad hoc Networks," *WPMC2003*, Yokosuka, Japan, 2003.
[20] Martikainen, O., "All-IP Trends in Telecommunications," *International Workshop NGNT*, 2002.
[21] Mohsin, M. and Prakash, R., "IP Address Assignment in a Mobile Ad Hoc Network," *MILCOM2002*, 2002.
[22] Internet Engineering Task Force, *RFC 792: Internet Control Message Protocol*, September 1981.
[23] Internet Engineering Task Force, *RFC draft 2461: Neighbor Discovery for IP version 6 (IPv6)*, December 1998.
[24] Shiflet, C. F., Belding-Royer, E. M., and Perkins, C. E., "Address Aggregation in Mobile Ad hoc Networks," Paris, France, June 2004.
[25] Balachandran, A., Voelker, G. M., and Bahl, P., "Wireless Hotspots: Current Challenges and Future Directions," *Mobile Networks and Applications 10*, Springer Science, 2005, pp. 265–274.
[26] "Evolution towards All-IP: the Service Layer," *Ericsson White Paper*, February 2005.

[27] Leggio, S., Manner, J., Hulkkonen, A., and Raatikainen, K., "Session Initiation Protocol Deployment in Ad-Hoc Networks: A Decentralized Approach," *IWWAN2005*, London, UK, May 2005.

[28] Barry, M. and McGrath, S., "QoS Techniques in Ad Hoc Networks," *ANWIRE*, Glasgow, UK, April 2003.

[29] Niyato, D. and Hossain, E., "Call Admission Control for QoS Provisioning in 4G Wireless Networks: Issues and Approaches," *IEEE Network Magzaine*, September/October 2005, pp. 5–11.

[30] Aguero, R., Berg, M., Choque, J., Hultell, J., Jennen, R., Markendahl, J., Munoz, L., Prytz, M. and Strandberg, O., "RRM Challenges for Non-Conventional and Low-Cost Networks in Ambient Networks," *WPMC'05*, Aalborg, Denmark, September 2005.

[31] Song, W., Jiang, H., Zhuang, W. and Shen, X., "Resource Management for QoS Support in Cellular/WLAN Interworking," *IEEE Network Magzaine*, September/October 2005, pp. 12–18.

[32] Narang, N. and Mittal, R., "Network Management for Next Generation Networks," *8th International Conference on Advanced Computing and Communications*, Cochin, India, December 2000.

[33] Zheng, R. and Kravets, R., "On-Demand Power Management for Ad Hoc Networks," *IEEE INFOCOM 2003*, San Francisco, USA, 2003.

[34] Iera, A., Molinaro, A., Polito, S. and Ruggeri, G., "End-to-End QoS Provisioning in 4G with Mobile Hotspots," *IEEE Network Magzaine*, September/October 2005, pp. 26–34.

[35] Haraszti, P., "Inter Operator Interfaces for End-to-End QoS," Applied IP and Multimedia Services Workshop, Heidleberg, October 2000.

[36] Bush, S., "A Simple Metric for AdHoc Network Adaptation," *IEEE Journal on Selected Areas in Communications*, 23(12), December 2005, pp. 2272–2287.

[37] Ahola, J., "Ambient Intelligence," *ERCIM (European Research Consortium for Information and Mathematics) NEWS*, No. 47, October 2001.

[38] Fuentes, L., Jimenez, D., and Pinto, M., "Towards the development of Ambient Intelligence Environments using Aspect-Oriented techniques," *ACP4IS*, Lancaster, UK, March 2004.

[39] Gustafsson, E. and Jonsson, A., "Always Best Connected," *IEEE Wireless Communications Magazine*, February 2003, pp. 49–55.

[40] Bylund, M. and Segall, Z., "Towards Seamless Mobility with Personal Servers," *INFO — The Journal of Policy, Regulation and Strategy for Telecommunications*, 6(3), 2004, pp. 172–179.

[41] Sasase, I., "Research Activities on the 4th Generation Mobile Communications and Ad-Hoc Networks," *International Workshop on Optical and Electronic Device Technology for Access Network*, Aalborg University — Keio University Joint Workshop for Broadband Wireless Communications, 2005.

[42] Garcia-Escalle, P. and Casares-Giner, V., "Location Management Strategies in Next Generation Personal Communications Services Network," *Wireless Networks — Telecommunications' New Age*, V(1), February 2004.

[43] Tafazolli, R., *Technologies for the Wireless Future: Wireless World Research Forum (WWRF)*, Wiley, October 2004.

Index